Nanocry
Titan

Nanocrystalline Titanium

Edited by

Halina Garbacz
Faculty of Materials Science and Engineering, Warsaw University of Technology, Warsaw, Poland

Irina P. Semenova
Institute of Physics of Advanced Materials, Ufa State Aviation Technical University, Ufa, Russia

Sergey Zherebtsov
Laboratory of Bulk Nanostructured Materials, Belgorod State University, Belgorod, Russia

Maciej Motyka
Faculty of Mechanical Engineering and Aeronautics, Rzeszow University of Technology, Rzeszow, Poland

ELSEVIER

Elsevier
Radarweg 29, PO Box 211, 1000 AE Amsterdam, Netherlands
The Boulevard, Langford Lane, Kidlington, Oxford OX5 1GB, United Kingdom
50 Hampshire Street, 5th Floor, Cambridge, MA 02139, United States

Library of Congress Cataloging-in-Publication Data
A catalog record for this book is available from the Library of Congress

British Library Cataloguing-in-Publication Data
A catalogue record for this book is available from the British Library

ISBN: 978-0-12-814599-9

Publisher: Matthew Deans
Acquisition Editor: Christina Gifford
Editorial Project Manager: Ana Claudia A. Garcia
Production Project Manager: Joy Christel Neumarin Honest Thangiah
Cover Designer: Christian J. Bilbow

Typeset by SPi Global, India

Contents

Contributors

Ana C. Crema de Almeida
Biomaterials Laboratory, Instituto Militar de Engenharia, Rio de Janeiro, Brazil

Emília dos Santos Monteiro
Biomaterials Laboratory, Instituto Militar de Engenharia, Rio de Janeiro, Brazil

Grigory S. Dyakonov
Institute of Physics of Advanced Materials, Ufa State Aviation Technical University, Ufa, Russia

Carlos N. Elias
Biomaterials Laboratory, Instituto Militar de Engenharia, Rio de Janeiro, Brazil

Daniel J. Fernandes
Biomaterials Laboratory, Instituto Militar de Engenharia, Rio de Janeiro, Brazil

José A. Galiza
Biomaterials Laboratory, Instituto Militar de Engenharia, Rio de Janeiro, Brazil

Halina Garbacz
Faculty of Materials Science and Engineering, Warsaw University of Technology, Warsaw, Poland

Witold Habrat
Faculty of Mechanical Engineering and Aeronautics, Rzeszow University of Technology, Rzeszow, Poland

Hiroyuki Kokawa
Department of Materials Processing, Graduate School of Engineering, Tohoku University, Sendai, Japan

Andrzej Królikowski
Faculty of Chemistry, Warsaw University of Technology, Warsaw, Poland

Terence G. Langdon
Materials Research Group, Faculty of Engineering and the Environment, University of Southampton, Southampton, United Kingdom

Angelos P. Markopoulos
Manufacturing Technology Section, School of Mechanical Engineering, National Technical University of Athens, Zografos, Greece

Sergey Mironov
Department of Materials Processing, Graduate School of Engineering, Tohoku University, Sendai, Japan

Maciej Motyka
Faculty of Mechanical Engineering and Aeronautics, Rzeszow University of Technology, Rzeszow, Poland

Georgy I. Raab
Institute of Physics of Advanced Materials, Ufa State Aviation Technical University, Ufa, Russia

Gennady Salishchev
Laboratory of Bulk Nanostructured Materials, Belgorod State University, Belgorod, Russia

Yutaka S. Sato
Department of Materials Processing, Graduate School of Engineering, Tohoku University, Sendai, Japan

Irina P. Semenova
Institute of Physics of Advanced Materials, Ufa State Aviation Technical University, Ufa, Russia

S. Lee Semiatin
Air Force Research Laboratory, Materials and Manufacturing Directorate, AFRL/RXCM, Wright-Patterson Air Force Base, Dayton, OH, United States

Jan Sieniawski
Faculty of Mechanical Engineering and Aeronautics, Rzeszow University of Technology, Rzeszow, Poland

Krzysztof Topolski
Faculty of Materials Science and Engineering, Warsaw University of Technology, Warsaw, Poland

Ruslan Z. Valiev
Institute of Physics of Advanced Materials, Ufa State Aviation Technical University, Ufa; Laboratory for Mechanics of Bulk Nanostructured Materials, Saint Petersburg State University, Saint Petersburg, Russia

Sergey Zherebtsov
Laboratory of Bulk Nanostructured Materials, Belgorod State University, Belgorod, Russia

Preface

Titanium is relatively "young" structural material; commercial production of titanium and titanium alloys was started in the late 1940s. Today, titanium alloys are a popular metallic material for a wide variety of commercial and industrial applications. Titanium and its alloys stand out primarily due to their excellent corrosion resistance and high specific strength. This explains their intensive use in the aerospace and other transportation structures, sports goods, and chemical installations. Titanium and selected titanium-based alloys have also excellent biomedical characteristics, such as biocompatibility and osseointegration, which makes them attractive for various medical applications. Solution-strengthened titanium alloys usually contain toxic alloying elements, such as Al and V, that, by dissolving away from the implant, might pose severe health risks to human health. For this reason, the use of unalloyed, high-strength titanium is a challenging and very important problem.

However, the application of titanium or low-alloyed titanium alloys is considerably limited due to relatively low yield stress. One of the promising ways to improve the mechanical properties of unalloyed titanium or low-alloyed titanium alloys is grain refinement to the nanocrystalline (NC) or ultrafine-grained (UFG) ranges (a brief discussion on terminology aspects can be found in the editorial article by R.Z. Valiev, H. Hahn, and T.G. Langdon for the special issue in *Adv. Eng. Mat.* (2010) p. 665). The formation of the NC or UFG structure in titanium using severe plastic deformation methods and enhanced mechanical properties (mainly strength and ductility) have been intensively studied and broadly discussed during the past decades. At the same time, there are some grain refinement techniques based on conventional deformation processing that still remain less known, in particular hydrostatic extrusion, rolling, and friction stir processing. Quite limited information can also be found on the other (than strength and ductility) mechanical and functional properties of nanostructured titanium, its workability, and possible applications.

This book provides a unique opportunity to bring together experts from different areas related to nanostructured titanium and gather a concise collection of modern materials on the main aspects of production, processing, and application of titanium of NC and UFG. It consists of two parts and starts with an overview to introduce and assess the main production techniques for nanostructured titanium, and discuss the most important fundamental aspects of grain refinement during large or severe deformation. The following chapters focus on mechanical and functional properties, performance attributes and demonstrate specific examples of nanostructured Ti application, predominantly in engineering and medicine. A combination of fundamentals, advanced knowledge, and intricate technological details can make this book useful for a broad audience of scientists and engineers from academia and industry.

We would like to thank all the contributors to this book for their hard and very fruitful work as well as our colleagues from the nanomaterials community for the collaboration and productive discussions. As editors, we would very much appreciate receiving any feedback and critical remarks on the contents of this book and, if appropriate, continue further discussion on the current state and future prospects of this challenging field.

Halina Garbacz, Irina P. Semenova, Sergey Zherebtsov, and Maciej Motyka

List of abbreviations

AES	Auger electron spectroscopy
AFM	atomic force microscopy
APT	atom probe tomography
ARB	accumulative roll bonding
AS	advancing side (in the FSP process)
bcc	body-centered cubic (crystal structure)
BSE	backscattered electrons (SEM mode)
CBN	cubic boron nitride
CD	compression direction
CDRX	continuous dynamic recrystallization
CG	coarse-grained
CP	commercially pure
CPE	constant phase element
CR	cold rolling
CV	coefficient of variation
CW	cold working
DDRX	discontinuous dynamic recrystallization
DRX	dynamic recrystallization
DSA	dynamic strain aging
EBSD	electron backscatter diffraction
ECAP	equal-channel angular pressing
ECAP-C	ECAP-conform
ECM	extracellular matrix
EDS	energy-dispersive X-ray spectroscopy
EIS	electrochemical impedance spectroscopy
fcc	face-centered cubic (crystal structure)
FSP	friction stir processing
GB	grain boundary
GNB	geometrically necessary boundary
HAGB	high-angle grain boundary
HCF	high-cycle fatigue
hcp	hexagonal close-packed (crystal structure)
HE	hydrostatic extrusion
HFRR	high frequency reciprocating rig
hMSCs	human mesenchymal stem cells
HPT	high-pressure torsion
HRTEM	high-resolution transmission electron microscopy
HV	hardness measured by Vickers's method

IDB	incidental dislocation boundary
IGC	inert gas condensation
JMAK	Johnson–Mehl–Avrami–Kolmogorov (equation)
LAGB	low-angle grain boundary
LCF	low-cycle fatigue
LSEM	large-strain extrusion machining
LSM	large-strain machining
LTR	rolling at subzero temperature (low temperature rolling)
MAF	multiaxial forging
MC	microcrystalline
MS	magnetron sputtering
NC	nanocrystalline
ND	direction normal to rolling plane
NG	nano-grained
OCP	open circuit potential
ODF	orientation distribution function
PBS	phosphate buffered saline
PCD	polycrystalline diamond
PDM	point defect model
PF	pole figure
RGD	arginylglycylaspartic acid
RS	retreating side (in FSP process)
RS	rotary swaging
RSM	response surface methodology
RT	room temperature
SAED	selected area electron diffraction
SBF	simulated body fluid
SE	secondary electrons (SEM mode)
SEM	scanning electron microscopy
SFE	stacking fault energy
SFT	sliding friction treatment
SMAT	surface mechanical attrition
S-N	stress vs the number of cycles to failure (curve)
SP	superplastic (deformation)
SPD	severe plastic deformation
SPF	superplastic forming
TD	transverse direction in rolled sheet
TEM	transmission electron microscopy
TMAZ	thermomechanically affected zone
TMP	thermomechanical processing
UFG	ultrafine-grained
UHMWPE	ultra-high-molecular weight polyethylene
USP	ultrasonic shot peening

UTS	ultimate tensile strength
WC	tungsten carbide
WR	warm rolling
XPS	X-ray photoelectron spectroscopy
XRD	X-ray diffraction
YS	yield strength (yield stress)

SECTION 1

Production of nanocrystalline titanium by large or severe plastic deformation

CHAPTER

1 High-pressure torsion and equal-channel angular pressing

Irina P. Semenova*, Ruslan Z. Valiev*,†, Terence G. Langdon‡

Institute of Physics of Advanced Materials, Ufa State Aviation Technical University, Ufa, Russia Laboratory for Mechanics of Bulk Nanostructured Materials, Saint Petersburg State University, Saint Petersburg, Russia† Materials Research Group, Faculty of Engineering and the Environment, University of Southampton, Southampton, United Kingdom‡*

1.1 NANOSTRUCTURING OF TITANIUM BY HIGH-PRESSURE TORSION

Studies performed over the last decades have demonstrated that the formation of the nanostructured state in a material is an efficient way to improve its physical and mechanical properties [1–6]. One of main approaches to producing nanostructure materials is based on refining the microstructure until nanograins in bulk samples using severe plastic deformation (SPD) methods [3–6]. Its principle is that during SPD materials are subjected to large deformations under conditions of high applied pressures [4]. As ultrafine-grained (UFG) metals and alloys fabricated using SPD methods have, as a rule, grain sizes in the range of 100–500 nm and usually contain different nanostructural elements inside the grains (nanotwins, nanoparticles, segregations, etc.), these elements significantly affect the material properties. As a result, the materials are assigned to the class of nanostructured bulk materials [6]. High-pressure torsion (HPT) and equal-channel angular pressing (ECAP) techniques were successfully applied to produce UFG structures in pure Ti. However, the structural features of UFG titanium strongly depend on the method and the treatment regime. The features of the formation of a nanostructured state in commercially pure titanium by the HPT method are discussed in this chapter.

Specimens from commercially pure (CP) titanium VT1-0 (Russia) were initially processed by HPT under an applied pressure of 5 GPa to a true logarithmic strain of about 7 at room temperature [7]. They had the shape of thin discs with a thickness of 0.3 mm and diameter of 10 mm. Fig. 1.1 shows typical transmission electron microscopy (TEM) images of the HPT-processed structure in CP Ti at different magnifications. It is seen that a homogeneous UFG structure with a mean grain size of about 100 nm is formed (Fig. 1.1A and B). This structure is characterized by the following features: first, by extinction contours which are clearly seen in the grain interiors

Nanocrystalline Titanium. https://doi.org/10.1016/B978-0-12-814599-9.00001-8

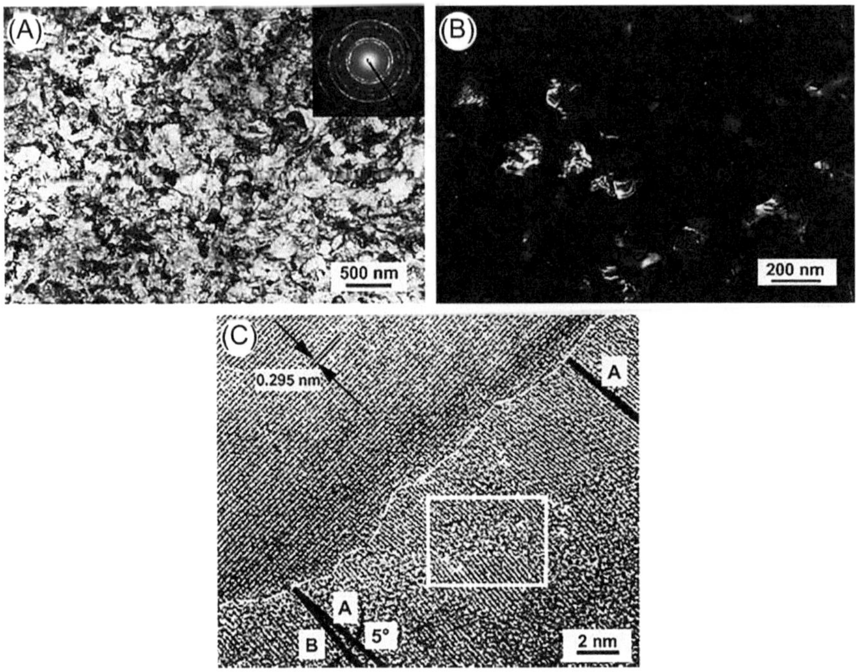

FIG. 1.1

Typical TEM bright field (with SAED as an *inset*) (A) and dark-field (B) micrographs and HRTEM image (C) of as-processed CP Ti [9] .

indicative of large internal stresses; second, grain boundaries appear to be of mostly high angle misorientation but they are not well-defined and there is no banded contrast at the boundaries that is characteristic of well-formed grain boundaries as usually observed by TEM [7,8]. The HRTEM images show that the grain boundaries in the processed material are typically wavy, curved, or corrugated (Fig. 1.1C) indicating their highly nonequilibrium character [4,9]. Moreover, a variation in the misorientation angle is observed along the same boundary by about 5 degrees, probably due to disclinations. A high level of defects is also observed in the grain interiors. In many zones of the grain interior the lattice fringes are poorly defined (area marked by the square in Fig. 1.1C) suggesting local distortions of the crystalline lattice.

In Ref. [9] an unusual aspect of low-temperature annealing on the boundary structure was observed. It was shown that during annealing, grain growth in the HPT-processed Ti starts at a temperature of 350°C and higher. An analysis of short-term annealing for 10 min at temperatures of 250°C and 300°C impacted the microstructure of the HPT-processed Ti and revealed important changes as compared to the as-produced state [9]. First, there is a significant decrease in the lattice distortions without visible grain growth during annealing. At a higher resolution it was shown that annealing at 250°C revealed a rearrangement of defects resulting in their movement from the grain interiors to near the grain boundary areas. An atomic resolution image of such a typical highly distorted grain boundary area is shown in Fig. 1.2A. An ordered array of dislocations [10] with periodicity of $\lambda = 0.7$ nm can be observed in some areas of the high angle grain boundary. The schematic illustration of these "defect" regions is presented in Fig. 1.2B. The estimated dislocation density is very high and is equal to about $10^{14}\,m^{-2}$, which is indicative of the nonequilibrium character of the grain boundaries after this annealing. However, as with any ordering phenomenon, this process lowers the internal stresses (and, therefore, the free energy) in the structure as a system that was confirmed by an analysis of XRD patterns which exhibited peak broadening after this heat treatment [9].

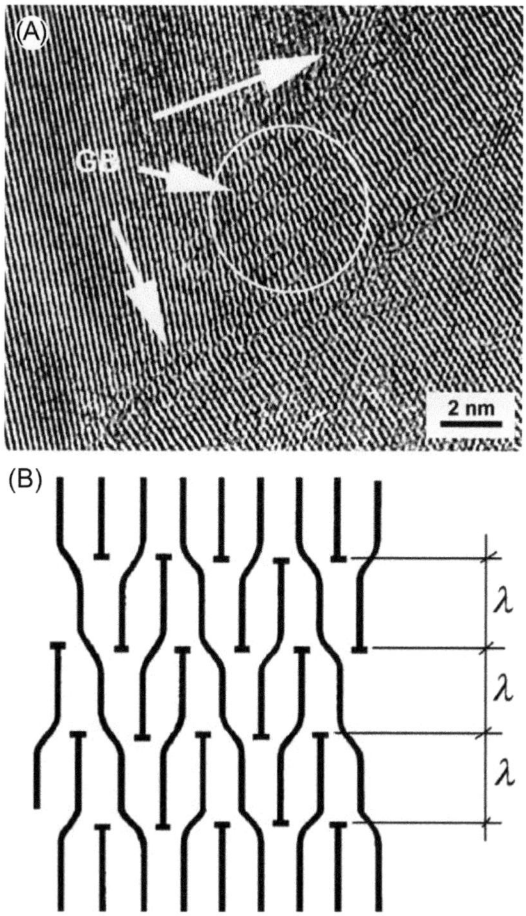

FIG. 1.2

(A) HRTEM of GB region in CP Ti after HPT and annealing at 250°C for 10 min; (B) scheme of the ordered defect structure in selected area [9].

In the late 1990s, researchers showed that the consolidation of metal powders via SPD techniques could be implemented to prepare high density samples with grain sizes smaller than 50 nm [11,12]. In a recent study [13] there was a report of a novel processing approach for producing nanocrystalline CP Ti with grain sizes smaller than 40 nm. The results support the hypothesis that it is possible to synthesize nanostructured CP Ti of high density by using a new approach involving HPT processing of powder after cryomilling.

Fig. 1.3 shows bright-field and dark-field TEM images of the microstructure at the peripheral zone (approximately 3 mm from the center) of the sample after HPT.

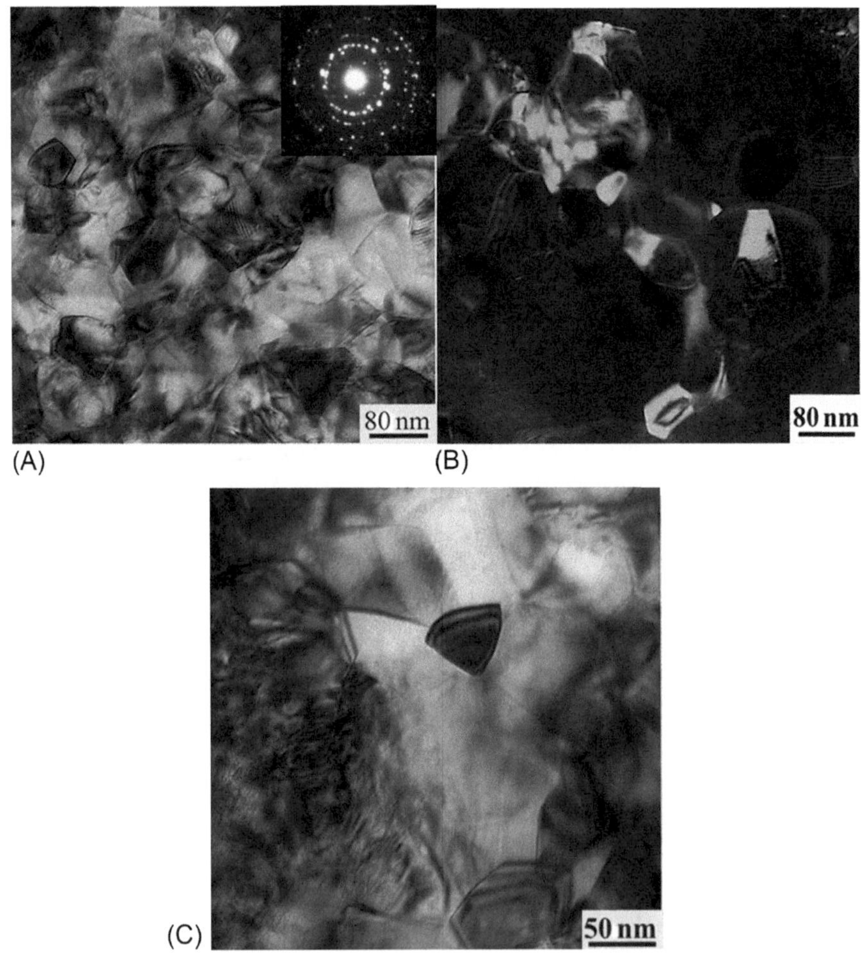

FIG. 1.3

(A, B) Bright-field and dark-field images of ultrafine grains formed in peripheral part of the sample; (C) ultrafine grains [13].

This region does not show any significant porosity and the microstructure of Ti consists of ultrafine grains with an average size ranging from ~25 to ~40 nm (see Fig. 1.3A). However, some coarser grains with an average size of ~120 ± 50 nm are also observed in the microstructure (see Fig. 1.3B). The fraction of the large grains in the microstructure does not exceed 7% and it is likely that the large grains in the Ti sample are formed during HPT through coalescence of the initial small grains in the cryomilled Ti powders. A similar phenomenon was observed in Cu powders subjected to HPT [14]. In this study a possible mechanism of nonuniform grain growth can be also associated with dynamic recrystallization, because SPD processing is performed at elevated deformation temperatures ($T \approx 300°C$) when grain boundary diffusional processes are activated. The appearance of a nonuniform contrast within the grains in the dark-field images indicates a high lattice distortion (see Fig. 1.3C). The total dislocation density in the grain interiors is estimated to be approximately $2 \times 10^{14} m^{-2}$. The spots on the selected area electron diffraction (SAED) patterns form concentric circles (see Fig. 1.3A) and this feature is consistent with the formation of high angle boundaries which are typical of nanostructured materials produced by HPT under the conditions that involve a high number of rotations [4]. The microstructure in the central region of the sample revealed a nonuniform distribution of grain sizes ranging from ~25 to ~240 nm (see Fig. 1.4). Such nonuniformity of the structure formed in the center of HPT samples can be attributed to the competing processes of deformation, dynamic recrystallization, and grain growth, which are likely to be activated at elevated temperatures (300°C).

Thus, HPT of CP Ti at room temperature may result in a nanostructured state formation characterized by a very small grain size from ~120 to ~40 nm, a high dislocation density, enhanced internal stresses, and high angle misorientations of the grain boundaries.

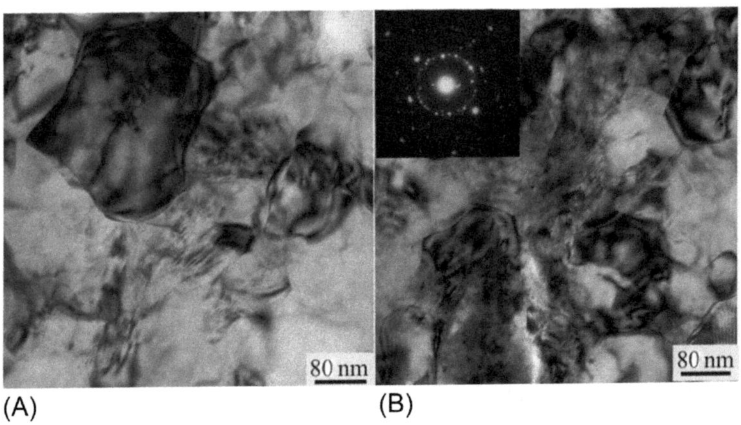

FIG. 1.4

(A) TEM micrographs coarse grains; (B) ultrafine grains in the center of the sample [13].

1.2 GRAIN REFINEMENT BY ECAP AND ITS MODIFICATION

The ECAP technique that involves deformation of massive samples by simple shear was developed by V.M. Segal with his colleagues in the 1970s. It subjected materials to plastic strains without changing their cross-section and therefore permitting repeated deformation [15]. The required condition for UFG structure formation in the billet is shear at the point of channel intersection at lowered temperatures (lower than the recrystallization temperature) ensuring an achievement of the accumulated strain degree of $\varepsilon \sim 4$–6. In the early 1990s, Valiev and coauthors developed this technique and applied it as an SPD technique to form structures with submicrocrystalline and nanometer grain sizes [4,16].

Usually at frequently used angles of straining of 90 degrees, each pass corresponds to the additional strain degree of about 1 [17,18]. Furthermore, the billet orientation when passing through the channels or the route is an important parameter of ECAP. In Refs. [19–21] different ECAP techniques were considered: the billet orientation remained unchanged at each pass (route A); after each pass the billet was rotated about its axis by 90 degrees (route B); after each pass the billet was rotated about its longitudinal axis by 180 degrees (route C). Later, the subtypes of the route B, called B_A and B_C, were introduced for consideration. These two subtypes differ in the rotation direction between the subsequent passes. Thus, route B_A consists of alternative rotation clockwise and counterclockwise by 90 degrees and in route B_C the rotation is performed only clockwise. These routes differ in the shear direction at multiple passes of the billet through the intersecting channels.

Thus, processing by ECAP is multifactorial. Its successful implementation depends on the die-set geometry (angle of channels intersection, their shape and sizes), the parameters of ECA pressing (velocity, number of passes, route, temperature, lubricant), and the nature of the deformed material (initial structure, ductility, strength). One should also take into account possible heating of samples during ECAP [22–24], which also affects the microstructure formation. As is known, depending on the oxygen content and other impurities, commercially pure Ti can be of Grade 2, Grade 3, or Grade 4, which can significantly impact the mechanical properties of Ti in coarse-grained and UFG states [25,26].

The earliest studies were performed on Ti Grade 2, VT1-0 in its Russian counterpart for the chemical composition. The authors [27] studied the microstructure peculiarities formed in Ti Grade 2 as a result of ECAP at 400°C in three different routes: A, B, and C after first and second passes. The TEM studies of the samples after the first pass showed that the deformation occurred mainly via twinning (Fig. 1.5). During the second pass the deformation mechanism changed to dislocation slip. The prismatic $\langle a \rangle$ and pyramidal $\langle c+a \rangle$ slip occurred in turn with the twin bands for route C. The prismatic $\langle a \rangle$ slip was the main deformation mechanism for route B. Deformation was controlled by basal $\langle a \rangle$ slip and microtwinning in intermittent twin bands for route A. Such differences in the deformation behavior of Ti were interpreted by the texture formed during the first pass and by the different values of the Schmid factor for slip in the subsequent deformation. The results assumed that

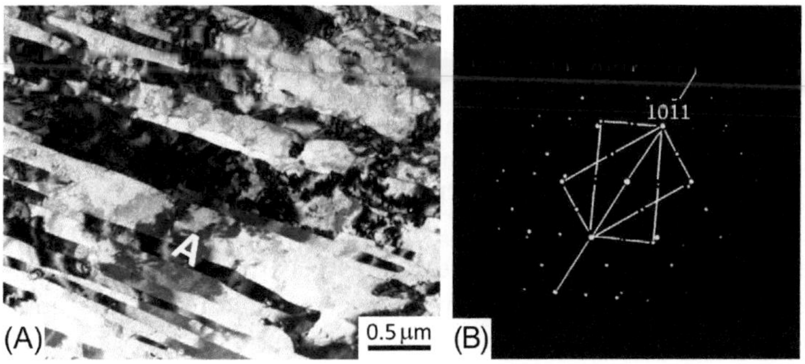

FIG. 1.5

The microstructure (A) and corresponding diffraction pattern (B) of Ti after one ECAP pass [27].

the texture formed during the first pass and determining the shear stresses for each slip system also affected the deformation mechanism during the second ECAP pass, and thus should be carefully considered during any analysis of structure formation in Ti during ECAP.

Another report [28] also studied the impact of the ECAP routes on the microstructure and properties of pure Ti VT1-0 (Ti Grade 2). The billets after eight ECAP passes at 450–400°C were examined. It was stated that route B_C was the most effective for refinement and formation of an equiaxed UFG structure in Ti whereas in routes B_A and C the grains had an elongated shape (Fig. 1.6B). As a result of the multipass ECAP processing, a structure with equiaxed grains of about 260nm was produced (Fig. 1.6A). On the whole, all the produced structures were characterized by a high dislocation density. The diffraction patterns displayed spots located about the circle which confirmed high angle misorientation of the grain boundaries. In addition, some separate spots were split, which was caused by low-angle grain boundaries.

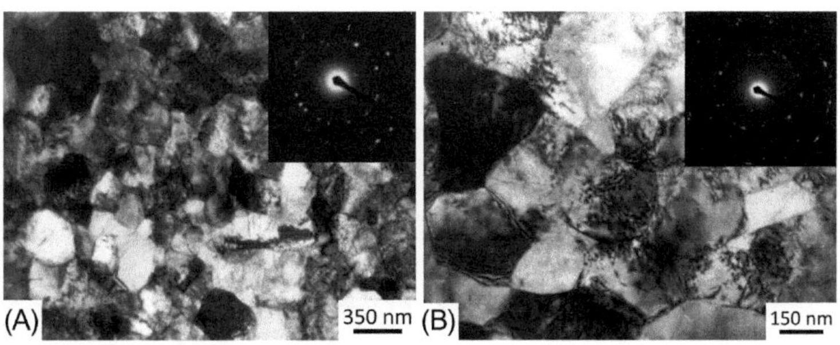

FIG. 1.6

The microstructure of VT1-0 after eight ECAP passes at different routes, transverse section: (A) the route B_C; (B) the route B_A [28].

In Ref. [29] the possibility of grain refinement in Ti Grade 2 to 0.2–0.25 μm was studied with a deformation temperature reduction to 300°C as a result of a channel intersection angle increase from 90 to 120 degrees. The increase of hydrostatic pressure in the deformation zone even with a temperature decrease enhanced the material deformability and resulted in an accumulated strain increase up to the sample failure. In Ref. [30] Zhao et al. showed that CP Ti can be processed by ECAP at room temperature by increasing the channel angle within the die and using a relatively slow ram speed. The microstructure of the pressed material contains shear bands and deformation twins as found in titanium pressed at significantly higher temperatures. Examples of transmission electron microscopy micrographs are shown in Fig. 1.7 taken after one pass with $\phi = 120$ degrees at room temperature. The structure in Fig. 1.7A is composed of elongated bands having widths of ~0.4 m and within these bands there are high densities of dislocations formed into irregular cell structures. Similar twins are visible in many regions of the pressed samples and are a characteristic feature of the microstructures produced by ECAP in CP Ti [27]. Accordingly, it is concluded that ECAP may be used to successfully press CP Ti at room temperature because the limited slip systems available in the hexagonal close-packed (hcp)

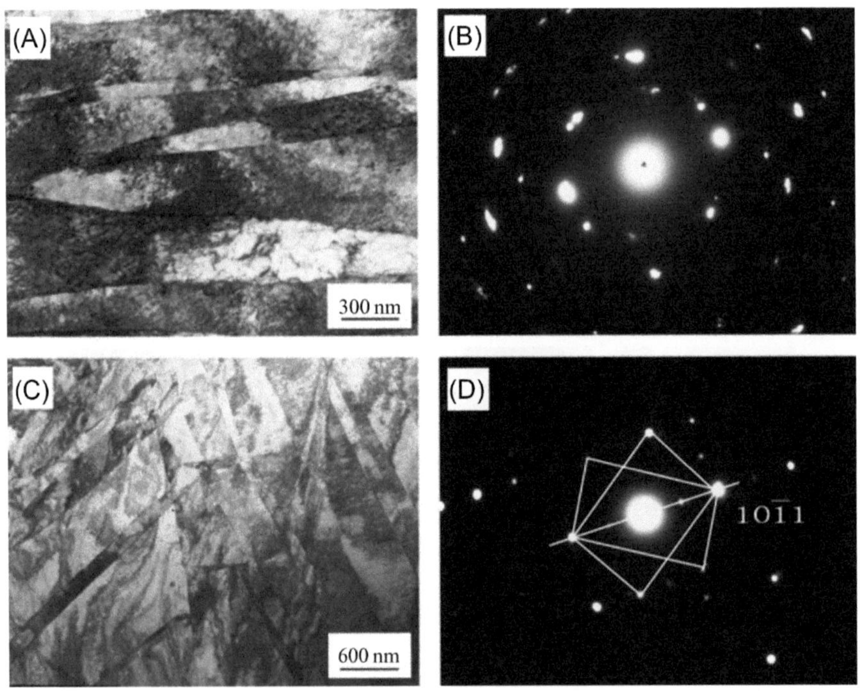

FIG. 1.7

Microstructure by TEM after one pass at room temperature showing (A) shear bands, (B) the corresponding SAED pattern, (C) twins, and (D) the corresponding SAED pattern [30].

metal are assisted by the development of $\{10\overline{1}1\}$ deformation twins and the microstructure after pressing through a single pass consisting of a mixture of shear bands and mechanical twins.

In Ref. [31] the dependence of the initial grain size and deformation velocity on the grain size after ECAP was demonstrated. The ECAP was carried out on the billets at 400°C with constant pressing speeds of 0.2 or 2.8 mm/s. A microstructure examination revealed that the accommodation mechanism of a large strain induced by ECAP changed from deformation twinning to dislocation slip by decreasing the grain size or the pressing speed. The results are expected to provide useful information on the optimization of ECAP conditions for producing UFG hcp metals and alloys.

Commercially pure Ti Grade 4 has a lower deformability due to the high content of oxygen (up to 0.34 wt%), which can impact the peculiarities of the UFG structure formation. For example, in Ref. [32] it was shown that in Ti Grade 4 billets the size of grains/subgrains of about 0.4 μm was achieved only after eight ECAP passes, whereas the achievement of an accumulated strain of 4 was a sufficient condition for UFG structure formation in Ti Grade 2 [33]. It is known that with the increasing content of impurities in Ti, the critical shear stresses increase in all the slip planes. Possibly the structure transformation in Ti Grade 4 requires higher shear stresses and, therefore, accumulation of higher strain degrees. Fig. 1.8A demonstrates that after four ECAP passes, the structure is not sufficiently homogeneous with nonuniform etching fields of microsections cut out from the central and peripheral billet sections, thereby providing direct evidence for the structural heterogeneity. The bright-field TEM images show the subgrain structure characterized by a high dislocation density with the subgrain size varying from 0.5 to 1 μm (Fig. 1.8B). The SAED patterns taken from these regions reveal considerable misorientations in the subgrain structure. With the increase of the ECAP passes to eight, there appear equiaxed grains of about 0.6 μm in size with thickness contours at their boundaries (Fig. 1.8B). Lattice dislocations can be found inside some grains with a density that is lower than after four ECAP passes. Azimuthal scattering of spots in the SAED patterns and a complex diffraction contrast in the bright-field TEM images testify to considerable internal stresses (Fig. 1.8B).

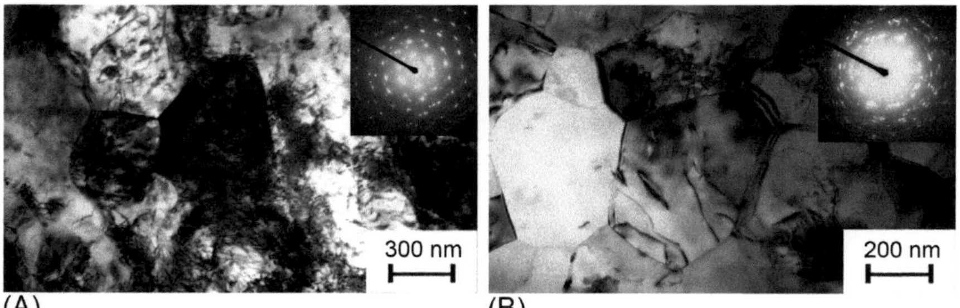

FIG. 1.8

Microstructure of Ti Grade 4 after ECAP, four passes (A) and eight passes (B), cross-section, TEM [32].

One of the important parameters of the UFG structure formed during SPD is the crystallographic texture. An experimental study on the character of the distribution of preferred crystallographic orientations was performed by X-ray analysis. In order to perform texture studies, samples with sizes of $1 \times 10 \times 10\,mm^3$ were cut in the plane of the longitudinal section X–Y that passes through the geometrical center of the cylinder billet (Fig. 1.9).

A study of the crystallographic texture was performed by X-ray analysis with the diffractometer DRON-3 equipped with a texture device. Cu–K radiation was used to scan the Bragg reflections in the radial angle range from 0 to 80 degrees and azimuthal range from 0 to 360 degrees. POPLA software was used to analyze the data. As a result, the pole figures (PF) were obtained for Ti in the initial CG state and the state after ECAP via route B_C with the number of passes equal to 1 and 8 (Fig. 1.10). Fig. 1.10 displays the general view of the crystalline texture in Ti before and after ECAP. This is testified by asymmetrically located high-level texture maxima. One may identify prismatic orientations $\langle 101 \rangle$ $\{110\}$ and twin orientations typical of coarse crystalline Ti.

During the analysis of PF for Ti after the first cycle of ECAP (Fig. 1.9B) it was stated that this structure can be described as an aggregate of the main texture components and the so-called orientations of the basal type. The texture maxima corresponding to these components are well observed on PF (0001). It is revealed that after the first pass, the basal orientation prevails over the prismatic and pyramidal orientations. The distribution of preferred crystallographic orientations is typical of Ti processed by ECAP [27] and is asymmetric which demonstrates the asymmetry of

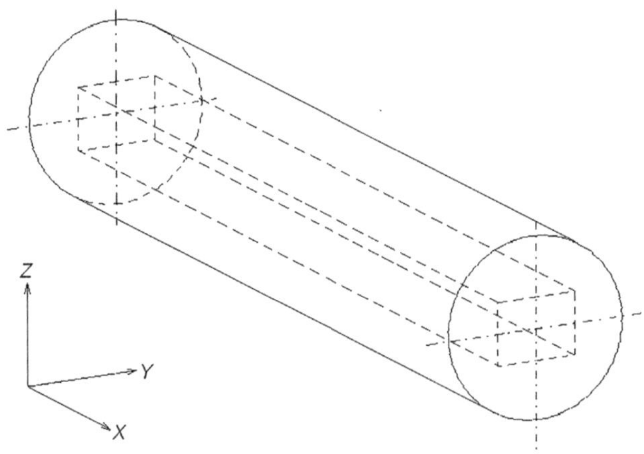

FIG. 1.9

The scheme of a billet cut for X-ray analyses.

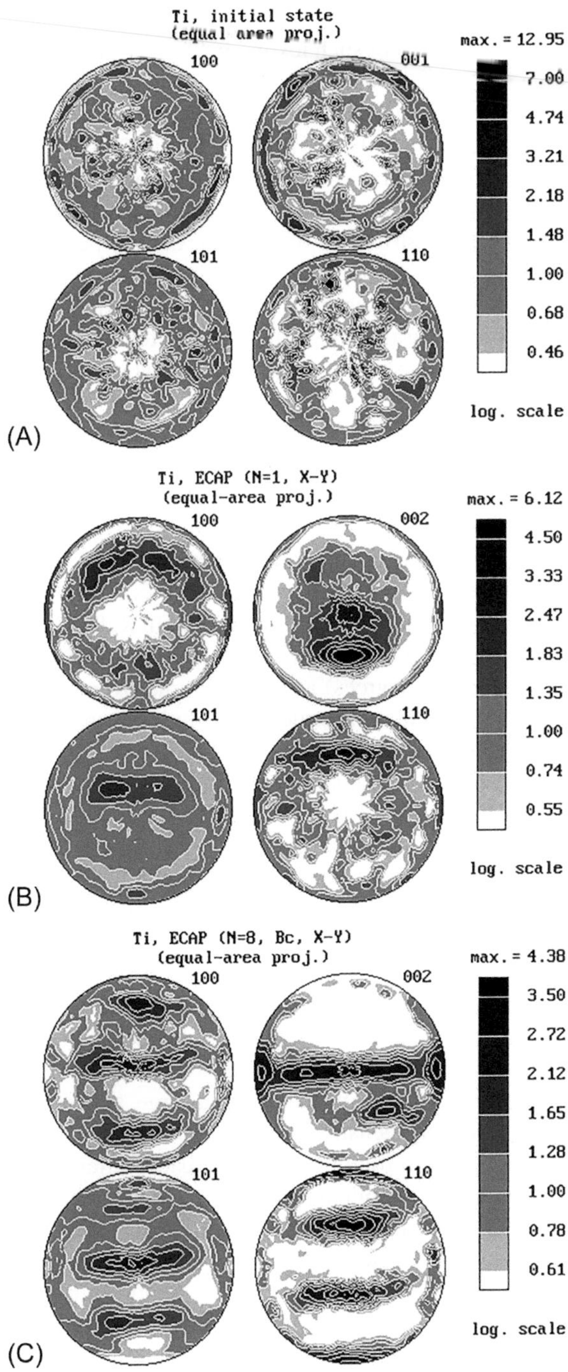

FIG. 1.10

Experimental pole figures Ti Grade 4 in various structural states for *X–Y* section in initial state (A), after one pass (B) and eight passes (C) of ECAP.

the deformation. The formation of such texture in Ti can be explained by the limited number of possible slip systems during ECAP.

An analysis of the texture states formed after eight passes (Fig. 1.10C) points to the main orientations: {0001} ⟨110⟩, {101} ⟨110⟩, and {100} ⟨110⟩. These two latter are on the whole the most typical ones for hcp Ti (with the ratio of lattice parameters $c/a = 1.587$) and testify to the occurrence of pyramidal and prismatic slip systems of dislocations. However, the distribution of the polar density points to the basal component of {0001} ⟨110⟩ as the strongest compared with other orientations. This fact characterizes the work of prevailing basal slip systems during ECAP via route B_C. Thus, a contribution of the basal component in the texture state produced by ECAP prevails, and the work of the basal slip systems of dislocations is important during the formation of this texture state.

One of the new modifications of ECAP techniques is ECAP-Conform (ECAP-C). This method permits fabrication of a UFG structure in long-length rods and provides high production capacity and high perspectives for industrial applications [34–36].

As was shown earlier, Ti Grade 4 and Grade 2 are usually processed by ECAP at a temperature of 400–450°C in the die set with an angle of channel intersection $\Phi = 90$ degrees because at lower temperatures there is an appearance of surface defects and cracking which leads to billet failure [28]. In Ref. [37] ECAP-C was performed at 200°C which was connected with a decrease of strain intensity (angle of channel intersection $\Phi = 120$ degrees) and with an improvement of the process tribology. TEM results showed that the structure evolution increased with the number of passes and this is typical also of conventional ECAP. For example, it is characterized by an increase of the total dislocation density and the formation of a larger fraction of high angle boundaries. Fig. 1.11C and D shows that a UFG structure of a granular type mostly forms in Ti after 10 passes, where this is equivalent to $e = 7.0$.

In Ref. [37] it is shown also that in the course of ECAP-C a crystallographic texture forms in a rod, which can be described as a combination of orientations of pyramidal {012} ⟨100⟩ and prismatic {100} ⟨010⟩ components (Fig. 1.12). During ECAP-C the slip along the basal planes is hindered, which can be associated with the lower processing temperatures of titanium compared to conventional ECAP. In addition, TEM is typically used to study the substructure but it cannot provide sufficient quantitative data. To obtain comprehensive information of the microstructural evolution and strengthening mechanisms, a large amount of information is needed to cover the boundary density, types, misorientation, and crystallographic textures. By contrast, high-resolution electron backscatter diffraction (EBSD) appears to be a better choice. Despite its limited angular accuracy (~2 degrees), this technique permits a quantification of microstructural features over a much larger scale than TEM and thus provides insights that may be more significant from a statistical point of view. An earlier report [38] demonstrated the feasibility of high-accuracy EBSD for UFG titanium deformed by ECAP-C. The results of these measurements were used to elucidate the contributions of various hardening mechanisms to the overall material strength. This aspect is described in more detail in Chapter 7.

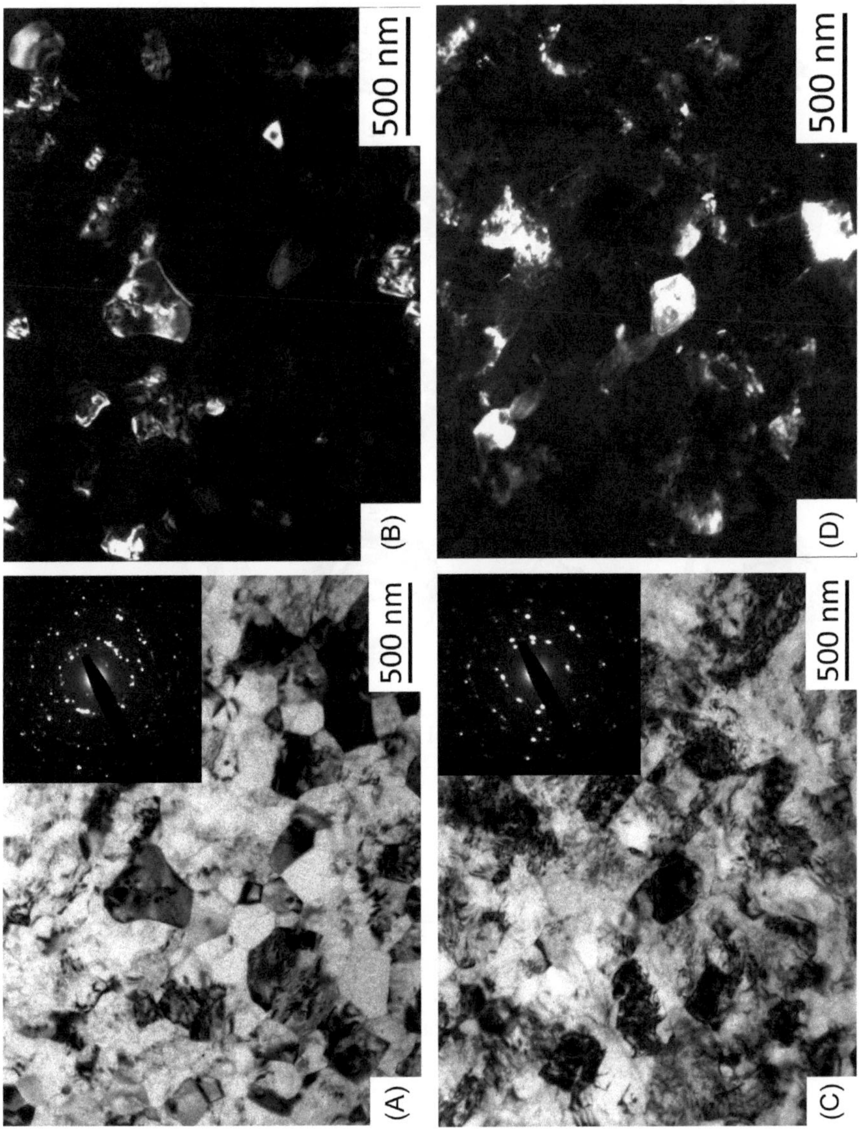

FIG. 1.11

TEM images of Ti Grade 4 after 8 (A, C) and 10 (B, D) ECAP-C passes in transverse sections [37].

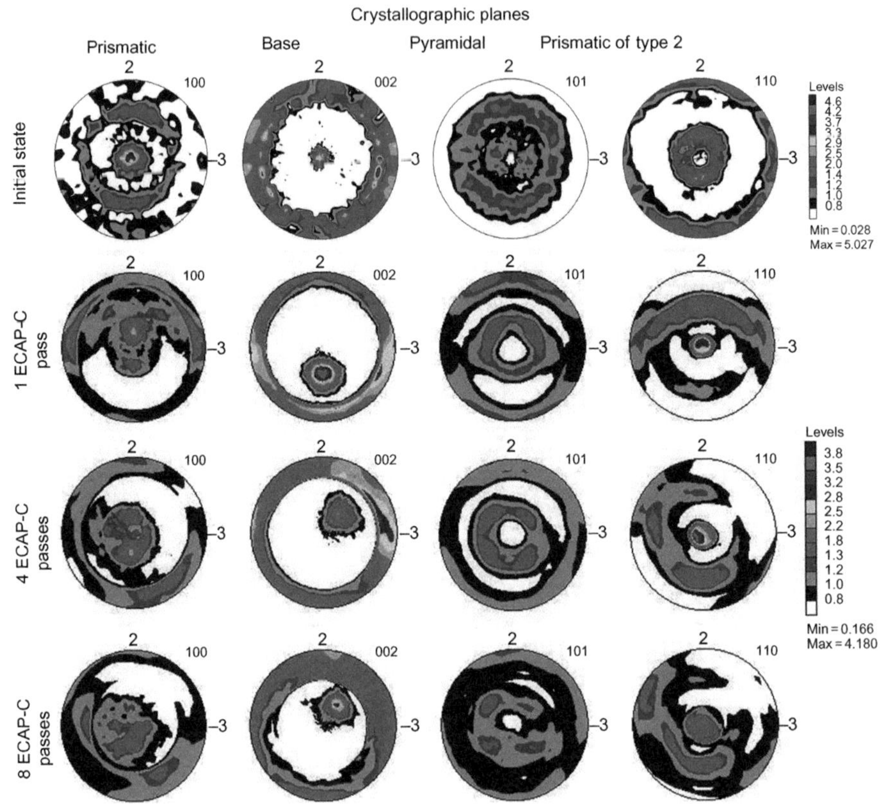

FIG. 1.12

Experimental pole figures (1010), (0002), (101), and (110) Ti Grade 4 in various structural states [37].

1.3 SUMMARY

In generalizing the above results, one can say that a principal possibility for UFG structure formation by HPT and ECAP in pure Ti is confirmed by many researchers. It is shown that the scheme, route, temperature, strain rate, applied pressures, initial grain size, and impurity content all affect the UFG structure parameters in Ti. Additionally:

1. HPT has an advantage over other SPD techniques due to the possibility of application of very high pressures at room temperature, which results in the formation of grains <80 nm. The UFG structure formed in Ti has features

common with other metals subjected to HPT such as high dislocation density, enhanced internal stresses, high angle misorientations of grain boundaries.

2. The ECAP and ECAP-C techniques have certain restrictions, as they require a temperature increase, which ensures a UFG structure formation with a grain size of over 250 nm in pure Ti. The transformation of low-angle boundaries into high angles is achieved when the accumulated logarithmic strain degree is over 4.0.
3. It is shown that in Ti with hcp lattice the dislocation density during plastic deformation is hindered, therefore, twinning plays a significant role in structure refinement especially at the early stages of deformation. While the strain degree further increases, twins play an important role in structure fragmentation and in the formation of new grain boundaries.

ACKNOWLEDGMENTS

The work of R.Z.V. was supported in part by Saint Petersburg State University in the framework of Call 3 project (id 26130576) and in part by the RF Ministry for Education and Science under Grant agreement No. 14.586.21.0061 (unique project identifier RFMEFI58618X0061). I.P.S. thanks the Ministry of Education and Science of the Russian Federation within the scopes of the basic part of the state assignment for educational institutions (Grant No. 16.5979.2017).

REFERENCES

[1] H. Gleiter, Nanocrystalline materials, Prog. Mater. Sci. 33 (1989) 223–315, https://doi.org/10.1016/0079-6425(89)90001-7.

[2] I.V. Alexandrov, Y.T. Zhu, T.C. Lowe, R.K. Islamgaliev, R.Z. Valiev, Microstructures and properties of nanocomposites obtained through SPTS consolidation of powders, Metall. Mater. Trans. A 29 (1998) 2253–2260, https://doi.org/10.1007/s11661-998-0103-4.

[3] R.Z. Valiev, R.K. Islamgaliev, I.V. Alexandrov, Bulk nanostructured materials from severe plastic deformation, Prog. Mater. Sci. 45 (2000) 103–189, https://doi.org/10.1016/S0079-6425(99)00007-9.

[4] R.Z. Valiev, Nanostructuring of metals by severe plastic deformation for advanced properties, Nat. Mater. 3 (2004) 511–516, https://doi.org/10.1038/nmat1180.

[5] R.Z. Valiev, Y. Estrin, Z. Horita, T.G. Langdon, M.J. Zehetbauer, Y.T. Zhu, Producing bulk ultrafine-grained materials by severe plastic deformation, JOM 58 (4) (2006) 33–39, https://doi.org/10.1007/s11837-006-0213-7.

[6] R.Z. Valiev, T.G. Langdon, The art and science of tailoring materials by nanostructuring for advanced properties using SPD techniques, Adv. Eng. Mater. 12 (8) (2010) 677–691, https://doi.org/10.1002/adem.201000019.

[7] A.A. Popov, I.Y. Pyshmintsev, S.I. Demakov, A.G. Illarionov, T.C. Lowe, R.Z. Valiev, Structural and mechanical properties of nanocrystalline titanium processed by severe deformation processing, Scr. Mater. 37 (1997) 1089–1094, https://doi.org/10.1016/S1359-6462(97)00210-8.

[8] A.V. Sergueeva, V.V. Stolyarov, R.Z. Valiev, A.K. Mukherjee, Advanced mechanical properties of pure titanium with ultrafine grained structure, Scr. Mater. 45 (2001) 747–752, https://doi.org/10.1016/j.prostr.2017.11.030.

[9] R.Z. Valiev, A.V. Sergueeva, A.K. Mukherjee, The effect of annealing on tensile deformation behavior of nanostructured SPD titanium, Scr. Mater. 49 (2003) 669–674, https://doi.org/10.1016/S1359-6462(03)00395-6.

[10] D.B. Williams, C.B. Carter, Transmission Electron Microscopy: A Textbook for Materials Science II, Plenum Press, New York, 1996, 260.

[11] A.V. Korznikov, I.M. Safarov, D.V. Laptionok, R.Z. Valiev, Structure and properties of superfine-grained iron compacted out of ultradisperse powder, Acta Metall. Mater. 39 (1991) 3193–3197, https://doi.org/10.1016/0956-7151(91)90054-5.

[12] R.Z. Valiev, Processing of nanocrystalline materials by severe plastic deformation consolidation, in: D.L. Bourell (Ed.), Synthesis and Processing of Nanocrystalline Powder, The Minerals, Metals and Materials Society, Warrendale, PA, 1996, pp. 153–161.

[13] I. Semenova, I. Timokhina, R. Islamgaliev, E. Lavernia, R. Valiev, Nanocrystalline Ti produced by cryomilling and consolidation by severe plastic deformation, Metals 5 (2015) 206–215, https://doi.org/10.3390/met5010206.

[14] H. Wen, Y. Zhao, Y. Liu, O. Torera, K.M. Nesterov, R.K. Islamgaliev, R.Z. Valiev, E. J. Lavernia, High-pressure torsion-induced grain growth and detwinning in cryomilled Cu powders, Philos. Mag. 90 (2010) 4541–4550, https://doi.org/10.1080/14786435.2010.514579.

[15] V.M. Segal, Materials processing by simple shear, Mater. Sci. Eng. A 197 (1995) 157–164, https://doi.org/10.1016/0921-5093(95)09705-8.

[16] R.Z. Valiev, N.A. Krasilnikov, N.K. Tsenev, Plastic deformation of alloys with submicron-grained structure, Mater. Sci. Eng. A 137 (1991) 35–40, https://doi.org/10.1016/0921-5093(91)90316-F.

[17] Y. Iwahashi, J. Wang, Z. Horita, M. Nemoto, T.G. Langdon, Principle of equal-channel angular pressing for the processing of ultrafine-grained materials, Scr. Mater. 35 (1996) 143–146, https://doi.org/10.1016/1359-6462(96)00107-8.

[18] R.Z. Valiev, T.G. Langdon, Principles of equal-channel angular pressing as a processing tool for grain refinement, Prog. Mater. Sci. 51 (2006) 881–981, https://doi.org/10.1016/j.pmatsci.2006.02.003.

[19] Y. Iwahashi, Z. Horita, M. Nemoto, T.G. Langdon, An investigation of microstructural evolution during equal-channel angular pressing, Acta Mater. 45 (11) (1997) 4733–4741, https://doi.org/10.1016/S1359-6454(97)00100-6.

[20] Y. Iwahashi, M. Furukawa, Z. Horita, M. Nemoto, T.G. Langdon, Microstructural characteristics of ultrafine-grained aluminum produced using equal-channel angular pressing, Metall. Mater. Trans. A 29 (9) (1998) 2245–2252, https://doi.org/10.1007/s11661-998-0102-5.

[21] Y. Iwahashi, Z. Horita, M. Nemoto, T.G. Langdon, The process of grain refinement in equal-channel angular pressing, Acta Mater. 46 (9) (1998) 3317–3331, https://doi.org/10.1016/S1359-6454(97)00494-1.

[22] F.Z. Utyashev, F.U. Enikeev, V.V. Latysh, Comparison of deformation methods for ultrafine-grained structure formation, Ann. Chim 21 (6–7) (1996) 379–389.

[23] D. Yamaguchi, Z. Horita, M. Nemoto, T.G. Langdon, Significance of adiabatic heating in equal-channel angular pressing, Scr. Mater. 41 (1999) 791–796, https://doi.org/10.1016/S1359-6462(99)00233-X.

[24] H.S. Kim, Prediction of temperature rise in equal channel angular pressing, Mater. Trans. 42 (2001) 536–538, https://doi.org/10.2320/matertrans.42.536.
[25] R. Boyer, G. Welsch, E. Collings (Eds.), Materials Properties Handbook: Titanium Alloys, ASM International, 1998, p. 1048.
[26] G. Purcek, G.G. Yapici, I. Karaman, H.J. Maier, Effect of commercial purity levels on the mechanical properties of ultrafine-grained titanium, Mater. Sci. Eng. A 528 (2011) 2303–2308, https://doi.org/10.1016/j.msea.2010.11.021.
[27] D.H. Shin, I. Kim, J. Kim, S.L. Semiatin, Microstructure development during equal channel angular pressing of titanium, Acta Mater. 51 (2003) 983–996, https://doi.org/10.1016/S1359-6454(02)00501-3.
[28] V.V. Stolyarov, Y.T. Zhu, I.V. Alexandrov, T.C. Lowe, R.Z. Valiev, Influence of ECAP routes on the microstructure and properties of pure Ti, Mater. Sci. Eng. A 299 (2001) 59–67, https://doi.org/10.1016/S0921-5093(00)01411-8.
[29] G.I. Raab, E.P. Soshnikova, R.Z. Valiev, Influence of temperature and hydrostatic pressure during equal-channel angular pressing on the microstructure of commercial-purity Ti, Mater. Sci. Eng. A 387–389 (2004) 674–677, https://doi.org/10.1016/j.msea.2004.01.137.
[30] X. Zhao, W. Fu, X. Yang, T.G. Langdon, Microstructure and properties of pure titanium processed by equal-channel angular pressing at room temperature, Scr. Mater. 59 (2008) 542–545, https://doi.org/10.1016/j.scriptamat.2008.05.001.
[31] I. Kim, J. Kim, D.H. Shim, K.T. Park, Effects of grain size and pressing speed on the deformation mode of commercially pure Ti during equal channel angular pressing, Metall. Mater. Trans A 34 (2003) 1555–1558, https://doi.org/10.1007/s11661-003-0267-x.
[32] I.P. Semenova, G.H. Salimgareeva, V.V. Latysh, R.Z. Valiev, Nanostructuring of Ti in long-sized rods by severe plastic deformation, Proceedings of the 8th ESAFORM Conference on Material Forming, Cluj-Napoca, Romania, vol. II, 2005, pp. 657–660.
[33] I.P. Semenova, V.V. Latysh, G.H. Sadikova, Y.T. Zhu, T.C. Lowe, R.Z. Valiev, Microstructure and mechanical properties of long, ultrafine-grained Ti rods, in: Y.T. Zhu, V. Varyukhin (Eds.), Proceedings of the NATO Advanced Study Institute on Nanostructured Materials by High-Pressure Severe Plastic Deformation, II. Mathematics, Physics and Chemistry212 2006, pp. 235–241, https://doi.org/10.1007/1-4020-3923-9_32.
[34] G.J. Raab, R.Z. Valiev, T.C. Lowe, Y.T. Zhu, Continuous processing of ultrafine grained Al by ECAP-Conform, Mater. Sci. Eng. A 382 (2004) 30–34, https://doi.org/10.1016/j.msea.2004.04.021.
[35] C. Xu, S. Schroeder, P.B. Berbon, T.G. Langdon, Principles of ECAP-Conform as a continuous process for achieving grain refinement: application to an aluminum alloy, Acta Mater. 58 (2010) 1379–1386, https://doi.org/10.1016/j.actamat.2009.10.044.
[36] Y.T. Zhu, T.C. Lowe, R.Z. Valiev, G.I. Raab, US Patent No. 7, 152, 448, USC 72/262, Continuous Equal-Channel Angular Pressing, Los Alamos National Security, LLC (Los Alamos, NM), 2006, p. 26. filed December 16, 2004, issued December.
[37] A.V. Polyakov, I.P. Semenova, G.I. Raab, V.D. Sitdikov, R.Z. Valiev, Peculiarities of ultrafine-grained structure formation in Ti grade-4 using ECAP-Conform, Rev. Adv. Mater. Sci. 31 (16) (2012) 78–84.
[38] G.S. Dyakonov, S. Mironov, I.P. Semenova, R.Z. Valiev, S.L. Semiatin, Microstructure evolution and strengthening mechanisms in commercial-purity titanium subjected to equal-channel angular pressing, Mater. Sci. Eng. A 701 (2017) 289–301, https://doi.org/10.1016/j.msea.2017.06.079.

CHAPTER

Combined processing ECAP + TMP

2

Irina P. Semenova*, Georgy I. Raab*, Ruslan Z. Valiev*,†

*Institute of Physics of Advanced Materials, Ufa State Aviation Technical University, Ufa, Russia**
Laboratory for Mechanics of Bulk Nanostructured Materials, Saint Petersburg State University, Saint Petersburg, Russia†

2.1 DRAWING AND ROLLING OF ECAPed-Ti

Billets and semi-items produced by high-pressure torsion (HPT) and equal-channel angular pressing (ECAP) usually are small in size and this restricts their application in industry. One of the promising approaches to production of semi-items and items with required sizes is the combination of ECAP and its modifications with conventional techniques of thermomechanical processing (TMP), for example, drawing and rolling. The formation of ultrafine-grained (UFG) structures in Ti can lead to qualitative changes in the structure formation processes during subsequent deformation. For example, the UFG structure formed during ECAP changes the structure evolution processes during subsequent drawing. In particular, it becomes possible for such deformation mechanisms as grain boundary sliding to operate, and this in turn promotes an increase in the material deformability and additional refinement of the UFG structure. The combination of ECAP with subsequent plastic deformation can be another procedure for controlling the material structure and its mechanical properties.

Numerous data on structure refinement by different deformation techniques show that the scale factor and deformation conditions considerably affect the UFG structure formation and final grain size. Thus, a structure with a grain size of 80–120 nm forms in thins plates after HPT, and in a bulk billet with a diameter of 20–40 mm after ECAP the grain size is 250–500 nm in the same material [1,2]. Moreover, the initial geometry actively affects the size of structural elements when using one processing technique. Thus, during ECAP with one level of accumulated strain and with all other conditions being equal, one cannot produce the structure with the same grain size for samples with diameters of 10 and 40 mm. It is established that the main reason of such behavior is differences in the geometrical parameters of the deformation zone, which are formalized via the specific surface parameter (Ay)

Nanocrystalline Titanium. https://doi.org/10.1016/B978-0-12-814599-9.00002-X

of the deformation zone and its increment (ΔAy) [2]. The specific surface of the deformation zone, calculated as the ratio between the initial areas of the zone to its volume, characterizes the scale factor, and its increment during plastic treatment characterizes the intensity of deformation impact and the intensity of the initial structure fragmentation. In practice, these criteria can be calculated and may be depicted as triple diagrams (Fig. 2.1). Using the known ratios, the structure parameters of billets after processing are subjected to estimation and prediction, in particular, using the Halt ratio, one can calculate the grain size after severe plastic deformation [4]. The analysis of the plotted diagrams leads to the conclusion that an enhancement of the absolute values of Ay and ΔAy, respectively, increases the strength and reduces the grain size. Therefore, if one controls the criteria Ay and ΔAy, during processing, one can affect the size of formed fragments (grains) and strength characteristics of the materials after processing. For example, such conditions can be implemented in combinations of different techniques such as ECAP+rolling and/or drawing.

A practical test of the proposed criteria was performed using the example of a combined processing of commercially pure Ti by ECAP+drawing at temperatures lower than the recrystallization temperature. During ECAP when ΔAy and Ay remain constant, the process of strength accumulation and structure refinement damping is observed. In case the criterion Ay grows considerably from 0.6 to 3.2, for example, and at the subsequent drawing an intensive increase in strength and further structural refinement are observed (Fig. 2.2).

As a result of the use of combined processing techniques that enable actively controlling the scale factor parameters, for example, using ECAP and subsequent deformation treatments, it is possible to produce high strength rod-shaped items with a diameter of 20–25 mm and lengths of up to 1.5–2 m (Fig. 2.3). Using the combined processing that includes ECAP or ECAP-Conform and subsequent drawing, rolling

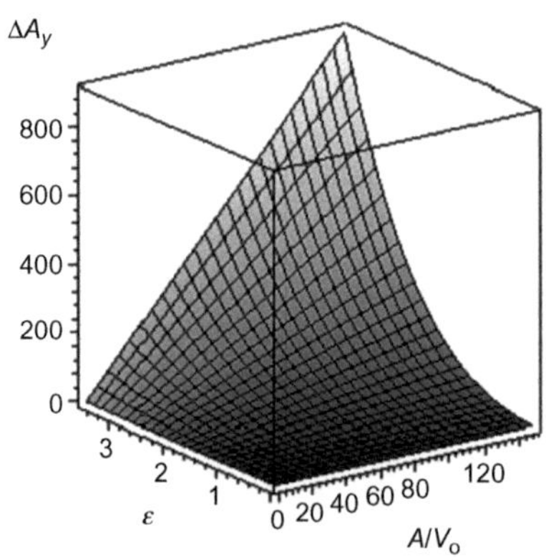

FIG. 2.1

Diagram of interrelations between the increment of the deformation nucleation specific surface area ΔAsp (in mm^{-1}), the deformation nucleation specific surface area (in mm^{-1}), and true deformation degree ε [3].

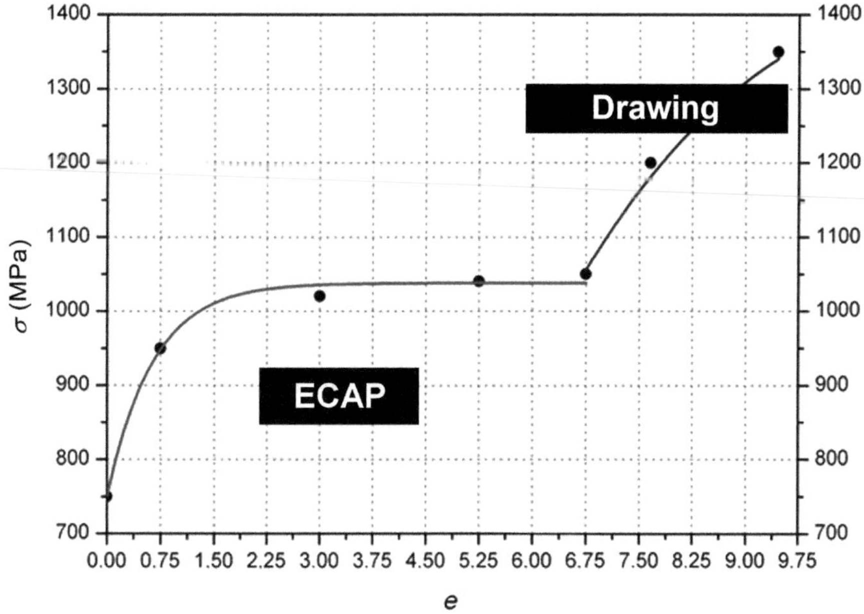

FIG. 2.2

Changes in the ultimate strength in the deformation process of the workpiece made of pure titanium by the ECAP ($\alpha = 120$ degrees, route B_C, A_{sp} const.) with consecutive drawing ($T = 140 \pm 10°C$, A_{sp} var.) with the level of accumulated deformation [3].

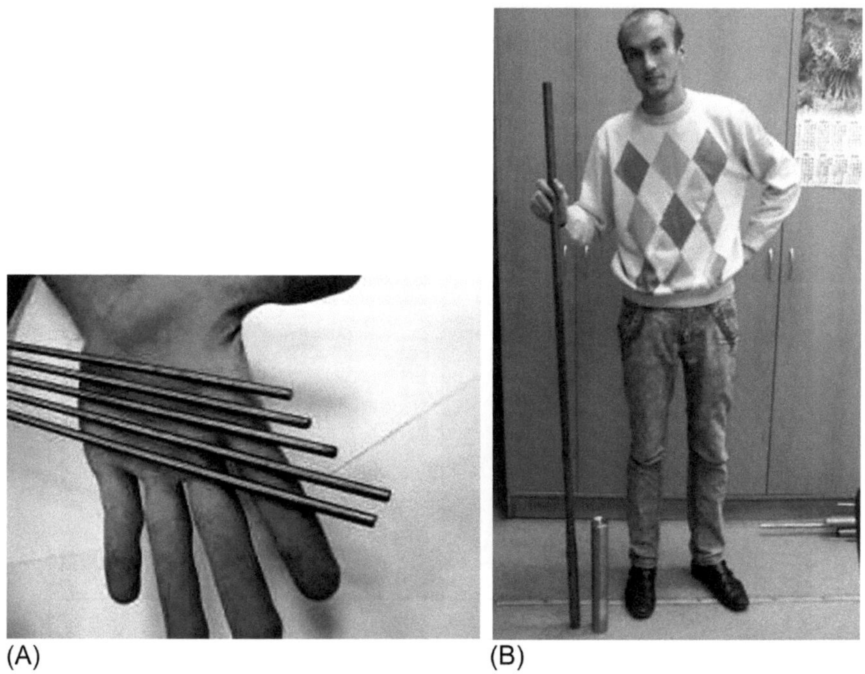

FIG. 2.3

Elongated rods made of (A) nanostructured pure titanium; (B) nanostructured alloy Ti–6Al–4V [3].

is much more promising for producing commercial batches of semi-items and items. Thus, an approach will allow the production of commercially successful rods of 5–12 m long and wire with high strength from Ti and its alloys [3].

2.2 MICROSTRUCTURE AND TEXTURE

2.2.1 MICROSTRUCTURE OF Ti AFTER ECAP AND DRAWING/ROLLING

In Refs. [5–7] the role of ECAP in combination with cold rolling is described. In particular, under cold rolling of conventional coarse-grained Ti, the structural refinement occurs mainly due to twinning (Fig. 2.4). The structure contains twins (over 500 nm wide) lying in the plane and with enhanced dislocation density. As the rolling degree increases (from 15% to 88%), the subgrain structure transforms and mainly low-angle grain boundaries are then formed. As a consequence, the structure refinement after cold rolling leads to a strength increase and considerable ductility reduction.

A combination of ECAP and cold rolling [6,7,8] in Ti Grade 2 (VT1-0) leads to more significant structural refinement at the stage of ECAP, and in the cross-section of the rod after rolling the grain size can achieve 120 nm, with a shape that is close to equiaxed. However, in the longitudinal section of the billet the grains are elongated along the rolling direction, they are fragmented and contain a large number of dislocations and other defects of the crystalline lattice, mainly low-angle grain boundaries. In the rods processed by ECAP and subsequent cold rolling, it is possible to achieve a very high strength which exceeds the strength of conventional Ti by almost two times but the ductility is very limited (9%) and this can negatively affect its structural properties and in particular the fatigue resistance.

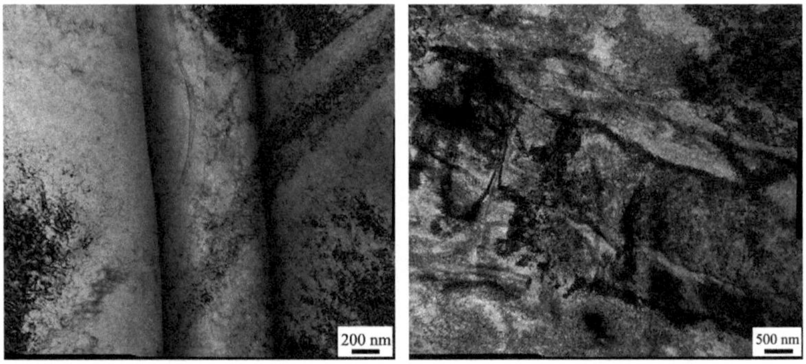

(A) (B)

FIG. 2.4

Microstructures of coarse grained Ti Grade 2 after rolling: 40% (A), 80% (B).

In Figs. 2.5 and 2.6 the peculiarities of UFG structure transformation are demonstrated using Ti Grade 4 with a diameter of 25 mm during subsequent drawing to a diameter of 7 mm. The drawing is performed at room temperature with intermediate annealing not higher than 300°C to ensure technological ductility. The total accumulated strain degree is about 75%. Fig. 2.5 displays the images of the rod microstructure in the longitudinal sections before (Fig. 2.5A) and after final drawing

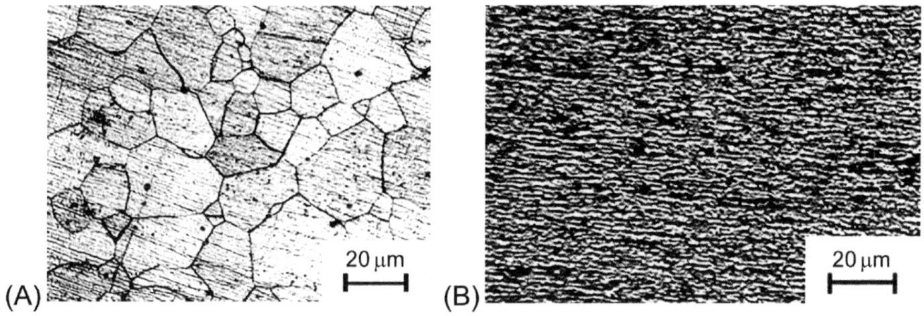

FIG. 2.5

(A) The initial microstructure of Ti Grade 4 (∅25 mm) and (B) after ECAP and drawing to ∅7. The longitudinal section of the rod.

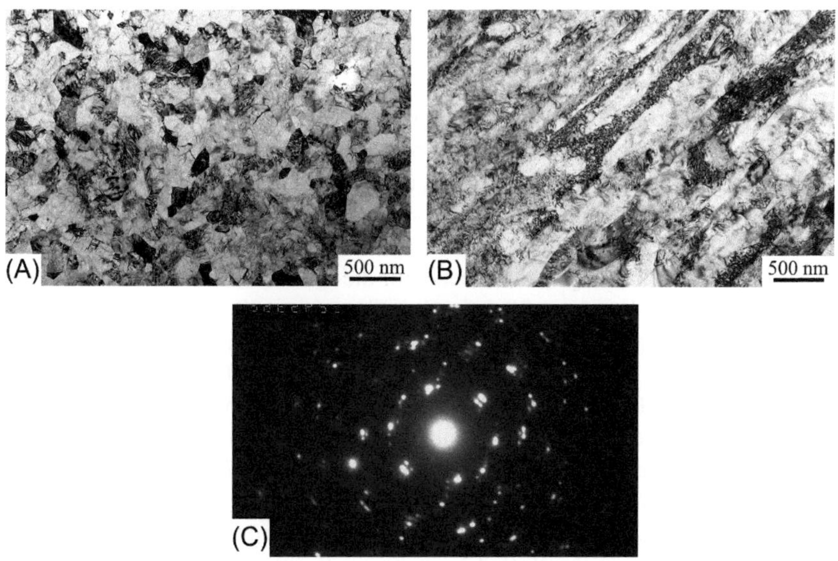

FIG. 2.6

Microstructure by TEM of the UFG Ti Grade 4 after ECAP and drawing with accumulated strain 75%; (A) transverse section; (B) longitudinal section; (C) diffraction pattern to (B).

(Fig. 2.5B). In the microstructure image (Fig. 2.5B), the metallographic texture formation along the deformation direction is well observed in the longitudinal section of the rod. The texture is characterized by a strong elongation of grains of the α-phase. Fig. 2.6 presents bright-field images of the Ti Grade 4 microstructure after ECAP +drawing (rods 7mm) characterizing the areas of the studied samples cut from the rod. A statistical analysis of the TEM images of the rod microstructure in the transverse section shows that almost 70% of the volume is occupied by grains with a size from 0.15 to 0.20 μm (Fig. 2.6A). Based on the diffraction patterns with many spots on concentric circles, the boundaries of grains/subgrains have high angle misorientations. The merging of separate spots about the diffraction circle, typical of the structure in the longitudinal section, points to low-angle boundaries of subgrains and fragments found inside elongated grains (Fig. 2.6B and C).

It is known that warm deformation contributes to the formation of a microstructure with more equilibrium boundaries and an internal low dislocation density. In Ref. [9] warm rolling (WR) is used in the temperature range 300–350°C without intermediate annealing to 87%. The average strain degree per pass is about 10% which ensures good structural processing of a Ti Grade 4 rod.

The TEM images of Ti microstructures after WR display grain boundaries and the spots located uniformly along the concentric circles on the diffraction patterns, this suggests the formation of grains with mainly high angle boundaries (Fig. 2.7A). There are also considerable differences in the microstructure when the deformation temperatures change. It was reported that a temperature increase to 350°C during rolling contributes to the formation of thinner equilibrium grain boundaries with a reduced density of lattice dislocations. On average, the size of grains and subgrains in the rod microstructure is 120nm. Judging by the misorientation spectra obtained by the EBSD technique after WR, the fraction of high angle grain boundaries (HAGBs) is 72% (Fig. 2.7C) [9]. A comparative analysis of the grain shape in the microstructure of the longitudinal section of the rod shows an increase in the fraction of grains with an elongation factor $f_g \geq 2$ up to 60% after WR as compared to 10% after drawing. Thus, the grain shape becomes more equiaxed.

A quantitative analysis of the structures produced via the two combined technologies shows that in the longitudinal section of the rod after ECAP and WR, the fraction of fragments and grains with a size L in the range 600–200nm increased from 10% to 60% as compared to the structure after ECAP and drawing (Table 2.1). The size of the structure elements was estimated with the use of dark-field images.

Measurements of texture formation processes were performed using the diffractometer DRON-3M equipped with an automated texture goniometer. When the pole figures (PFs) are exposed, filtered X-ray radiation with Cu $K1$ (0.1540598nm) was used and the exposure in the reflection mode was performed within the change of the radial angle from 0 to 75 degrees and azimuthal angle from 0 to 360 degrees. As a result, a set of intensities of reflected X-ray beams $P_{hl}(y)$ was obtained. The diameter of the radiated area was 0.6mm. The study was performed in the geometric center of the cross-section of the billet (plane 2(−3)) (Fig. 2.8). The orientation distribution

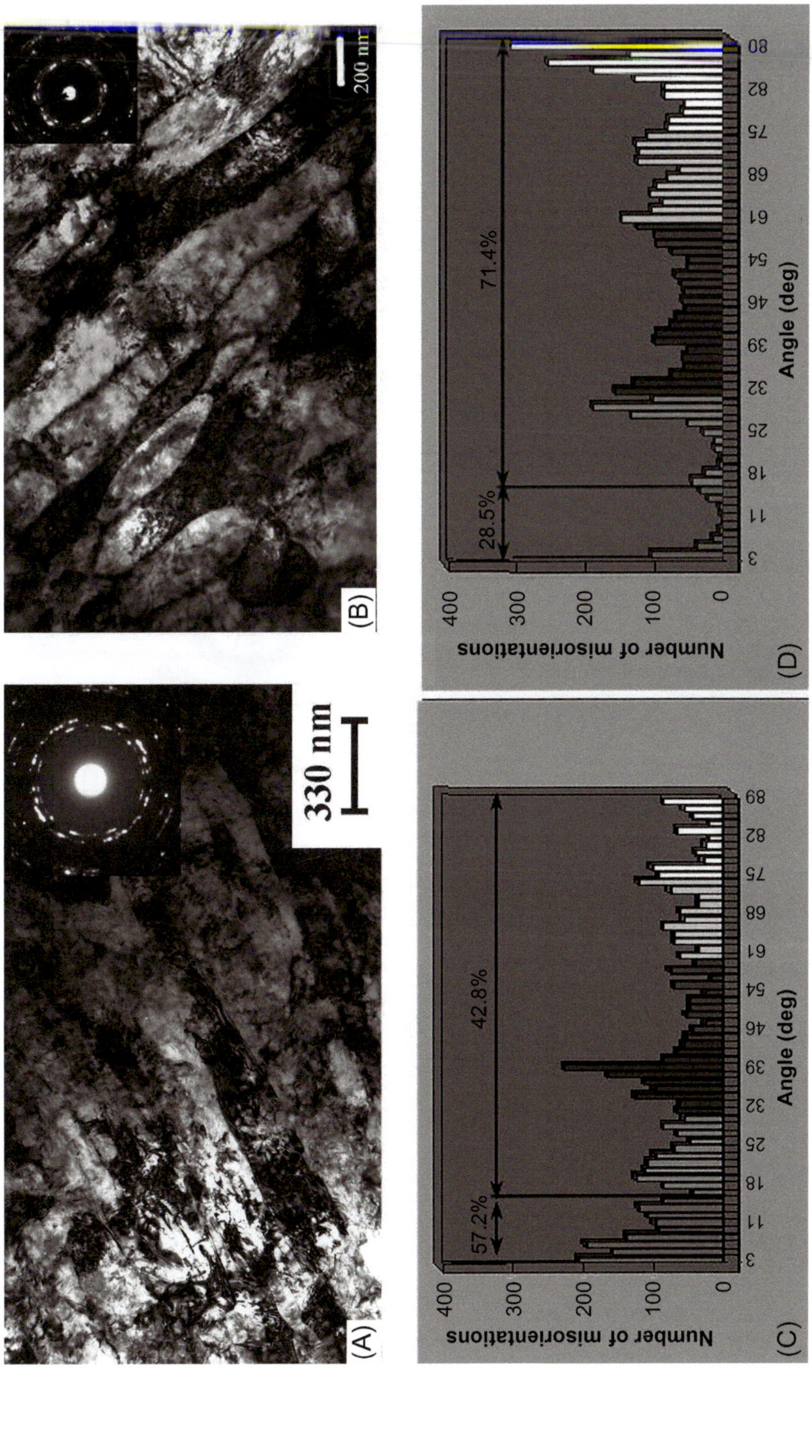

FIG. 2.7

Rod microstructures in the longitudinal section after ECAP + D (A, C) and after ECAP + WR (B, D). A, B—TEM, light-field image; C, D—SEM, EBSD analysis [9].

Table 2.1 Distribution of Fraction of Elongated Grains Over Sizes

L (nm)	ECAP and WR (%)	ECAP and Drawing (%)
800–600	30	83
600–200	60	10
<200	10	7

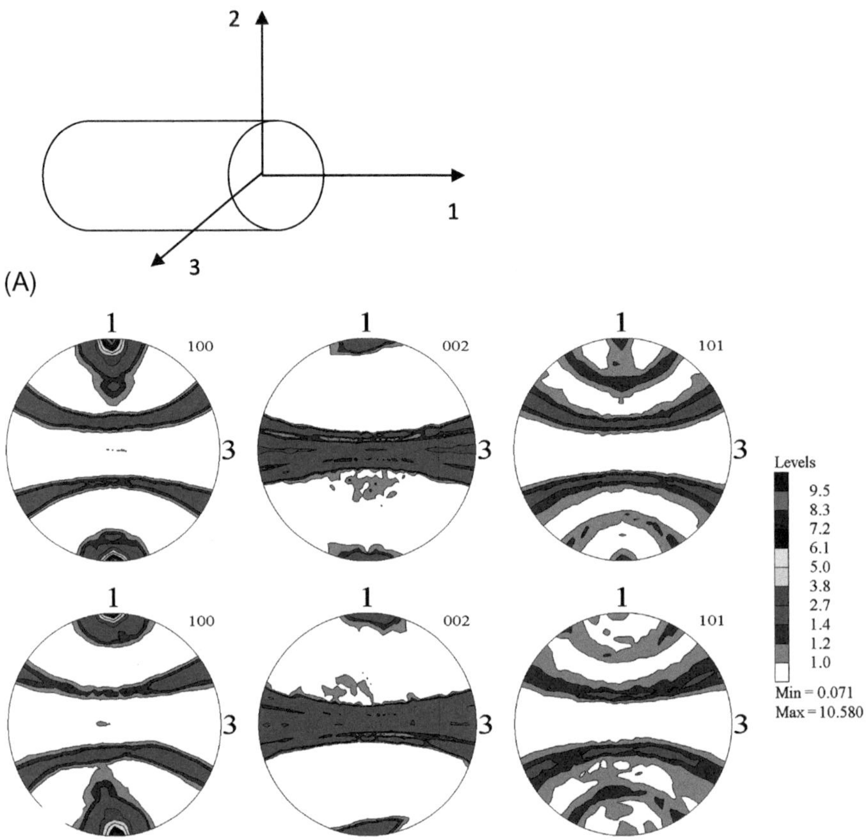

FIG. 2.8

(A) Scheme of direction at X-ray analyses; (B) experimental PF (1010), (0002), and (1011): *above*—ECAP + CR and *below*—ECAP + D in plane 1–3.

function (ODF) and complete PF are plotted as a result of the exposure of incomplete PF for planes (101), (100), and (0002). The results of the experimental studies were presented as complete PF and ODF calculated with the use of the LaboTEX software. Ti Grade 4 samples produced by ECAP+drawing and ECAP+WR were used for comparative studies. The experimental complete PF of Ti in both states is displayed in Fig. 2.8B. The crystallographic texture formed in Ti as a result of ECAP+WR and ECAP+drawing are quite similar. This is confirmed by the positioning of the texture maxima on the PF (100), (0002), and (101).

In order to more clearly interpret the experimental data, the PF were turned by 90 degrees against axis 3 counterclockwise so that axis 1 coincided with the direction of rolling or drawing. The turned PF are displayed in Fig. 2.8B in plane 1–3.

Based on the pole figures $(10\overline{1}0)$, (0002), and $(10\overline{1}1)$, during implementation of the mentioned deformation schemes, the highest pole density is observed in the texture components corresponding to the prismatic systems $\{10\overline{1}0\}\ \langle 0\overline{1}11\rangle$. Apparently, the dislocation slip activity is hindered along the basal and pyramidal systems which are supported by the lower density of their texture maxima.

In the process of implementation of the above deformation schemes, the prismatic slip systems have the highest volume fraction of orientations and are more favorably oriented for the deformation to take place along axis 1 (Fig. 2.8A). In addition, these systems have the lowest critical shear stresses. This apparently testifies to the highest activity of prismatic systems $\{100\}\ \langle 011\rangle$ during implementation of the deformation schemes. At the same time, the dislocation slip activity is hindered along the basal and pyramidal slip systems but nevertheless it is implemented. This is supported by the estimated values of their volume fractions.

In our laboratory in Ufa we developed a new approach to enhance the strength and ductility by controlling the grain boundary (GB) structure that can be considered as GB engineering of UFG metals and alloys [10,11]. In particular, for UFG Ti one of the effective approaches to ductility enhancement, as shown in Refs. [12,13], is formation of a high fraction of HAGBs in the UFG structure. High -angle boundaries promote the mechanisms of grain boundary sliding and/or contribute to strain hardening during plastic deformation, which results in ductility enhancement. Such studies have been recently carried out using HPT and ECAP [11–13].

In work [14] this approach is developed for Ti Grade 4 involving nanostructuring via combined techniques that include ECAP, subsequent drawing (Fig. 2.9A), and annealing. In this work it was revealed that after a 6-h annealing at 350°C, which leads to a maximum UTS with an improved uniform elongation, no significant grain growth was observed in UFG Ti. TEM images displayed the GB structure as thin and equilibrium, which can be seen by the typical banded contrast observed at many grain boundaries (Fig. 2.9B). At the same time, the GB structure was stated to be affected by oxygen and carbon segregations, observed with the help of atom probe tomography (APT) (Fig. 2.10).

(A) (B)

FIG. 2.9

Microstructure of CP Ti Grade 4 processed by ECAP and subsequent drawing in the transverse section of a rod seen by TEM: (A) before annealing; (B) after annealing at 350°C for 6h [14].

Therefore, based on the TEM, X-ray analysis, and three APT results, it was stated that such unusual mechanical behavior of the UFG Ti Grade 4 can be associated with recovery processes of nonequilibrium boundaries, which are accompanied with ordering of their structure and with impurity segregation formation at grain boundaries. Changes in the concentration of alloying elements and formation of their segregations near boundaries can influence the stress of plastic deformation development through the material. An increase of critical stresses necessary for nucleation of new dislocations of the boundaries in UFG metals promotes enhancement of the flow stress and additional strengthening.

2.2.2 MICROSTRUCTURE OF Ti AFTER ECAP-CONFORM AND DRAWING/ROLLING

One of the recent advanced modifications of ECAP is an ECAP-Conform technique which combined with subsequent drawing can be a cost-effective way to produce long-length rods with UFG structure [15,16]. Such rods are suitable for industrial applications, including medical implants [17].

The results of studies on the effect of the number of ECAP-C passes at 200°C on the rod microstructure after additional drawing are described in Ref. [18]. The research material was Ti Grade 4 with an initial diameter of 12mm. After ECAP

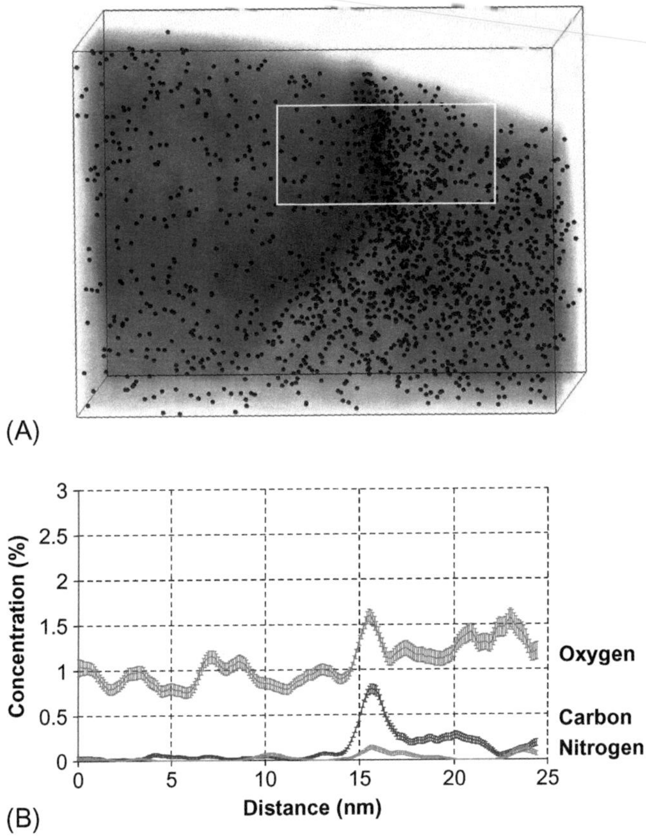

FIG. 2.10

(A) 3D distribution of carbon atoms (*red*; *gray* in the print version) in the vicinity of a grain boundary; (B) composition profiles of interstitial impurities across the grain boundary shown by the *rectangle* in (A). Values are reported together with the standard deviation [14].

75% reduction by drawing to a diameter of 6mm was performed at a temperature of $T = 200°C$. The microstructures within the Ti Grade 4 rod subjected only to drawing consisted of subgrains elongated in the form of "fibers" parallel to the drawing axis (Fig. 2.11A). This type of deformation forms mostly low-angle subgrain boundaries.

A different situation was observed when the billets were subjected to ECAP prior to drawing. An increase of the number of passes from 1 to 10 has a significant impact on the microstructures formed in the rod. However, even after one ECAP pass the longitudinal section of the billet showed grains elongated along the direction of deformation, and some of these grains had finer subgrain structure and increased dislocation density (Fig. 2.11B). The microstructure of the rod processed by six ECAP passes followed by drawing showed grains elongated along the direction of deformation, but the inner areas were divided by low-angle dislocation boundaries and the

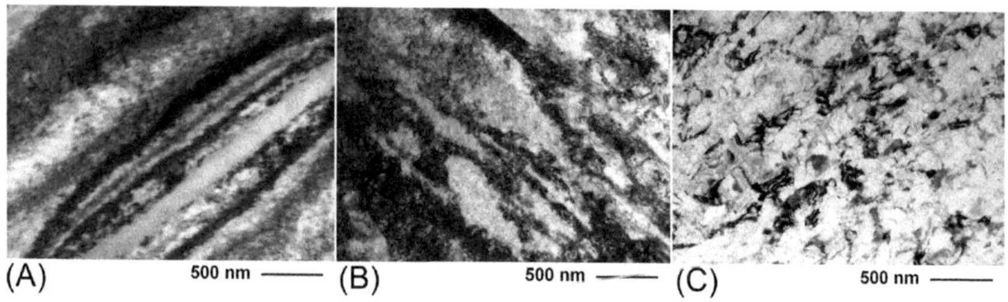

FIG. 2.11

TEM of longitudinal sections showing the microstructure of rods (A) after drawing without ECAP; (B) after one pass of ECAP+drawing; and (C) after six passes of ECAP+drawing with $\varepsilon \sim 75\%$ [18].

size of the subgrains reached 200 nm (Fig. 2.11C). Similar structures were reported earlier due to ECAP and subsequent cold rolling or drawing [5–7]. The increase in the number of ECAP passes to 10 leads to additional reduction of the average sizes of grains/subgrains to 170 nm and an enhancement of the fraction of grains with an equiaxed shape in both transverse (Fig. 2.12A) and longitudinal (Fig. 2.12B) sections. The uniform distribution of spots on the concentric circles in the diffraction patterns indicates an increase in the fraction of HAGBs (Fig. 2.12A and B).

The texture evolution in ECAP-C and subsequent drawing is shown in Ref. [19]. In particular Ti Grade 4 samples were studied after six ECAP-C passes and drawing at 200°C from 12 to 6 mm in diameter. Fig. 2.13 demonstrates experimental PFs for CP Ti after six ECAP-C passes and drawing. The texture can be described as (1010) fiber and such texture is typical of CP Ti after forging and/or drawing [20]. The sharpness of the texture maxima increases dramatically in comparison with the ECAP-C (see Fig. 2.12). The activity of slip systems in hcp pure Ti depends on

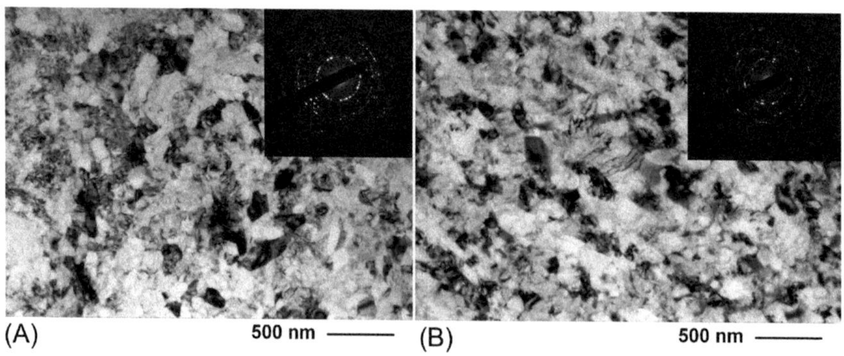

FIG. 2.12

Microstructure of Grade 4 Ti rods processed via 10 passes of ECAP-C at 200°C and drawing with $\varepsilon \sim 75\%$ at 200°C: (A) transverse section, (B) longitudinal section [18].

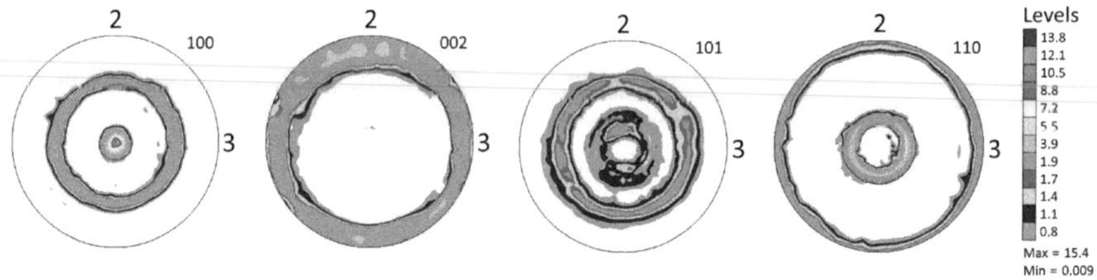

FIG. 2.13

Experimental PFs (1010), (0002), (1011), and (1120) for CP Ti Grade 4 after six ECAP-C passes and subsequent drawing [19].

the value of the critical resolved shear stress for these slip systems. For CP Ti, the critical resolved shear stress values are 90 MPa for the prismatic slip system, 105 MPa for the pyramidal $\langle a\rangle$ slip system, 115 MPa for the basal slip system, and 280 MPa for the pyramidal $\langle a+c\rangle$ slip systems [21]. During drawing the basal and prismatic slip systems of the second type are most favorably oriented for slip of dislocations during straining (Fig. 2.13).

The results of the microstructure and texture characterization assisted by the EBSD technique are described in Ref. [22]. After ECAP-C at 250°C to an accumulated true strain of 4.2 (six passes), the material was drawn at 200°C to a total area reduction of 76%. The results indicated that a strong texture consisting of $\langle 10\bar{1}0\rangle$ - drawing axis (i.e., $a\langle 10\bar{1}0\rangle$-fiber texture) had been formed.

This observation agrees well with X-ray measurements of the same material [19].

2.3 SUMMARY

According to the results of the performed studies, a significant impact of thermomechanical processing by drawing and/or rolling, and annealing after ECAP on the UFG structure evolution in commercially pure Ti can been stated:

(a) additional grain refinement and its size reduction up until 0.1 μm;
(b) depending on the deformation temperature in formation of different grain boundary types (low angle and high angle), different shape of grains (equiaxed, elongated, fragmented);
(c) as a result of drawing and rolling, a sharp texture forms with the highest fraction of orientations of prismatic slip systems;
(d) after low-temperature annealing of UFG Ti the elongated grains transform into equiaxed ones, the fraction of high angle boundaries increases to 70%, and the state of grain boundaries changes due to segregation of impurities.

The modification of the ECAP technique is the ECAP-Conform and subsequent drawing is the advanced way for production of UFG Ti rods for industrial applications, including medical implants (see Chapter 12).

ACKNOWLEDGMENTS

The work of R.Z.V. was supported by the Saint Petersburg State University Grant No. 6.37.204.2016. I.P.S. thanks the Russian Scientific Foundation Grant No. 16-19-10356.

REFERENCES

[1] R.Z. Valiev, T.G. Langdon, Principles of equal-channel angular pressing as a processing tool for grain refinement. Prog. Mater. Sci. 51 (2006) 881–981, https://doi.org/10.1016/j.pmatsci.2006.02.003.

[2] F.Z. Utyashev, G.I. Raab, Impact of scale factor on grain refinement during severe plastic deformation. J. KSHP. OMD. 11 (2008) 13–20. (in Russian).

[3] I.P. Semenova, G.I. Raab, R.Z. Valiev, Nanostructured titanium alloys: new developments and application prospects. Nanotechnol. Russ. 9 (5–6) (2014) 320–333, https://doi.org/10.1134/S199507801403015X.

[4] G.I. Raab, F.Z. Utyashev, Development of SPD technologies of production line for rod-shaped semiproducts out of nanostructured titanium for medical application. Mater. Sci. Forum 667–669 (2011) 1159–1164, https://doi.org/10.4028/www.scientific.net/MSF.667-669.1159.

[5] V.V. Stolyarov, Y.T. Zhu, T.C. Lowe, R.Z. Valiev, Microstructure and properties of pure Ti processed by ECAP and cold extrusion. Mater. Sci. Eng. A 303 (2001) 82–89, https://doi.org/10.1016/S0921-5093(00)01884-0.

[6] V.V. Stolyarov, Y.T. Zhu, I.V. Alexandrov, T.C. Lowe, R.Z. Valiev, Grain refinement and properties of pure Ti processed by warm ECAP and cold rolling. Mater. Sci. Eng. A 343 (2003) 43–50, https://doi.org/10.1016/S0921-5093(02)00366-0.

[7] T. Zhu, Y.R. Kolobov, G.P. Grabovetskaya, V.V. Stolyarov, N.V. Girsova, R.Z. Valiev, Microstructure and mechanical properties of ultrafine-grained Ti foil processed by equal-channel angular pressing and cold rolling. J. Mater. Res. 18 (4) (2003) 1011–1016, https://doi.org/10.1557/JMR.2003.0138.

[8] G.H. Salimgareeva, I.P. Semenova, V.V. Latysh, I.V. Kandarov, R.Z. Valiev, Combined SPD techniques to fabricate nanostructured Ti rods for medical application. Solid State Phenom. 114 (2005) 183–188, https://doi.org/10.4028/www.scientific.net/SSP.114.183.

[9] I.P. Semenova, E.B. Yakushina, V.V. Nurgaleeva, R.Z. Valiev, Nanostructuring of Ti-alloys by SPD processing to achieve superior fatigue properties. Int. J. Mater. Res. 100 (12) (2009) 1691–1696, https://doi.org/10.3139/146.110234.

[10] R.Z. Valiev, Nanostructuring of metals by severe plastic deformation for advanced properties. Nat. Mater. 3 (2004) 511–516, https://doi.org/10.1038/nmat1180.

[11] R.Z. Valiev, I.V. Alexandrov, N.A. Enikeev, M.Y. Murashkin, I.P. Semenova, Towards enhancement of properties of UFG metals and alloys by grain boundary engineering using SPD processing, Rev. Adv. Mater. Sci. 25 (2010) 1–10.

[12] R.Z. Valiev, A.V. Sergueeva, A.K. Mukherjee, The effect of annealing on tensile deformation behavior of nanostructured SPD titanium. Scr. Mater. 49 (2003) 669–674, https://doi.org/10.1016/S1359-6462(03)00395-6.

[13] I.P. Semenova, G.H. Salimgareeva, V.V. Latysh, T. Lowe, R.Z. Valiev, Enhanced fatigue strength of commercially pure Ti processed by severe plastic deformation. Mater. Sci. Eng. A503 (2009) 92–95, https://doi.org/10.1016/j.msea.2008.07.075.

[14] I.P. Semenova, G.H. Salimgareeva, G. Da Costa, W. Lefebvre, R.Z. Valiev, Enhanced strength and ductility of ultra-fine grained Ti processed by severe plastic deformation. Adv. Eng. Mater. 12 (8) (2010) 803–807, https://doi.org/10.1002/adem.201000059.

[15] G.I. Raab, R.Z. Valiev, T.C. Lowe, Y.T. Zhu, Continuous processing of ultrafine grained Al by ECAP-Conform. Mater. Sci. Eng. A 382 (2004) 30–34, https://doi.org/10.1016/j.msea.2004.04.021.

[16] Y.T. Zhu, T.C. Lowe, R.Z. Valiev, G.I. Raab, Continuous Equal-Channel Angular Pressing, Los Alamos National Security, LLC (Los Alamos, NM), 2006 US Patent No. 7,152,448, USC 72/262, filed December 16, 2004, issued December 26.

[17] R.Z. Valiev, I.P. Semenova, V.V. Latysh, H. Rack, T.C. Lowe, J. Petruzelka, L. Dluhos, D. Hrusak, J. Sochova, Nanostructured titanium for biomedical applications. Adv. Eng. Mater. 8 (2008) B15–B17, https://doi.org/10.1002/adem.200800026.

[18] I.P. Semenova, A.V. Polyakov, G.I. Raab, T.C. Lowe, R.Z. Valiev, Enhanced fatigue properties of ultrafine-grained Ti rods processed by ECAP-Conform. J. Mater. Sci. 47 (22) (2012) 7777–7781, https://doi.org/10.1007/s10853-012-6675-9.

[19] D.V. Gunderov, A.V. Polyakov, I.P. Semenova, G.I. Raab, A.A. Churakova, E.I. Gimaltdinova, I. Sabirov, J. Segurado, V.D. Sitdikov, I.V. Alexandrov, N.A. Enikeev, R.Z. Valiev, Evolution of microstructure, macrotexture and mechanical properties of commercially pure Ti during ECAP-conform processing and drawing. Mater. Sci. Eng. A 562 (2013) 128–136, https://doi.org/10.1016/j.msea.2012.11.007.

[20] S. Nemat-Nasser, W.G. Guo, J.Y. Cheng, Mechanical properties and deformation mechanisms of a commercially pure titanium. Acta Mater. 47 (1999) 3705–3720, https://doi.org/10.1016/S1359-6454(99)00203-7.

[21] E.C. Oliver, M.R. Daymond, J. Quinta Da Fonseca, P.J. Withers, Intergranular stress evolution in titanium studied by neutron diffraction and self-consistent modelling. J. Neutron Res. 12 (2004) 33–37, https://doi.org/10.1080/10238160410001734423.

[22] G.S. Dyakonov, E. Zemtsova, S. Mironov, I.P. Semenova, R.Z. Valiev, S.L. Semiatin, An EBSD investigation of ultrafine-grain titanium for biomedical applications. Mater. Sci. Eng. A 648 (2015) 305–310, https://doi.org/10.1016/j.msea.2015.09.080.

CHAPTER

Hydrostatic extrusion

3

Halina Garbacz*, Krzysztof Topolski*, Maciej Motyka†

Faculty of Materials Science and Engineering, Warsaw University of Technology, Warsaw, Poland Faculty of Mechanical Engineering and Aeronautics, Rzeszow University of Technology, Rzeszow, Poland†*

3.1 INTRODUCTION

The history of hydrostatic extrusion (HE) is relatively long reaching as far back as the end of the 19th century. However for almost 100 years it has been used in industry, mainly for the forming of metal parts. The necessity of finding new bulk nanocrystalline metal production methods was induced by the development of nanotechnology. Among all promising techniques, HE was revealed to be very effective. Quite intensive research on HE applications for the manufacture of ultrafine-grained (UFG) and nanocrystalline (NC) metals have been conducted in Poland, by assistance from the Institute of High Pressure Physics UNIPRESS of Polish Academy of Science and the Faculty of Materials Science and Engineering of Warsaw University of Technology. At the beginning of 21st century, the first attempts to obtain NC titanium were successfully conducted by a multipass HE process. Since that time, further research on NC titanium have been comprehensively carried out, mainly in the areas of grain refinement enhancement, increasing product size, and its functional properties (particularly essential for biomedical applications). A brief description of HE methods enabling the severe plastic deformation (SPD) of titanium and the analysis of process conditions significant for its nanostructuring will be discussed in this chapter.

3.2 A BRIEF HISTORY OF HE

The method of HE was patented in 1893 by James Robertson, however the first documented experiments were conducted by Percy W. Bridgman in the 1950s [1]. Intensive research on HE development was conducted in the latter part of the 20th century. Leading researchers in that field were Avitzur (1960s, 1970s) [2,3], Pugh (1970s) [1], and Lange [4], Inoue with Nishihara [5] (1980s). Their impressive achievements

Nanocrystalline Titanium. https://doi.org/10.1016/B978-0-12-814599-9.00003-1

are documented in a few monographs and numerous publications. These are still the most valuable source of theoretical knowledge as well the documented results of HE processes and their industrial applications. All of these works have concerned all basic aspects of HE, such as the different properties of extruded metals, selection of strain degree, pressure medium, the angle of die, lubricants, sealing system, process tribology, and device construction and geometry of produced products (e.g., pipes, sections). Exemplary results of HE process, taking into consideration the effect of material hardness on extrusion pressure, are presented in Fig. 3.1.

The HE technology was first used on an industrial basis in the 1960s and 1970s by Western Electric Company (USA) and Kobe Steel (Japan). Near the end of 20th century the HE process was applied by ASEA (Sweden), Hitachi Cable Ltd. (Japan), National Standard (UK) and Sumitomo Light Metals Ltd. (Japan) [1,6]. However in the late 1990s industrial application of HE was significantly limited, mainly dominated by the use of conventional plastic working methods. At the beginning of 21st century HE technology was used by Dutch company Hydrex Materials BV (presently UTB Industry).

It should be pointed out that Poland has also had meaningful contributions to the development of the HE process. At the beginning of 1970s, in the Institute of High Pressure Physics UNIPRESS of Polish Academy of Science, W. Pachla and L. Styczynski started to work on HE and in 1984 it was industrially used for manufacturing of multicore wires, fine wires, and capillaries. Additionally in the 1990s, at IHPP UNIPRESS, the process was used to manufacture copper light wall tubes and high-temperature superconductive ribbons [6,7].

At present HE technology is investigated by only a few research centers in the world. For instance, there are some literature data on the extrusion of polymers

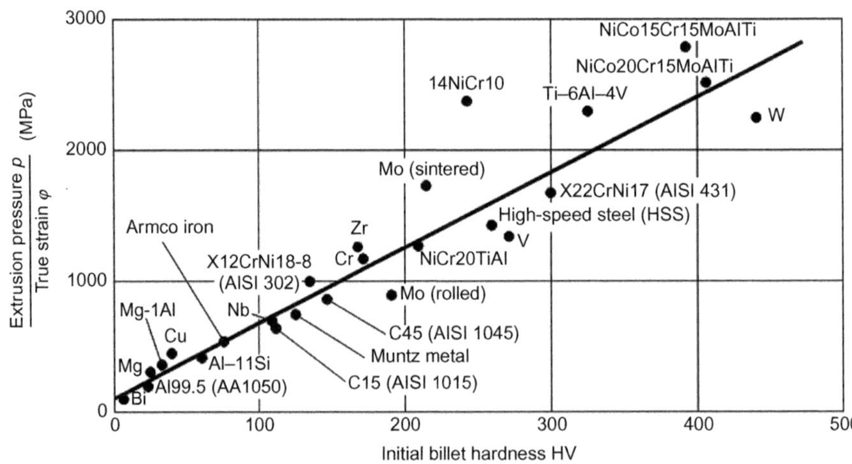

FIG. 3.1

Correlation between extrusion pressure (p), limit strain (φ), and initial billet hardness (H) obtained for various materials during its hydrostatic extrusion [4].

and polymer matrix composites for increasing their strength [8–10]. Other examples of HE applications are the extrusion of magnesium alloys [11], production of aluminum and copper fine wire bundles [12] and consolidation of aluminum UFG powders [13]. Moreover, HE can be used for casts made of hypereutectoid Al–Si alloys [14] for the production of bars and tubes with additional refinement of hard silicon precipitations. HE has the potential to assist in the formation of hard-deformable materials. Limited publications detail equipment and process parameters used (e.g., deformation way, die geometry, extrusion pressure) [15–17]. A significant contribution in HE development was made by research and development center of the Technical University of Berlin who conducted advanced research on devices and extruded materials [18,19].

All the documented research mentioned above preceded the use of HE as an SPD method in the field of nanotechnology.

3.3 BASIC CHARACTERIZATION OF THE HE TECHNIQUE

The HE process is a particular direct extrusion method and due to the crucial role of a liquid medium, is sometimes called "hydraulic direct extrusion." The device for HE (Fig. 3.2) is composed of few elementary parts: die (6) that processed material (1) is extruded through, a moving piston (4) and a cylindrical operational chamber (5) filled by pressure medium, e.g., oil (3).

The medium fills the operational chamber and surrounds the billet simultaneously. The movement of the piston starts the extrusion process and the direct contact billet-die plays the role of a sealed system. When the piston moves, the medium is gradually compressed and thus the hydrostatic pressure in the chamber increases. The plastic deformation process begins when the stress in the billet exceeds the value of material's yield strength and frictional resistance is overcome. Then the billet is extruded through the die hole, usually having a conical geometry.

The fundamental characteristic feature of HE process, distinguishing it from conventional direct extrusion, is the presence of a liquid medium, usually oil. It eliminates the direct contact of the billet with the chamber walls and piston. On the other hand it requires an efficient seal, and therefore the geometry and surface quality of the die hole and the beginning of billet should be designed and prepared properly. Due to the presence of the medium between the chamber's wall and billet, the friction

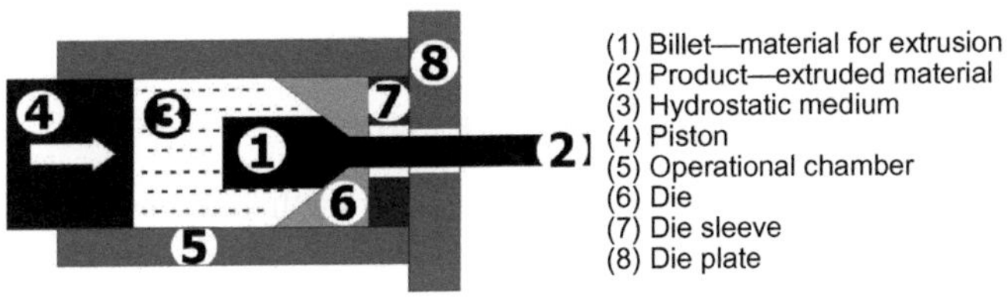

FIG. 3.2

Scheme of HE device [20].

forces are maximally reduced. That is, during the HE process, the extrusion pressure (force) is meaningfully lower compared to conventional extrusion. As a result the HE is characterized by two crucial advantages: (a) possibility of higher reduction ratio application, and (b) allowing the lower billet temperatures application. In a conventional extrusion method, radial stresses and friction forces are much higher, precluding the deformation process in many cases. However, the presence of liquid medium in HE does not eliminate frictional resistance completely. Friction forces are present between billet's and die's surfaces and contribute to an increase of extrusion pressure and this affects the surface quality of product.

Another advantageous feature of HE method is the triaxial state of compressive stresses. During the process, in the deformation zone (in the die hole) it is changed to triaxial nonuniform compression. Such a deformation mode favors plasticity of processed material (even brittle ones) and it is considered as the most beneficial for plastic working processes. It precludes the formation and propagation of microcracks in the material. Therefore the HE technique enables unusually high plastic deformation at room temperature, previously inaccessible by the conventional extrusion method. This aspect is particularly important for production of NC metals by SPD methods—grain refining requires a very high deformation degree and relatively low deformation temperature [20].

The next feature of the HE method, distinguishing it from other SPD ones, is the possibility of material compacting during the process. Moreover this technique allows processing of the material volumes (billet and product) larger than other SPD methods, such as equal-channel angular pressing (ECAP) or high-pressure torsion (HPT). The HE method can be realized using various sizes of die and their replacement is quite easy. It ensures a great freedom in the selection of deformation degree for individual extrusion passes, as well total strain. Free use of dies makes multipass extrusion easier for the gradual reduction of product diameters, which is usually necessary for the production of NC metals.

Some SPD methods, such as ECAP or multiaxial forging (MAF) allows the use of arbitrary high total strain, theoretically "infinite," which is dependent upon material properties. Among those techniques the HE is limited by the decreasing diameter of the product. Therefore, during the HE process the infinitely high plastic deformation cannot be achieved. Other disadvantages of HE are the energy loss for liquid medium compression and high requirements for a sealed system, the necessity of the injection and removal of the fluid for every extrusion cycle, and some additional complications when extrusion is realized at the elevated temperature.

3.4 HE PROCESS AS AN SPD METHOD

The HE process was considered as a possible SPD method at the beginning of 21st century in Poland. The first pioneer attempts of using this technique for UFG and NC metals, including pure titanium, were made cooperatively by researchers at the Faculty of Materials Science and Engineering of Warsaw University of Technology and the Institute of High Pressure Physics UNIPRESS of Polish Academy of Science [21–24]. Thanks to many cooperative experiments the suitability of HE for grain refinement of many metals and alloys were confirmed. It was revealed that the HE process can be successfully used for development of microstructure and properties of various materials, such as aluminum and its alloys, titanium, nickel, steel, copper, and its alloys. The first experiments on titanium were conducted in 2005–06 [21,25].

The results of numerous investigations confirmed that the HE process enables the development of NC structure in metals and their significant strengthening by a relatively high plastic deformation degree [20–28]. These features are in accordance with the generally accepted criteria for SPD methods.

Plastic deformation process in extrusion operations are realized by a significant change of billet dimensions. This aspect was discussed several times by scientists who tried to discredit using HE as an SPD method. Usually the base of such reasoning is V.M. Segal's patent [29], which emphasizes the invariability of sample shape before and after deformation. This condition means theoretically "infinite" deformation of processed material. However, the constant shape of deformed sample is not neither crucial condition nor requirement of SPD methods, it is a characteristic feature of particular technique only. Documented effects of HE for grain refinement in many metals states that the invariability of sample shape is not a requisite criterion for SPD methods and it is not the aim of such process. The most important criterion is the achievement of unconventional high plastic deformation in material, leading to its strengthening and nanostructuring.

3.5 GRAIN REFINEMENT BY HE

Specific deformation conditions during multipass HE ensures a very high efficiency of grain refinement. The microstructure developed by HE in various metals (such as Al, Cu, Ti) can be diversified, from equiaxed grains having low dislocation density in Al (Fig. 3.3A) and dislocation cells in Cu (Fig. 3.3B), to nanometric grains with high dislocation density in Ti [22,30–33]. The grain size of all mentioned metals hydrostatically extruded with the strain of $\varepsilon = 3.82$ was lower than 1 μm, but nanostructure was achieved in titanium only.

The analysis of the microstructural evolution during the HE process revealed that the first stage of extrusion leads to a meaningful increase of dislocation density in the material. Accumulated dislocations form lamellar structures (elongated bands) separated by boundaries, which successively evolve during the deformation, to the formation of high angle grain boundaries (HAGBs). Shape "twins" can be formed at the beginning stages of plastic deformation.

The formation of lamellar structures due to cold severe plastic deformation is quite well described in the literature [34–36]. Dislocation boundaries can be transformed to geometrically necessary boundaries (GNBs), where deformation is realized in various slip systems, and incidental dislocation boundaries, which form the cell structures between GNBs. The change of GNBs' orientation takes places

FIG. 3.3

Microstructure (TEM) of hydrostatically extruded metals: (A) Al, (B) Cu [33].

gradually, together with the increase of deformation degrees. The next passes of HE lead to the ordering and homogenization of material microstructure, as well as changing in the geometry of the grains and increasing of volume fraction of HAGBs. Zherebtsov and coauthors [34] have indicated that during HE process, the volume fraction of HAGBs increases from 26% for true strain of $\varepsilon = 0.72$ to 38% for $\varepsilon = 1.8$.

In case of commercially pure titanium—Grade 2—its nanostructuring by HE process is possible for true strain of $\varepsilon = 3.82$ [20]. HE with lower value of strain—e.g., $\varepsilon = 2.54$ (which corresponds to five passes)—leads to significant changes in the titanium structure but is not sufficient for its nanostructuring (Fig. 3.4). The microstructure on a transverse section of NC Ti bars (after seven passes of HE) is uniform and

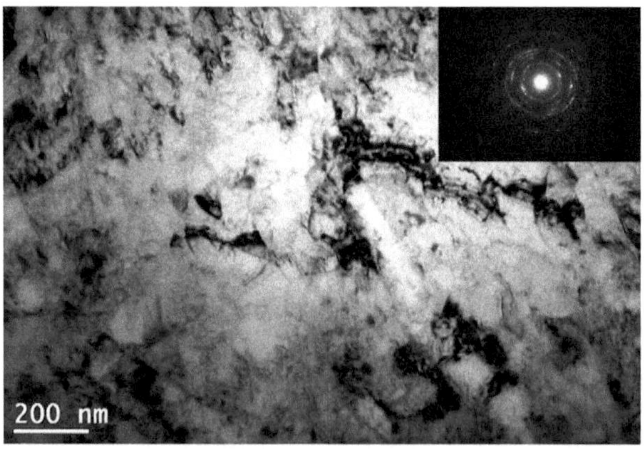

FIG. 3.4

Microstructure (TEM) of titanium (Grade 2) produced by HE method in five passes ($\varepsilon = 2.54$) (cross-section of bar).

composed of equiaxed grains with high dislocation density (Fig. 3.5). For instance, grains in NC titanium produced by HE had the following parameters: equivalent diameter $d_2 = 87$ nm and shape factor of $\alpha = 1.31$. It should be pointed out that more than 40% of grains have a size in the range of 60–100 nm, while the smallest ones just 20 nm. High grain refinement of such NC titanium was confirmed by electron diffraction patterns (Fig. 3.5A and C). The microstructure on a longitudinal section of extruded Ti Grade 2 bars is characterized by morphological texture and is composed of elongated lamellae 10–100 nm in width, parallel to extrusion direction (Fig. 3.5C and D). Inside the lamellae, the dislocations, subgrains, and nanometric grains are present. A characteristic feature of such HE processed titanium is the lack of twin boundaries in its nanostructure. According to Stolyarov and coauthors [37,38], twinning in titanium is replaced by dislocation slip, which is a more effective mechanism during SPD.

After the HE process, titanium samples have a sharp fiber texture, typical for extruded hexagonal materials [39,40]. The texture is characterized with the main maximum on the (10 − 10) pole figure in the position of extrusion direction, and the (0001) pole distributed in the surface perpendicular to this direction. Such texture

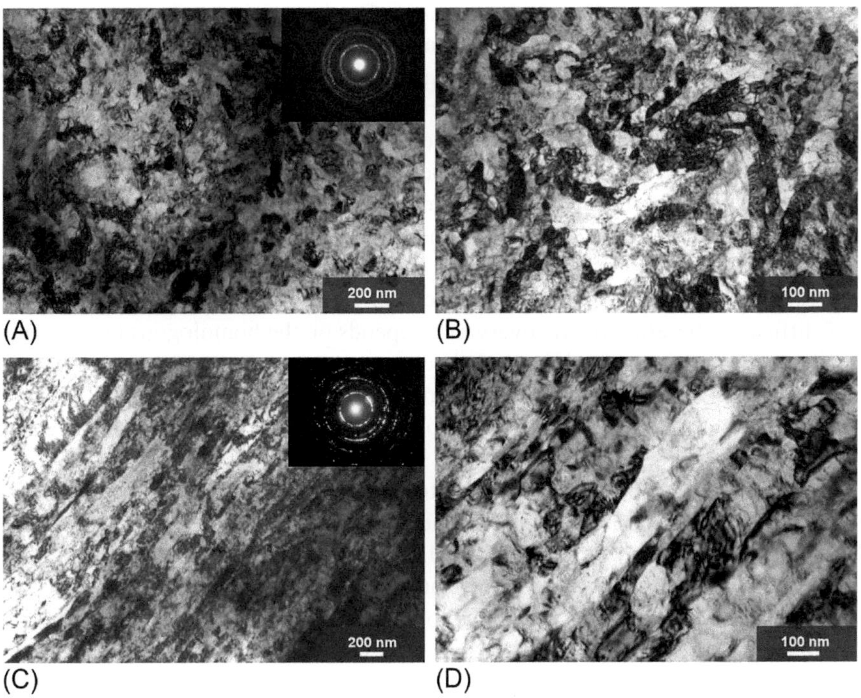

FIG. 3.5

Microstructure (TEM) of NC titanium (Grade 2) produced by HE method ($\varepsilon = 3.82$): (A, B) cross-section, (C, D) longitudinal section [33].

results in very different crystallographic planes observed on transversal and longitudinal cross-sections. On the transversal cross-section {10 − 10} planes of grains are almost exclusively observed, while on the longitudinal cross-section the grain orientation is distributed between the (0001) and {11 − 20} planes, with the maximum shifted towards the (0001) position.

The microstructure and final grain size of HE processed metals depends on the rate of accumulation, annihilation, and rearrangement of defects that are generated during the process. In general, all factors that enhance defects accumulation and suppress annihilation are beneficial for grain refinement. In this aspect, high strain rates usually applied in the HE process is very beneficial. It makes HE process adiabatic [27]. It was found, based on our own results and literature data, that the rise of temperature (ΔT) caused by the HE of titanium is in the range of 200–500°C—estimated according to Eq. (3.1) [39,40]:

$$\Delta T = 0.9\frac{p}{c \cdot \rho} \tag{3.1}$$

where p = extrusion pressure [Pa], c = specific heat [J/kg K], ρ = density [kg/m^3], 0.9 = conversion factor of plastic deformation work into heat.

Due to the high strain rate, the thermal effect is short and does not cause the exceeding of the recrystallization temperature for titanium.

The minimal grain size, feasible to achieve by the HE process for the same strain, depends on melting temperature, stacking fault energy and crystal structure of processed material, and also on the presence of alloying elements limiting the rate of thermally activated processes, i.e., recovery, recrystallization, and grain growth [41]. The effect of melting temperature on the average grain diameter of metals processed by HE: Al, Cu, Ni (fcc), Fe (bcc), and Ti (hcp) is shown in Fig. 3.5. The smaller effect of grain refinement in metals with a low melting temperature is explained by slight increase of dislocation density, limited by a fast recovery process. The energy of recovery is close to the activation energy of self-diffusion, therefore the recovery rate depends on the homologous temperature of deformation. In metals characterized by low melting temperature, e.g., Al, the recovery occurs even at room temperature. Any thermal effect related to HE pass intensifies that phenomena, simultaneously decreasing the capacity of grain refinement.

The temperature increase can be limited by pressure minimization—the reduction of strain in one deformation pass—and by using a cooling system. For example, UFG rods 3 mm in diameter made from Al and Cu were produced by the HE method in three passes with strain of 1.39, 1.39, and 1.02, respectively. NC Ti is hydrostatically extruded in more passes and due to the thermal effects, only during the first pass the strain is 1.39, then no more than 0.42 during following passes. Obviously the thermal effect during an HE pass is also affected by the mechanical properties of the material. Titanium deformed using lower strain caused similar thermal effects because it possesses much higher strength than Al and Cu. Ti has the highest melting temperature among the metals that were examined and has the most effective grain

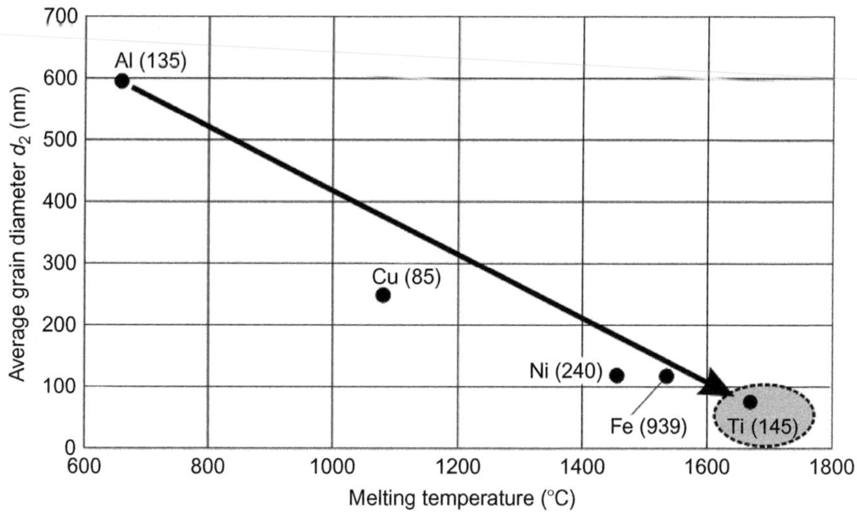

FIG. 3.6

Grain size of hydrostatically extruded metals (the value of SFE [mJ/m^2] in *brackets*).

refinement caused by the HE process (Fig. 3.6). The finalized nanostructure and high dislocation density in grains determines its high mechanical properties.

The morphology of structures formed in hydrostatically extruded metals also depends on stacking fault energy (SFE). In materials that possess low SFE value, the cross slip and dislocation climb proceeds with difficulty, so lower dislocation mobility limits the HAGBs formation. It is considered that HAGBs can be formed by the monotonic increase of grain boundaries misorientation due to the absorption of dislocation during follow-up extrusion passes [41]. Therefore the increase of grain boundaries misorientation requires certain minimal dislocation mobility necessary for the redevelopment of their structure. An example of metal characterized by a low value of SFE is Cu, which, when it has a microstructure developed by the HE process, composes of dislocation cells with a small misorientation angle.

In summary, the HE process at room temperature is an efficient method for microstructure development of quite a large group of metals. Its efficiency in terms of grain refinement is higher for metals possessing high melting temperature, such as Ti.

3.6 PRODUCTION OF NC TITANIUM BY THE HE METHOD

For the production of NC titanium using the HE method, some characteristic features of Ti should be mentioned, especially those that influence the extrusion process, both its course and effects. It should be emphasized that hydrostatic extrusion of titanium is a specific and demanding process, distinguishing it from the extrusion of other materials.

During the HE process, the adhesion of titanium to die material (usually steel) occurs and a mutual abrasion of die and billet takes place. It is widely known that titanium, together with steel (and almost all metals) form an unfavorable tribological pair with a high friction factor. During an HE pass, despite the discrete reduction of friction forces, the billet–die contact causes the relatively large adhesive wear of the die, as well as the billet, and generates friction force being a resistance of extrusion. It results in four aspects, all of which are crucial to consider:

(1) the increase of pressure needed for the deformation (extrusion), which restricts the possibility of extrusion with larger deformation during one pass of extrusion;
(2) the presence of "stick slip" phenomena (a material's discrete displacement during the extrusion due to adherence of titanium to die, with the next dynamic restart of extrusion causing the pressure reduction in a chamber);
(3) the necessity of a die's regeneration and replacement;
(4) resultant defects on the product's surface and increasing its roughness.

One of the possible solutions is surface treatment of the steel die or making it from other material. Another solution is a finishing (grinding, polishing) of die and/or billet surfaces for meaningful reduction of their roughness. Surface topography and development influence on the values of frictional resistance forces. Also, a surface treatment of billet can be used, e.g., modification by metallic coatings characterized by relatively good plasticity and adhesion. Such coatings are an effective lubricating medium, separating steel die and titanium billet.

The original results obtained show that by modification of the titanium surface with 10 μm aluminum coatings, these problems can be practically eliminated [20]. Titanium rods were covered with aluminum coatings using the magnetron sputtering method (Fig. 3.7A). During the extrusion process, the Al coating and Ti substrate

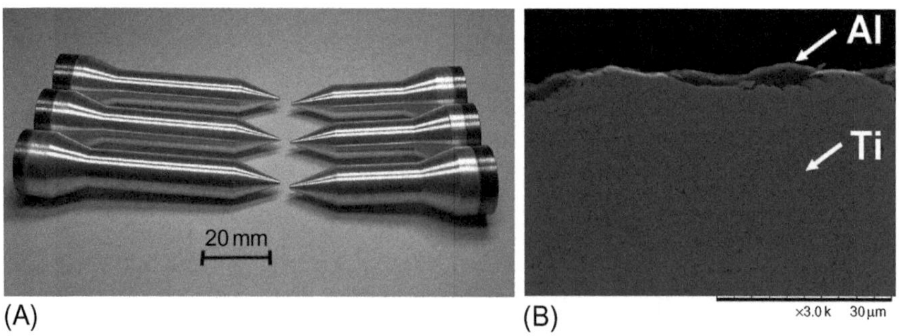

FIG. 3.7

Titanium billets with aluminum coating 10 μm in thickness intended for HE process (A) and surface layer of Ø3 mm rod after two HE passes: Ø12 → Ø6 → Ø3 mm (B).

were subjected to plastic deformation. Despite the strain, the Al coatings maintained good adhesion to the titanium substrate with no visible cracking or spalling (Fig. 3.7B). The results proved that the coatings functioned as an excellent lubricant which separated the titanium rods from the die. It allowed a reduction of the friction and protected the titanium from adhering to the die. This caused the reduction of the frictional wear of the die as well as the product. Also it was found, based on the macroscopic observations, a decrease of the surface roughness of the extruded rods. Another important advantage is a significant decrease of the extrusion pressure which permits increasing the strain applied in a single extrusion pass and deforming titanium with a higher reduction of the rod diameter. This can result in a decrease in the number of the extrusion passes needed for grain refinement and necessary to produce nanocrystalline Ti. This effect is demonstrated in Fig. 3.8, which shows the reduction of the diameter (*D*) of Ti rods from Ø12 to Ø3 mm. This reduction requires from 5 to 11 extrusion passes (points on the diagram) when the titanium is not coated, whereas the Al-coated rods can be extruded in only two HE passes.

Other characteristic features of titanium, making the HE process unique, are:

- fast work hardening and high value (close to 1) of "yield strength (YS)/ultimate tensile strength (UTS)" ratio. It results in small range of plastic deformation which can be applied. Low plasticity of titanium is also related to small numbers of easy slip systems (hcp structure).
- relatively high yield strength—from 350 to 1000 MPa.

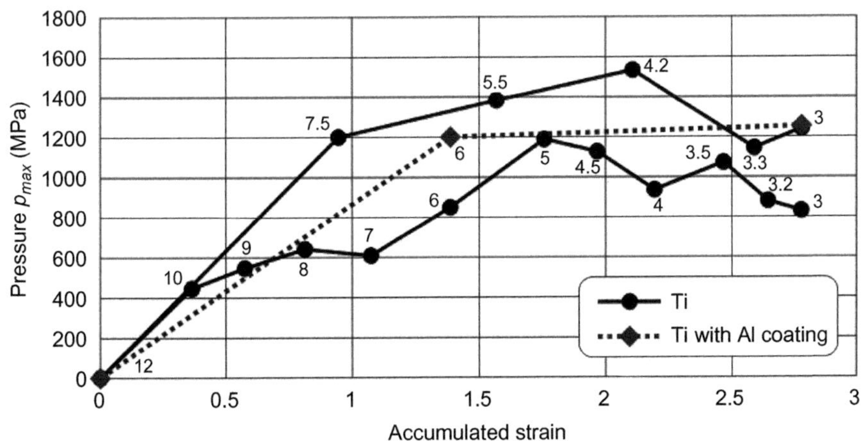

FIG. 3.8

The maximum extrusion pressure as a function of accumulated strain during successive extrusion passes (where numbers are the diameters of rods, *D*, 12, initial (billet); 3, final (product)). Demonstration of the influence of Al coatings resulted in reduction of the number of extrusion passes.

Reproduced from K. Topolski, W. Pachla, Halina Garbacz: progress in hydrostatic extrusion of titanium, J. Mater. Sci. 48 (2013) 4543–4548.

Titanium, during an HE pass, has a susceptibility for fast work hardening and high yield strength, and requires high pressure application, therefore the required accumulative strain in the HE process must be divided into lower partial strains. In other words, NC titanium can be produced by a multipass HE method only. Every extrusion pass is related to the reduction of billet/product diameters. Due to the limited length of the chamber and length increments of the product, it is necessary to cut semiproducts for billets used in the next pass of the HE process.

There are some factors and process parameters influencing the course and effects of NC titanium production using the HE method. They can be classified as a technological ones (related to HE technology and specification) and material ones (referring to properties of the titanium billet). Technological parameters are: accumulated total strain (ε),reduction of cross-section area (R), number of extrusion passes (n), maximal pressure of extrusion (p_{max}), strain rate ($\dot{\varepsilon}$), angle of conic die (2α), cooling of product behind the die, the presence of lubricants and the type of liquid medium. Basic material factors are: initial microstructure and properties of billet, its surface condition, the temperature of billet, and its geometry.

Obviously, similarly to other SPD methods, the crucial parameter for successful production of NC titanium is a suitably high value of accumulated strain ε [20,42]. Due to the multistage characteristic of hydrostatic extrusion, the true strain can be considered as an accumulated (total) or partial one (for one extrusion pass). During each pass the plastic deformation is induced into material. This deformation can be calculated using the formula:

$$\varepsilon = 2 \ \ln \frac{D_s}{D_f} \tag{3.2}$$

where D_s = starting rod diameter before extrusion, D_f = final rod diameter after extrusion.

On the basis of the research results it was revealed that NC titanium can be produced by the HE method, irrelevant to the initial microstructure of billet [43]. The main factor is a suitably high value of accumulated strain used, enabling the generation of high dislocation density in the processed material, which is necessary for partition of existing grains for smaller ones and formation of new HAGBs. In the case of hydrostatically extruded titanium Grade 2, the minimal total strain ensuring its nanostructuring is 3.0 [20,42,44]. Such value is relatively high for extrusion process; it forces a high reduction of the cross-section area. Due to very high pressure being required for extrusion, obtaining a strain of 3.0 in one operation is practically impossible. Additionally, such a deformation process would be characterized by very high thermal effect. Therefore NC titanium is produced by several or even more than 10 HE passes.

Billets for HE process can be initially heated up for reduction of its yield strength and the pressure required for its extrusion. The main advantage of such solution is a partial strain enhancement in the extrusion pass. However it is not used

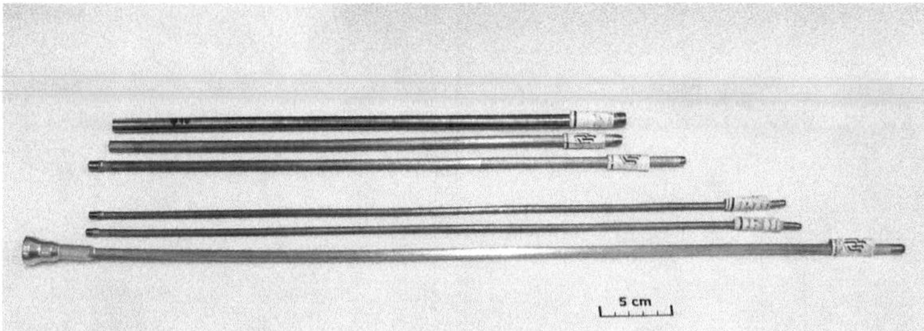

FIG. 3.9

Bulk NC titanium in the form of the rods with diameter of Ø5, Ø7, Ø8, and Ø10 mm.

for grain refinement of titanium because it does not allow the generation of enough dislocations (due to thermally activated processes being characteristic for hot deformation).

The maximal accumulated strain used for hydrostatically extruded titanium Grade 2 was 5.6. It allowed the production of NC titanium with an equivalent grain size of 47 nm [45]. It can be stated, based on other results, that nanostructuring of titanium Grade 2 at room temperature by HE process is possible for accumulated total strain of 3.0–5.6 [20,21,42,44,45].

Some other issues in the creation of NC titanium have been also analyzed. One important aspect was increasing the volume of the NC titanium that was produced. It was found the product could be scaled by increasing the billet diameter (Fig. 3.9) [42,44]. The microstructural effects of scaling are presented in Fig. 3.10. For example, a nanostructure was developed by the HE process with the total strain of 3.8 in rods: Ø3 mm in diameter from Ø20 mm billet and Ø5 mm in diameter from Ø33 mm billet. However it should be noted that an increase of billet diameter is limited by the construction and energetic parameters of the press that is used. Therefore, an alternative way to increase the diameter of NC titanium rods is the application of the minimal value of total strain (3.0) in HE process. Finally it was found that titanium Grade 2 can be hydrostatically extruded to the form of NC rods with diameters from Ø3 to Ø10 mm [20,42,44].

The increase of material volume produced by plastic working methods is usually related to uniformity of its microstructure and mechanical properties. Research results obtained for NC titanium rods, on their cross-section, indicate that microstructure and microhardness is similar in whole product's volume, independent to a rod's diameter (Table 3.1) [46]. Microstructural investigations using a transmission electron microscope confirmed the presence of nanometric grains with similar size and shape in hydrostatically extruded rods at both the beginning and end of the process.

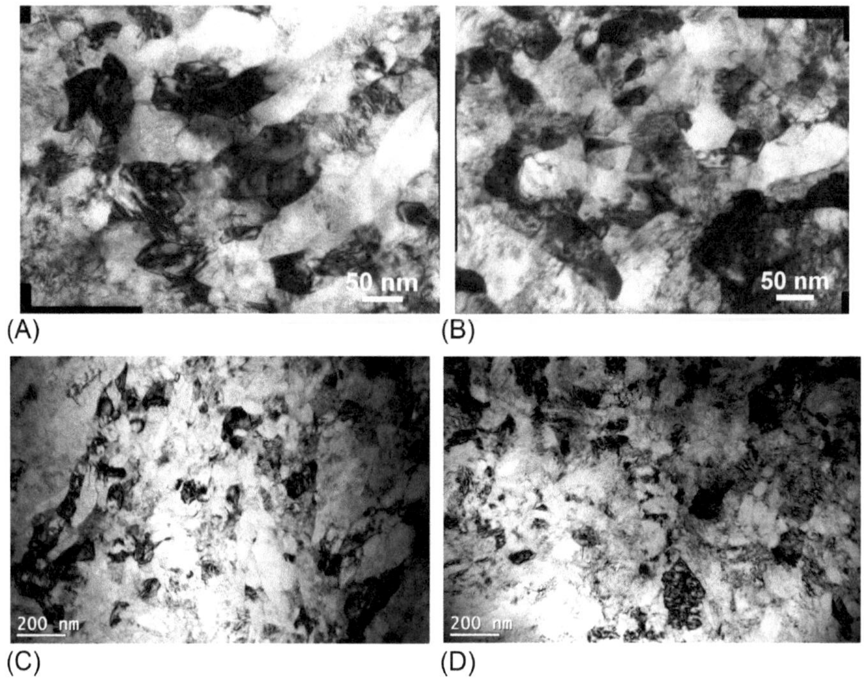

FIG. 3.10

Microstructure (TEM) of hydrostatically extruded Ti rods with diameter of Ø3 (A), Ø5 (B), Ø7 (C), and Ø10mm (D).

Table 3.1 Stereological Parameters of Microstructure (Equivalent Grain Diameter $E(d_2)$ With Coefficient of Variation $CV(d_2)$) and Hardness (HV0.2) of NC Titanium Rods (Cross-Sections) Produced by HE Method

	Diameter of Rod (mm)			
	Ø7		Ø10	
Parameter	**Beginning**	**End**	**Beginning**	**End**
$E(d_2)$, nm	62	67	73	65
$CV(d_2)$	0.39	0.34	0.34	0.28
HV0.2	256	266	266	265

3.7 SUMMARY

Nowadays, HE is considered as a part of SPD methods and enables the fabrication of bulk NC titanium in the form of rods. On the basis of the results of numerous experiments, it was proved that titanium Grade 2 can be HE processed at room temperature

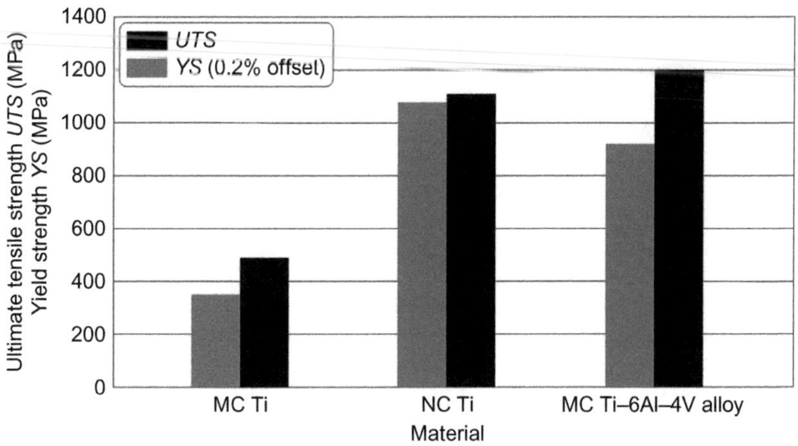

FIG. 3.11

Mechanical properties of NC titanium comparing with microcrystalline titanium and Ti–6Al–4V alloy [33].

into NC rods with a grain size range of 50–90 nm. These titanium rods possess extraordinary mechanical properties (comparable with Ti–6Al–4V alloys) with values of yield strength, ultimate tensile strength and hardness much higher than that of microcrystalline (MC) titanium (Fig. 3.11). NC titanium can be produced by the HE method in the form of rods with diameter from Ø3 to Ø10 mm. Their microstructure and mechanical properties in cross-section is uniform.

For NC titanium production, the biggest advantage of HE method is its triaxial stress state. It increases the plasticity of deformed materials and precludes the nucleation and propagation of microcracks. Additionally, the frictional resistance during the HE process is almost completely reduced, being beneficial for unconventional SPD. It should be also emphasized that HE technology can be combined with other SPD methods (e.g., ECAP) for further grain refinement in titanium or/and formation of products made from NC titanium.

REFERENCES

[1] H.L.D. Pugh, Mechanical Behaviour of Materials Under Pressure, Elsevier Publishing Company Limited, 1970.

[2] B. Avitzur, Hydrostatic extrusion, J. Eng. Ind. Trans. ASME 87 (1965) 487–494.

[3] B. Avitzur, H.C. Sortais, Experimental study of hydrostatic extrusion, J. Eng. Ind. Trans. ASME 88 (1966) 659–668.

[4] K. Lange, Handbook of Metal Forming, McGraw-Hill Book Company, 1985 USA.

[5] N. Inoue, M. Nishihara (Eds.), Hydrostatic Extrusion: Theory and Applications, Elsevier Applied Science Publishers, London, 1985.

[6] www.hydroextrusion.pl.

[7] W. Pachla, L. Styczynski, Thermally activated structural changes during hydrostatic extrusion of copper and aluminium, in: PhD Thesis, Warsaw University of Technology, Warsaw, 1978 (in Polish).
[8] M. Bonner, L.S. Saunders, I.M. Ward, G.W. Davies, M. Wang, K.E. Tanner, W. Bonfield, Anisotropic mechanical properties of oriented HAPEX™, J. Mater. Sci. 37 (2002) 325–334.
[9] M. Wang, N.H. Ladizesky, K.E. Tanner, I.M. Ward, W. Bonfield, Hydrostatically extruded HAPEX™, J. Mater. Sci. 35 (2000) 1023–1030.
[10] J.Kurpiewski Bielefeldt, Wyciskanie hydrostatyczne jako metoda modyfikacji mechanicznych właściwości polimerów, Polimery R.32 (7) (1987) 284–287.
[11] W.H. Sillekens, J.A.F.M. Schade van Westrum, A.J. den Bakier, P.J. Vet, Hydrostatic extrusion of magnesium: process mechanics and performance, Mater. Sci. Forum 426–432 (2003) 629–636.
[12] S. Lee, H.J. Park, S.S. Kim, T.H. Choi, E.Z. Kim, G.A. Lee, The potentiality of micro-scaled multi-filament wire forming using repetitive hydrostatic extrusion process, Adv. Mater. Res. 26–28 (2007) 77–80.
[13] S. Lee, C.Y. Yoon, H.J. Park, S.S. Kim, E.J. Kim, T.H. Choi, K.Y. Rhee, A study of hydrostatic extrusion as a consolidation process for fabricating ultrafine-grained bulk Al-Mg alloy, J. Mater. Process. Technol. 191 (2007) 396–399.
[14] H.G. Jeong, D.J. Yoon, E.Z. Kim, H.J. Park, K.H. Na, The influence by hydrostatic extrusion on the microstructure and extrudability of the IM processed hypereutectic Al–Si–X alloys, J. Mater. Process. Technol. 130–131 (2002) 438–443.
[15] J.-C. Hung, C. Hung, The design and development of a hydrostatic extrusion apparatus, J. Mater. Process. Technol. 104 (2000) 226–235.
[16] Y.M. Hwang, M.T. Yang, Study of hydrostatic extrusion processes with extra-high extrusion ratio, Key Eng. Mater. 233–236 (2003) 311–316.
[17] A.H. Elkholy, Parametric optimization of power in hydrostatic extrusion, J. Mater. Process. Technol. 70 (1997) 111–115.
[18] K. Muller, et al., Fundamentals of Extrusion Technology, Giesel Verlag, 2004.
[19] http://www.fzs.tu-berlin.de.
[20] K. Topolski, W. Pachla, H. Garbacz, Progress in hydrostatic extrusion of titanium, J. Mater. Sci. 48 (2013) 4543–4548.
[21] W. Pachla, M. Kulczyk, A. Swiderska-Sroda, M. Lewandowska, H. Garbacz, A. Mazur, K.J. Kurzydlowski, Proc. Conf. ESAFORM 2006, (2006) pp. 535–538. Glasgow, United Kingdom.
[22] K.J. Kurzydlowski, H. Garbacz, M. Richert, Effect of severe plastic deformation on the microstructure and mechanical properties of Al and Cu, Rev. Adv. Mater. Sci. 8 (2004) 129–133.
[23] M. Lewandowska, H. Garbacz, W. Pachla, A. Mazur, K.J. Kurzydlowski, Hydrostatic extrusion and nanostructure formation in an aluminium alloy, Solid State Phenom. 101–102 (2005) 65–68.
[24] K.J. Kurzydlowski, M. Richert, B. Leszczynska, H. Garbacz, W. Pachla, Microstructural refinement under high plastic strain rates during hydrostatic extrusion, Solid State Phenom 114 (2006) 117–122.
[25] H. Garbacz, M. Lewandowska, W. Pachla, K.J. Kurzydlowski, Structural and mechanical properties of nanocrystalline titanium and 316LVM steel processed by hydrostatic extrusion, J. Microsc. 223 (2006) 272–274.
[26] M. Lewandowska, K.J. Kurzydlowski, Recent development in grain refinement by hydrostatic extrusion, J. Mater. Sci. 43 (2008) 7299–7306.

[27] K.J. Kurzydlowski, M. Lewandowska, Fabrication of nanostructured materials by hydrostatic extrusion: advantages and limitations, Mater. Sci. Forum 561–565 (2007) 913–916.
[28] K.J. Kurzydlowski, Hydrostatic extrusion as a method of grain refinement in metallic materials, Mater. Sci. Forum 503–504 (2006) 341–348.
[29] V.M. Segal, USSR Patent No. 575892, (1977).
[30] H. Garbacz, W. Pachla, T. Wierzchon, K.J. Kurzydlowski, Processing by hydrostatic extrusion of titanium coated with aluminides, J. Solid State Phenom 114 (2006) 63–68.
[31] Report from the research grant no PBZ-KNB-096/2003 on "Technologies of nanocrystalline metals and alloys production", (2003). Warsaw.
[32] M. Kulczyk, W. Pachla, A. Mazur, R. Diduszko, H. Garbacz, M. Lewandowska, W. Wojkowski, K.J. Kurzydlowski, Microstructure and mechanical properties in nickel deformed by hydrostatic extrusion, Mater. Sci. Pol. 23 (2005) 839–846.
[33] H. Garbacz, Microstructure and Properties of Nanocrystalline Titanium, Warsaw University of Technology Press, Warsaw, 2010 (in Polish).
[34] S. Zherebtsov, W. Lojkowski, A. Mazur, G. Salishchev, Structure and properties of hydrostatically extruded commercially pure titanium, Mater. Sci. Eng. A 527 (2010) 5596–5603.
[35] D.A. Hughes, N. Hansen, Microstructure and strength of nickel at large strains, Acta Mater. 48 (2000) 2985–3004.
[36] A. Belyakov, K. Tsuzaki, Y. Kimura, Y. Mishima, Tensile behavior of submicrocrystalline ferritic steel processed by large-strain deformation, Philos. Mag. Lett. 89 (2009) 201–212.
[37] V.V. Stolyarov, Y.T. Zhu, I.V. Alexandrov, T.C. Lowe, R.Z. Valiev, Grain refinement and properties of pure Ti processed by warm ECAP and cold rolling, Mater. Sci. Eng. A343 (2003) 43–50.
[38] V.V. Stolyarov, Y.E. Beigel'zimer, D.V. Orlov, R.Z. Valiev, Refinement of microstructure and mechanical properties of titanium by twist extrusion and subsequent rolling, Phys. Met. Metall. 99 (2005) 204–211.
[39] A. Chojnacka, J. Kawalko, H. Koscielny, J. Guspiel, A. Drewienkiewicz, M. Bieda, W. Pachla, M. Kulczyk, K. Sztwiertnia, E. Beltowska-Lehman, Corrosion anisotropy of titanium deformed by the hydrostatic extrusion, Appl. Surf. Sci. 426 (2017) 987–994.
[40] E.C. Moreno-Valle, W. Pachla, M. Kulczyk, B. Savoini, M.A. Monge, C. Ballesteros, I. Sabirov, Anisotropy of uni-axial and bi-axial deformation behavior of pure titanium after hydrostatic extrusion, Mater. Sci. Eng. A 588 (2013) 7–13.
[41] M. Lewandowska, Development Microstructure and Properties of Aluminium Alloys Hydrostatically Extruded, Warsaw University of Technology Press, Warsaw, 2006 (in Polish).
[42] K. Topolski, H. Garbacz, K.J. Kurzydlowski, Nanocrystalline titanium rods processed by hydrostatic extrusion, Mater. Sci. Forum 584–586 (2008) 777–782.
[43] K. Topolski, H. Garbacz, W. Pachla, K. Kurzydlowski, The influence of the initial state on microstructure and mechanical properties of hydrostatically extruded titanium, Solid State Phenom. 140 (2008) 191–196.
[44] K. Topolski, H. Garbacz, W. Pachla, K.J. Kurzydlowski, Bulk nanostructured titanium fabricated by hydrostatic extrusion, Phys. Status Solidi C7 (5) (2010) 1391–1394.
[45] W. Pachla, M. Kulczyk, M. Sus-Ryszkowska, A. Mazur, K.J. Kurzydlowski, Nanocrystalline titanium produced by hydrostatic extrusion, J. Mater. Process. Technol. 205 (2008) 173–182.
[46] K. Topolski, H. Garbacz, W. Pachla, K.J. Kurzydlowski, Homogeneity of bulk nanostructured titanium obtained by hydrostatic extrusion, Mater. Sci. Forum 674 (2011) 47–51.

CHAPTER

4 Friction-stir processing

Sergey Mironov, Yutaka S. Sato, Hiroyuki Kokawa

Department of Materials Processing, Graduate School of Engineering, Tohoku University, Sendai, Japan

4.1 INTRODUCTION

Friction-stir processing (FSP) is relatively new grain-refinement technique invented in 1999 [1]. In fact, this approach is a derivative of well-known friction-stir welding technology [2] which has been upgraded for microstructural modifications. The basic concept of FSP is illustrated in Fig. 4.1. This technique requires a specially-designed nonconsumable tool, which typically consists of two principal parts: the larger upper shoulder and the smaller bottom probe, as shown in Fig. 4.1. The tool is rotated with a high rotational rate and then is plunged into a workpiece until the shoulder comes into intimate contact with the workpiece surface. The tool is traversed in a direction of interest under the condition where frictional heating and plastic deformation raise the local temperature of the material so it can be plastically deformed at low flow stress. The hot material, which is constrained by the colder surrounding material, the tool shoulder and a backing plate, flows around the rotating probe, and then fills the cavity at the rear of the translating tool (Fig. 4.1). The combination of the translation and the rotational motions of the tool creates a characteristic asymmetry between two sides of the processed zone: the advancing side (AS), where the direction of the tool rotation is the same as that of the tool translation, and the retreating side (RS), where they are opposite. The processing conditions at these two sides are believed to be different and thus the evolved microstructure may also differ. The material that flows around the tool undergoes extreme levels of plastic deformation and thermal exposure, which normally leads to a significant microstructure refinement in the center of the processed zone (the so-called stir zone).

Currently, FSP is well-proven to be a highly efficient technique, used to produce fine-grained sheet materials with excellent mechanical properties. However, due to high requirements for processing conditions, FSP of commercial-purity titanium is relatively difficult, and, until recently, works in this area were seldom [3,4]. Nevertheless, the progress in the tool manufacturing technology in the past few years have given rise to a growing interest in FSP of this material. With the aim to provide a

Nanocrystalline Titanium. https://doi.org/10.1016/B978-0-12-814599-9.00004-3

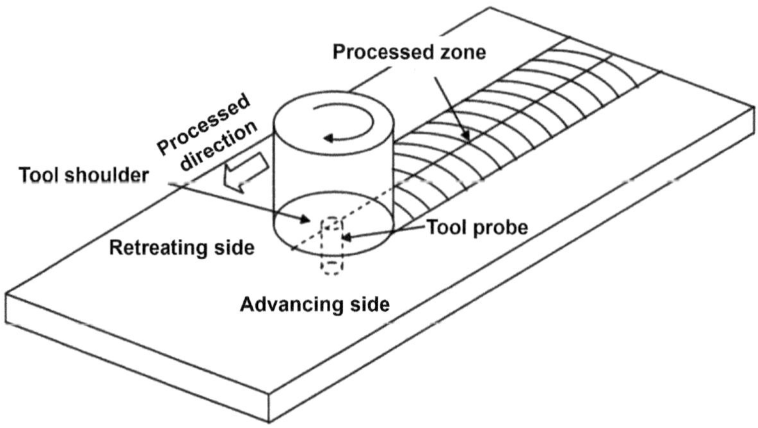

FIG. 4.1

Schematic illustration of friction-stir processing.

useful guide in this area, this chapter attempts to summarize the current state of understanding and development of this process.

4.2 CRITICAL ISSUES OF PROCESS

At typical FSP temperatures, commercial-purity titanium is characterized by relatively high strength, low thermal conductivity, and enhanced reactivity to oxygen. This imposes strict demands to the processing conditions. In this section, the critical aspects of this process are briefly addressed, including tool materials, working temperatures, and processing window.

Considering severe working conditions, the tools for friction-stirring of commercial-purity titanium should be manufactured from a high-temperature-resistant material which has no chemical interaction with titanium. Currently, the tools are typically fabricated from refractory materials including tungsten-based [5–11] or molybdenum-based [12,13] alloys. Importantly, all these tool materials are expensive and therefore FSP of commercial-purity titanium is a relatively costly process.

The processing temperature is one of the key issues in friction-stirring. In commercial-purity titanium, the temperature is of particular importance because this material may experience phase transformations. To avoid this undesirable effect, the working temperature is usually kept below the β-transus. It should be emphasized, however, that the FSP temperature is a complex function of tool rotation speed, tool travel speed, tool material and design, etc. and therefore it is difficult to control precisely. In some cases, FSP temperature may exceed the β-transus [14]. The typical measured temperature profiles are characterized by a steep rise to the peak temperature followed by a relatively slow cool down, as shown in Fig. 4.2A. The latter

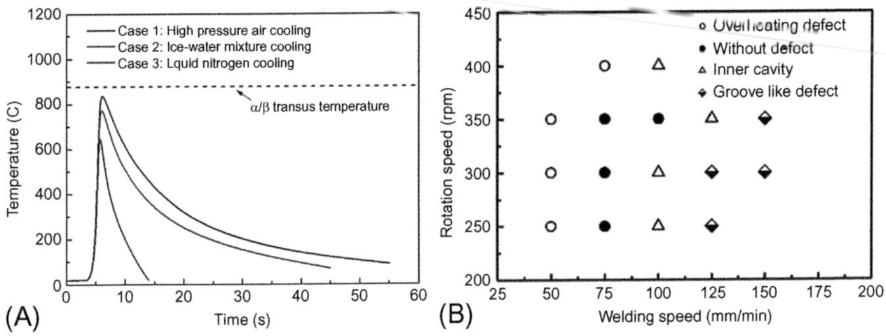

FIG. 4.2

Measured temperature profiles (A) and FSP processing window (B).

Panel (A): After N. Xu, Q. Song, Y. Bao, Y. Jiang, J. Shen, X. Cao, Twinning-induced mechanical properties modification of CP-Ti by friction stir welding associated with simultaneous backward cooling, Sci. Technol. Weld. Join. 22 (2017) 610–616, https://doi.org/10.1080/13621718.2017.1286444. Panel (B): After F.C. Liu, J. Liao, Y. Gao, N. Nakata, Influence of texture on strain localization in stir zone of friction stir welded titanium, J. Alloys Compd. 626 (2015) 304–308, https://doi.org/10.1016/j.jallcom.2014.12.011.

effect is most likely attributable to the relatively low thermal conductivity of titanium. To minimize the postprocessing microstructure coarsening, the cooling rate is sometimes enhanced by using a specially designed cooling systems, as exemplified in Fig. 4.2A.

The characteristic of FSP of commercial-purity titanium is a very narrow processing window (Fig. 4.2B). The subtle changes of process variables may result in macroscale defects and thus have dramatic effects on the quality of the processed material. As a result, a very tight process control is necessary. The limited processing window is thought to be associated with low heat conductivity of titanium which presumably results in large temperature gradients within the stir zone. Considering that the tool shoulder is well accepted to be the main heat source during friction-stirring [4], a relatively low temperature is expected at the stir zone root. Indeed, the macroscale defects are often observed in this location (see, e.g., Ref. [12]).

In addition to the above issues, the processing machine should be capable to sustain the axial forging load as high as 50 kN because of the relatively high strength of commercial-purity titanium at working temperatures. Moreover, a shield of inert gas must be applied to minimize material oxidation.

4.3 MACROSCALE STRUCTURE

Typical low-magnification overview of transverse cross-section of the friction-stir processed material is shown in Fig. 4.3. Three characteristic microstructural zones may readily be identified: the stir zone, the base material, and a narrow transition region—the so-called thermomechanically affected zone (TMAZ).

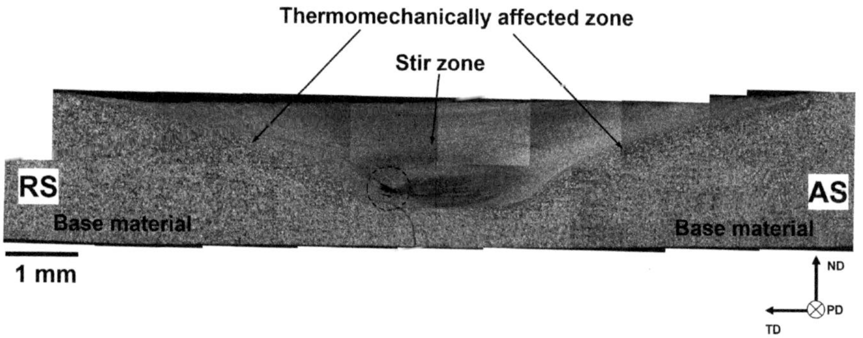

FIG. 4.3

Low-magnification optical overview of the transverse cross-section of the friction-stir processed material. Note: RS and AS are retreating side and advancing side, respectively; PD, ND, and TD are processing direction, normal direction, and transverse direction, respectively. The selected area shows tunnel-type defect in the bottom section of stir zone.

After S. Mironov, Y.S. Sato, H. Kokawa, Development of grain structure during friction stir welding of pure titanium, Acta Mater. 57 (2009) 4519–4528, https://doi.org/10.1016/j.actamat.2009.06.020.

In the stir zone, the material is entrapped by the rotating tool thus being "stirred." Accordingly, the material in this microstructural region experiences the largest strain and the highest temperature which typically give rise to drastic grain refinement. In the TMAZ, the material also undergoes deformation at elevated temperature but it does not directly contact with the rotating tool. As a result, the microstructural changes here are less severe and the evolved microstructure is transient in nature between the base material and the stir zone.

The stir zone is usually characterized by substantial variation in optical contrast (Fig. 4.3) thus evidencing inhomogeneous microstructure distribution. The darkest contrast (i.e., the finest microstructure) is normally found at the stir zone root.

The microhardness profiles measured across the processing material typically reveals material strengthening in the stir zone [5–7,9], as exemplified in Fig. 4.4. This effect is usually attributed to the underlying grain refinement [5–7,9]. In more detail, this issue is considered in the following section.

4.4 MICROSTRUCTURE OF THE STIR ZONE

4.4.1 STRUCTURE MORPHOLOGY

The typical microstructure developed in the stir zone is characterized by a relatively high proportion of low-angle boundaries (Fig. 4.5) which may consist of ~50% of total grain boundary area [12]. The evolved microstructure is very specific and is difficult to describe in conventional terms of grains, as shown in high-magnification inserts in Fig. 4.5. Instead, it is rather dominated by irregular mixture of low- and high-angle boundaries. In more detail, this microstructure morphology is discussed in Section 4.5.

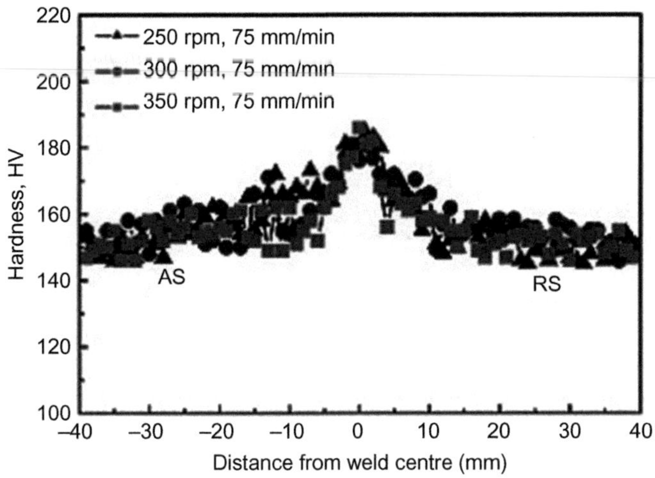

FIG. 4.4

Typical microhardness profiles measured across friction-stir processed material. Note: The remark in the *upper left corner* indicates processing conditions in terms of tool rotational speed and tool translation rate.

After F.C. Liu, J. Liao, Y. Gao, N. Nakata, Influence of texture on strain localization in stir zone of friction stir welded titanium, J. Alloys Compd. 626 (2015) 304–308, https://doi.org/10.1016/j.jallcom.2014.12.011.

The typical grain intercept length is reported to be ~5 μm [12] thus being beyond the ultrafine-grained range. At least partially, this effect is thought to be attributable to the high-temperature character of the FSP process (Fig. 4.2A). Importantly, the microstructure distribution is sometimes reported to be inhomogeneous within a stir zone with the finest grain size found in its bottom section [6]. This observation suggests a heterogeneous character of temperature distribution, as noted in Section 4.2.

Usually, the stir zone microstructure contains limited fraction of mechanical twins, presumably due to relatively high processing temperature. On the other hand, dislocation density is reported to be relatively high [7,8] (Fig. 4.6) thus indicating an incompleteness of the recrystallization process.

4.4.2 TEXTURE

It is well accepted that grain refinement during large plastic deformation is closely linked to texture development (see, e.g., Ref. [15]). Therefore, texture analysis may provide a useful insight into the grain structure evolution during FSP.

Although material flow during FSP appears to be very complicated, the predominant deformation mode is expected to be simple shear [16]. Ideal orientations of hexagonal close-packed metals (including titanium) under simple shear are given in Fig. 4.7A and Table 4.1 [17,18]. Importantly, the shear plane and direction during FSP are not stationary but follow the rotating tool [16]. The geometrical arrangement of the possible shear directions within the stir zone appears to be a circle, as

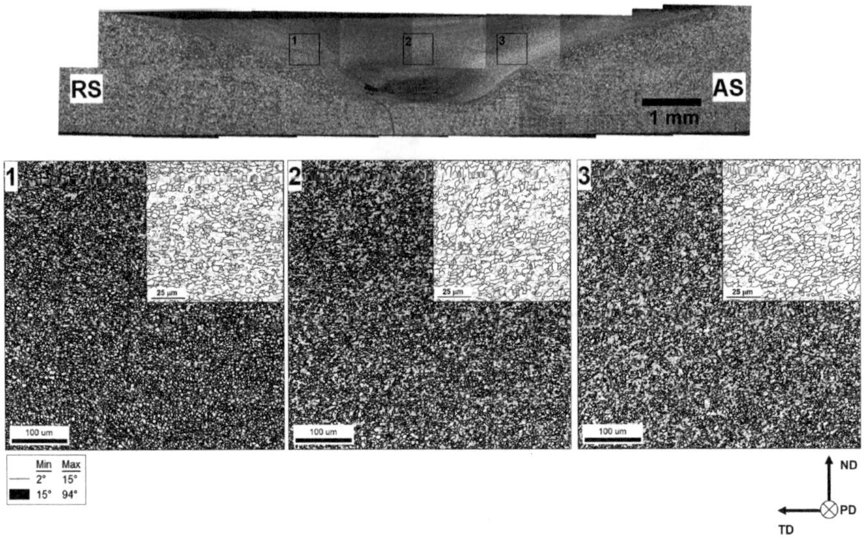

FIG. 4.5

EBSD grain boundary maps showing microstructure distribution in stir zone. Note: High-resolution EBSD maps of the microstructure are given in the *top-right corners* of the low-resolution maps. In the maps, low-angle boundaries are depicted as *thin red lines* (*gray* in the print version) and high-angle boundaries as *solid black lines*, respectively.

After S. Mironov, Y.S. Sato, H. Kokawa, Development of grain structure during friction stir welding of pure titanium, Acta Mater. 57 (2009) 4519–4528, https://doi.org/10.1016/j.actamat.2009.06.020.

illustrated in Fig. 4.7B. Orientation of the shear plane is generally not so evident, but is often assumed to be nearly parallel to the stir zone borders. If so, the shear surface may be expected to be close to the truncated cone, as shown in Fig. 4.7B.

The typical textures measured in friction-stirred titanium are shown in Fig. 4.8A. However, considering a specific orientations of shear surface and shear line (Fig. 4.7B), the measured textures typically require an appropriate rotation before analysis. The appropriately rotated textures are usually dominated by the partial P-fiber $\{hkil\}\ \langle\bar{2}110\rangle$ texture [5–7,9,11–13], as exemplified in Fig. 4.8B. This indicates a prevalence of prism slip, thus being in excellent agreement with well accepted concept of plastic strain of α-titanium. In some cases, however, evidence of B-fiber $\{0001\}\ \langle uvtw\rangle$ textures are also found [7,9] thus indicating an additional activation of the basal slip.

4.4.3 MISORIENTATION DISTRIBUTION

Typical misorientation distribution measured in the stir zone is shown in Fig. 4.9. To illustrate the relationship between the misorientation distribution and texture, the so-called uncorrelated (or texture-derived) distribution is also given in the figure. In contrast to the measured distribution displaying the misorientation data between

FIG. 4.6

TEM images of stir zone microstructure produced at processing speed of 50 mm/min (A), 100 mm/min (B), 200 mm/min (C), 300 mm/min (D).

After H. Fujii, Y. Sun, H. Kato, K. Nakata, Investigation of welding parameter dependent microstructure and mechanical properties in friction stir welded pure Ti joints, Mater. Sci. Eng. A 527 (2010) 3386–3394, https://doi.org/10.1016/j.msea.2010.02.023.

neighboring pixels in an EBSD map, the uncorrelated distribution is calculated assuming no spatial correlation between the pixels. In other words, it accounts for all possible misorientations between the sampled pixels (including noncontiguous ones).

Typically, the experimental misorientation-angle distributions are characterized by a sharp low-angle peak, a much broader irregular maximum in the angular range of 5–60 degrees, and a wide gap in the misorientation interval of ~60–94 degrees [6,9,12], as shown in Fig. 4.9. In the case of the relatively low-temperature FSP, however, it may additionally contain a twin-induced peak near 90 degrees [9]. The measured misorientation-axis distributions usually exhibit dense clusters near the $\langle 0001 \rangle$ and $\langle 2\bar{1}\bar{1}0 \rangle$ poles (Fig. 4.9), these clusters being produced mainly by the 5–30 degrees and 5–60 degrees boundaries, respectively [12].

Of particular importance is a good match between the experimental and texture-derived distributions (Fig. 4.9). This gives the confidence that the development of deformation-induced boundaries during FSP was significantly influenced by the texture evolution. Indeed, the formation of the partial P-fiber texture in the stir zone implies an alignment of the $\langle 11\bar{2}0 \rangle$ slip direction with macroscale shear direction.

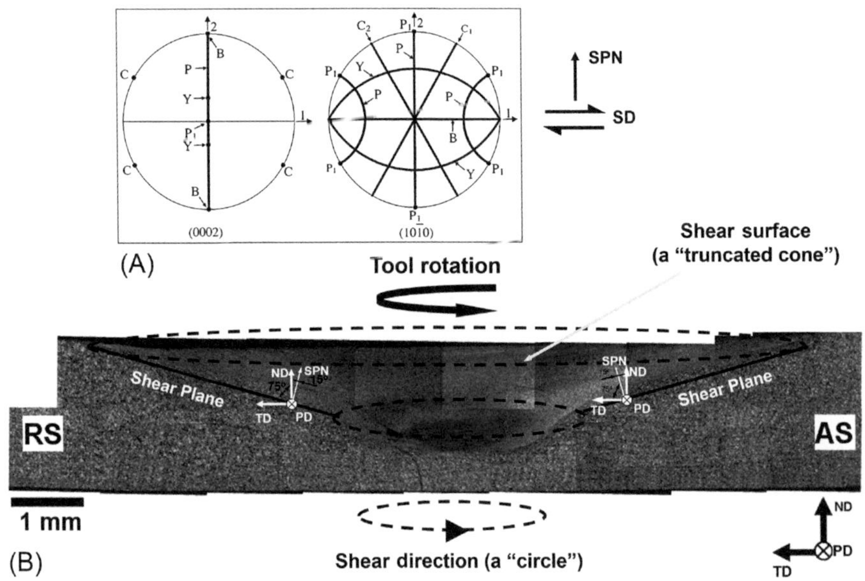

FIG. 4.7

(A) Ideal orientations of hcp metals under simple shear as they appear in {0001} and {1010} pole figures and (B) schematic representation of typical orientations of shear plane and shear directions within the stir zone. Note: SPN and SD are shear plane normal and shear direction, respectively.

Panel (A): After B. Beausir, L.S. Toth, K.W. Neale, Ideal orientations and persistence characteristics of hexagonal close packed crystals in simple shear, Acta Mater. 55 (2007) 2695–2705, https://doi.org/10.1016/j.actamat.2006.12.021. Panel (B): After S. Mironov, Y.S. Sato, H. Kokawa, Development of grain structure during friction stir welding of pure titanium, Acta Mater. 57 (2009) 4519–4528, https://doi.org/10.1016/j.actamat.2009.06.020.

If so, the grains are arbitrarily rotated around this axis with the maximal intensity in the angular interval $\sim\pm50$–60 degrees, as shown in Fig. 4.8B. Therefore, the broad irregular maximum in the angular range of ~5–60 degrees (and the gap beyond ~60 degrees) as well as the dense cluster near the $\langle 2\bar{1}\bar{1}0\rangle$ pole in Fig. 4.9 simply resulted from the partial P-fiber texture developed in the stir zone.

4.5 MICROSTRUCTURAL EVOLUTION IN TRANSITION ZONE

The transition nature of the microstructure in the thermomechanically affected zone is a key issue for understanding how the original grain structure was transformed into the stir zone microstructure during FSP. Therefore, this microstructural region is often used for analysis of microstructural evolution. Fig. 4.10A shows an EBSD map taken from this zone with selected areas given at higher

Table 4.1 Main Ideal Orientations in Simple Shear Deformation of hcp Metals

Notation	Euler Angles (φ_1, Φ, φ_2) (degrees)	{*hkil*}⟨*uvtw*⟩	Fiber Axis and Its Alignment	Associated Slip Modes
B-fiber	(0, 90, 0–60)	{0001}⟨*uvtw*⟩	⟨0001⟩// shear plane normal	$(0002)\langle 1\bar{2}10\rangle$, $\{10\bar{1}1\}\langle 1\bar{2}10\rangle$, $\{10\bar{1}1\}\langle \bar{1}\bar{1}23\rangle$
P-fiber	(0, 0–90; 0)	$\{hkil\}\langle \bar{2}110\rangle$	$\langle 2\bar{1}\bar{1}0\rangle$// shear direction	$\{10\bar{1}0\}\langle 1\bar{2}10\rangle$, $(0002)\langle 1\bar{2}10\rangle$, $\{10\bar{1}1\}\langle 1\bar{2}10\rangle$
Y-fiber	(0, 30, 30–60)	—	—	$\{10\bar{1}1\}\langle 1\bar{2}10\rangle$
C-fiber	(60, 90, 0–60)	—	—	$\{10\bar{1}1\}\langle \bar{1}\bar{1}23\rangle$, $\{11\bar{2}2\}\langle \bar{1}\bar{1}23\rangle$

After B. Beausir, L.S. Toth, K.W. Neale, Ideal orientations and persistence characteristics of hexagonal close packed crystals in simple shear, Acta Mater. 55 (2007) 2695–2705, https://doi.org/10.1016/j.actamat.2006.12.021; S. Li, Orientation stability in equal channel angular extrusion. Part II: hexagonal close-packed materials, Acta Mater. 56 (2008) 1031–1043, https://doi.org/10.1016/j.actamat.2007.11.003.

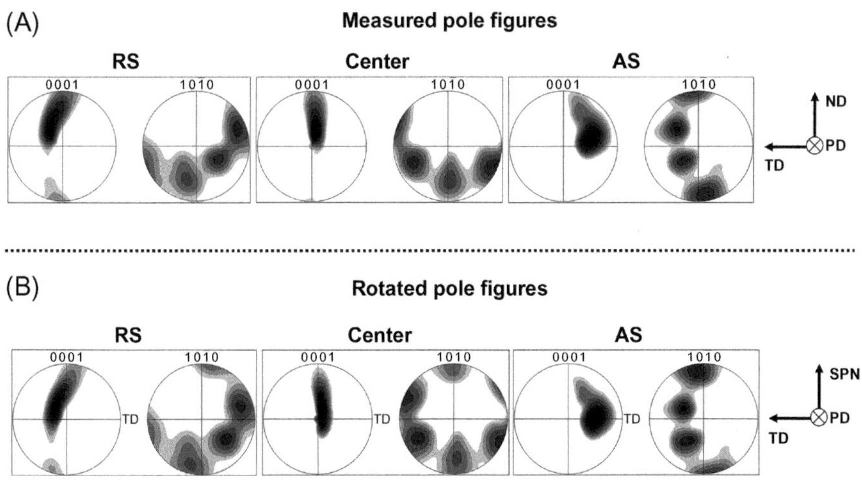

FIG. 4.8

Typical textures developed in stir zone.

After S. Mironov, Y.S. Sato, H. Kokawa, Development of grain structure during friction stir welding of pure titanium, Acta Mater. 57 (2009) 4519–4528, https://doi.org/10.1016/j.actamat.2009.06.020.

magnifications in Fig. 4.10B–D. For simplicity, the large EBSD map was divided into three microstructural areas that correspond to relatively low strain, moderate strain, and high strains, as indicated in Fig. 4.10A. Microstructural changes in these areas are considered in the following three subsections.

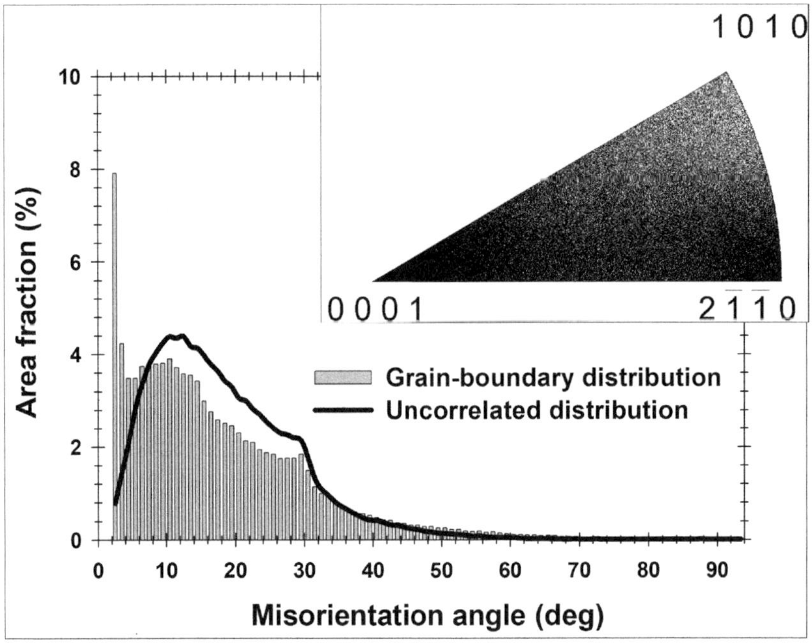

FIG. 4.9

Typical misorientation distribution measured in stir zone. Note: Misorientation-axis distribution is shown in the *top-right corner.*

After S. Mironov, Y.S. Sato, H. Kokawa, Development of grain structure during friction stir welding of pure titanium, Acta Mater. 57 (2009) 4519–4528, https://doi.org/10.1016/j.actamat.2009.06.020.

4.5.1 LOW STRAINS

Near the outer edge of the transition zone, original grains tended to be aligned in a common direction (Fig. 4.10A). This effect is often observed in friction-stirred materials, being usually attributable to geometrical effect of strain. Within the grains, an extensive development of deformation-induced low-angle boundaries was observed (Fig. 4.10B). Frequently they extended across entire grains. The extended deformation-induced boundaries were reasonably straight in a grain scale and their traces were often close to the traces of the $\{10\overline{1}0\}$ or $\{11\overline{2}0\}$ prism slip planes. A typical example of the crystallographic alignment of the boundaries is shown in the high-magnification insert in the bottom-right corner of Fig. 4.10B.

An important observation was the clustering of the misorientation axes of the deformation-induced low-angle boundaries near $\langle 0001 \rangle$ pole (Fig. 4.11A). The rotation axes of dislocation boundaries provide constraints on the slip systems which are potentially active during deformation. It is well accepted that the slip activity of a particular system rotates a crystalline lattice around the axis $\overline{w} \parallel (\overline{b} \times \overline{n})$, where $\overline{b}$ is the Burgers vector and $\overline{n}$ is the slip plane normal. Among all possible slip modes available in α-titanium (prism, basal, and pyramidal), the prism slip *only* provides a

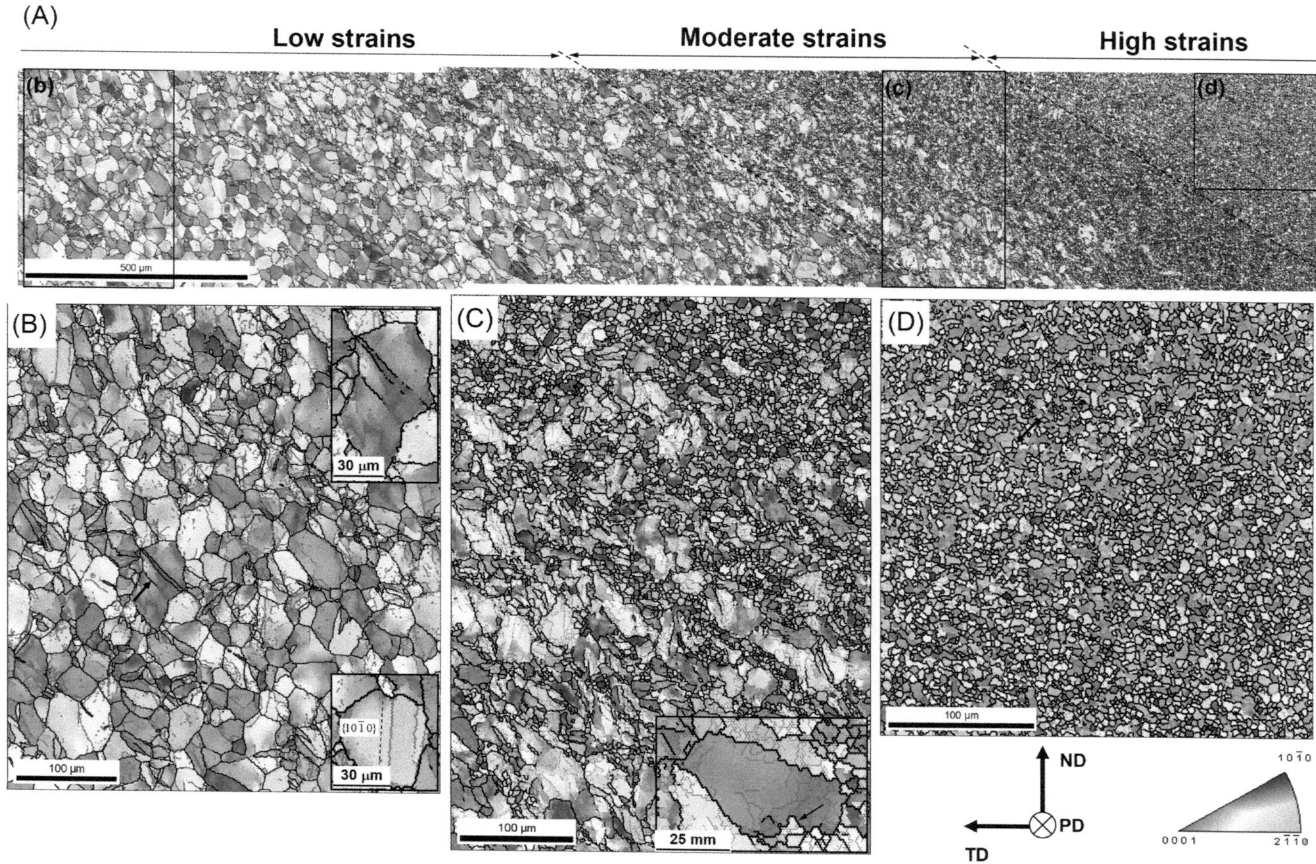

FIG. 4.10

Composite EBSD map showing microstructure evolution in thermomechanically affected zone (A) with selected areas shown in higher magnifications in (B)–(D). The individual grains are *colored* according to their crystallographic directions relative to PO; *color* (*different gray shades* in the print version) *code triangle* is shown in the *bottom right corner*. Low-angle boundaries and high-angle boundaries are depicted as *gray* and *black lines*, respectively.

After S. Mironov, Y.S. Sato, H. Kokawa, Development of grain structure during friction stir welding of pure titanium, Acta Mater. 57 (2009) 4519–4528, https://doi.org/10.1016/j.actamat.2009.06.020.

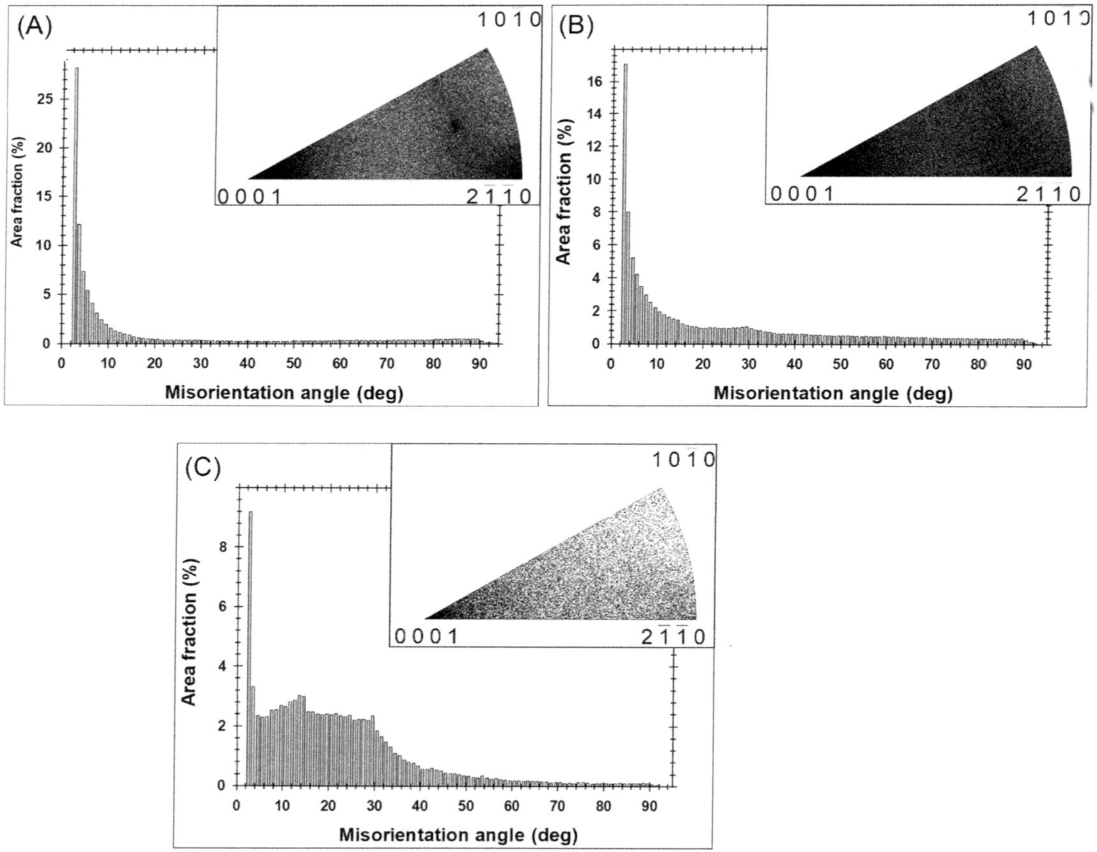

FIG. 4.11

Evolution of misorientation distributions in thermomechanically affected zone: low strains (A), moderate strains (B), and high strains (C). Note: The appropriate misorientation-axis distributions are shown in the *top-right corners*.

After S. Mironov, Y.S. Sato, H. Kokawa, Development of grain structure during friction stir welding of pure titanium, Acta Mater. 57 (2009) 4519–4528, https://doi.org/10.1016/j.actamat.2009.06.020.

lattice rotation around the $\langle 0001 \rangle$. Thus, the observed proximity of the misorientation axes of the low-angle boundaries to $\langle 0001 \rangle$ as well as their alignment with the prism planes indirectly indicates enhanced activity of the prism slip.

The deformation-induced boundaries within the grains were usually low-angle in nature, but some their segments have already accumulated misorientation exceeding 15 degrees (an example shown with an arrow in Fig. 4.10B). This indicates a gradual transformation of low-angle boundaries into high-angle ones.

Sporadic twins were observed in some grains. Remarkably, misorientations across the twin boundaries often deviated from the exact twin-matrix relationship. This process is illustrated in the high-magnification insert shown in the top-right corner of Fig. 4.10B; in the figure, the boundaries of the $\{10\bar{1}2\}$ twins (within 5-degree tolerance) are highlighted in red. It is seen that the twin boundaries gradually transformed into random (not twin) grain boundaries (shown in black). This is a well-known phenomenon which is usually attributable to the deformation-induced crystallographic rotations of the twins and matrix from their initial orientations (see, e.g., Ref. [19]).

4.5.2 MODERATE STRAINS

Approaching the stir zone, original grains became thinner with higher aspect ratios, thus reflecting an increase in strain (Fig. 4.10C). Moreover, the boundaries of the parent grains became wavy, thereby indicating local grain-boundary migration at elevated temperature. The details of this process are illustrated in the high-magnification insert shown in the bottom-right corner of Fig. 4.10C. It is seen that the segments of the original grain boundaries were bowed at the triple points created by transverse low-angle boundaries. This phenomenon may be interpreted in terms of local imbalance between different mobility of low-angle boundaries and high-angle boundaries [20]. The development of this process gave rise to grain-boundary bulges (an example shown with an arrow in the high-magnification insert in Fig. 4.10C), which eventually transformed into fine grains. These fine grains typically have equiaxed morphology and contained almost no substructure. Closer to the stir zone, they increase in volume fraction, creating a "necklace" structure decorating the elongated parent grains (Fig. 4.10C).

Remarkably, the misorientation distribution in this microstructural area was characterized by a substantial randomization of the misorientation axes (Fig. 4.11B). Together with the grain-boundary bulging mentioned earlier, this observation indirectly indicated that the fine grains presumably originated from *discontinuous* recrystallization.

4.5.3 HIGH STRAINS

At the stir zone extremity, microstructure morphology drastically changed (Fig. 4.10D). The original grains were broken down and the evolved microstructure was very specific. To a first approximation, it could be described in terms of a "matrix," which consisted of irregular mixture of low- and high-angle boundaries, and fine equiaxed grains. The microstructure was evidently directional (Fig. 4.10D), tending to align roughly parallel to the border between the transition zone and the stir zone (Fig. 4.10A), i.e., the presumed macroscale shear plane (Fig. 4.7B).

The formation of the "matrix" structure was most likely attributable to extensive development of deformation-induced boundaries. Moreover, the formation of the characteristic misorientation distribution (Fig. 4.11C) indicates a close relationship between the grain-boundary development and texture, as discussed in Section 4.4.3. Therefore, the underlying microstructural process appears to resemble continuous recrystallization. On the other hand, the fine equiaxed grains presumably originate from grain boundary bulging, as discussed in the previous section, and, if so, they reflect a concurrent development of discontinuous recrystallization. Finally, the clear alignment if the grain structure with geometry of macroscale shear evidences geometrical effect of strain.

4.6 SUMMARY

In this chapter, the current understanding of FSP commercial-purity titanium was briefly reviewed. The main conclusions are as follows.

Considering strict requirements to processing conditions, this technique does not currently appear to straightforward for widespread applications. Moreover, accounting for relatively high-temperature character of FSP process, it is unlikely readily applicable for production of ultrafine-grained structure. Further optimization of this technology is required.

The final microstructure produced in the stir zone is characterized by a mean grain size of ~5 μm, proportion of low-angle boundaries of ~50% and relatively strong P-fiber $\{hkil\}$ $\langle\bar{2}110\rangle$ texture.

Microstructural evolution during processing appears to be a complex process involving continuous recrystallization, discontinuous recrystallization, and geometrical effect of strain.

REFERENCES

[1] R.S. Mishra, M.W. Mahoney, S.X. McFadden, N.A. Mara, A.K. Mukherjee, High strain rate superplasticity in a friction-stir processed 7075 Al alloy. Scr. Mater. 41 (1999) 163–168, https://doi.org/10.1016/S1359-6462(99)00329-2.

[2] W.M. Thomas, Friction Stir Butt Welding, Int. Patent o. PCT/GB92/02203(1991).

[3] R. Nandan, T. DebRoy, H.K.D.H. Bhadeshia, Recent advances in friction-stir welding—process, weldment structure and properties. Prog. Mater. Sci. 53 (2008) 980–1023, https://doi.org/10.1016/j.pmatsci.2008.05.001.

[4] R.S. Mishra, Z.Y. Ma, Friction stir welding and processing. Mater. Sci. Eng. R 50 (2005) 1–78, https://doi.org/10.1016/j.mser.2005.07.001.

[5] F.C. Liu, J. Liao, Y. Gao, N. Nakata, Influence of texture on strain localization in stir zone of friction stir welded titanium. J. Alloys Compd. 626 (2015) 304–308, https://doi.org/10.1016/j.jallcom.2014.12.011.
[6] H. Liu, K. Nakata, N. Yamamoto, J. Liao, Grain orientation and texture evolution in pure titanium lap joint produced by friction stir welding. Mater. Trans. 51 (2010) 2063–2068, https://doi.org/10.2320/matertrans.M2010242.
[7] H. Fujii, Y. Sun, H. Kato, K. Nakata, Investigation of welding parameter dependent microstructure and mechanical properties in friction stir welded pure Ti joints. Mater. Sci. Eng. A 527 (2010) 3386–3394, https://doi.org/10.1016/j.msea.2010.02.023.
[8] W.-B. Lee, C.-Y. Lee, W.-S. Chang, Y.-M. Yeon, S.-B. Jung, Microstructural investigation of friction stir welded pure titanium. Mater. Lett. 59 (2005) 3315–3318, https://doi.org/10.1016/j.matlet.2005.05.064.
[9] N. Xu, Q. Song, Y. Bao, Y. Jiang, J. Shen, X. Cao, Twinning-induced mechanical properties modification of CP-Ti by friction stir welding associated with simultaneous backward cooling. Sci. Technol. Weld. Join. 22 (2017) 610–616, https://doi.org/10.1080/13621718.2017.1286444.
[10] H. Liu, K. Nakata, N. Yamamoto, J. Liao, Friction stir welding of pure titanium lap joint. Sci. Technol. Weld. Join. 15 (2010) 428–432, https://doi.org/10.1179/136217110X12731414740031.
[11] S. Bahl, P.L. Nithilaksh, S. Suwas, S.V. Kailas, K. Chatterjee, Processing-microstructure-crystallographic texture-surface property relationships in friction stir processing of titanium. JMEPEG 26 (2017) 4206–4216, https://doi.org/10.1007/s11665-017-2865-6.
[12] S. Mironov, Y.S. Sato, H. Kokawa, Development of grain structure during friction stir welding of pure titanium. Acta Mater. 57 (2009) 4519–4528, https://doi.org/10.1016/j.actamat.2009.06.020.
[13] Y. Zhang, Y.S. Sato, H. Kokawa, S.H.C. Park, S. Hirano, Grain structure and microtexture in friction stir welded commercial purity titanium. Sci. Technol. Weld. Join. 15 (2010) 500–505, https://doi.org/10.1179/136217110X12785889549624.
[14] Y. Zhang, Y.S. Sato, H. Kokawa, S.H.C. Park, S. Hirano, Stir zone microstructure of commercial purity titanium friction stir welded using pcBN tool. Mater. Sci. Eng. A 488 (2008) 25–30, https://doi.org/10.1016/j.msea.2007.10.062.
[15] D.A. Hughes, N. Hansen, High angle boundaries formed by grain subdivision mechanism. Acta Mater. 45 (1997) 3871–3886, https://doi.org/10.1016/S1359-6454(97)00027-X.
[16] R.W. Fonda, K.E. Knipling, Texture development in friction stir welds. Sci. Technol. Weld. Join. 16 (2011) 288–294, https://doi.org/10.1179/1362171811Y.0000000010.
[17] B. Beausir, L.S. Toth, K.W. Neale, Ideal orientations and persistence characteristics of hexagonal close packed crystals in simple shear. Acta Mater. 55 (2007) 2695–2705, https://doi.org/10.1016/j.actamat.2006.12.021.
[18] S. Li, Orientation stability in equal channel angular extrusion. Part II: hexagonal close-packed materials. Acta Mater. 56 (2008) 1031–1043, https://doi.org/10.1016/j.actamat.2007.11.003.
[19] S.Y. Mironov, G.A. Salishchev, M.M. Myshlyaev, R. Pippan, Evolution of misorientation distribution during warm "abc" forging of commercial-purity titanium. Mater. Sci. Eng. A 418 (2006) 257–267, https://doi.org/10.1016/j.msea.2005.11.026.
[20] P.B. Prangnell, J.S. Hayes, J.R. Bowen, P.J. Apps, P.S. Bate, Continuous recrystallisation of lamellar deformation structures produced by severe plastic deformation. Acta Mater. 52 (2004) 3193–3206, https://doi.org/10.1016/j.actamat.2004.03.019.

CHAPTER

Production of bulk nanocrystalline mill products by conventional metalforming methods

5

Sergey Zherebtsov*, Gennady Salishchev*, S. Lee Semiatin[†]

*Laboratory of Bulk Nanostructured Materials, Belgorod State University, Belgorod, Russia**

Air Force Research Laboratory, Materials and Manufacturing Directorate, AFRL/RXCM, Wright-Patterson Air Force Base, Dayton, OH, United States[†]

5.1 INTRODUCTION

Typically, the formation of nanocrystalline (NC) or ultrafine-grained (UFG) microstructures in metallic materials is associated with various methods of severe plastic deformation (e.g., equal-channel angular pressing or high-pressure torsion; see Chapters 1 and 2). However, these methods require specialized equipment and pose limitations with regard to product assortment, workpiece/perform size, tool material/tool life, etc. By contrast, microstructure refinement during large (or severe) deformation was studied initially using conventional processes such as rolling and extrusion [1–3]. Because the processing of titanium alloys often involves metalworking operations such as rolling, forging, and extrusion, it can be hypothesized that NC titanium and titanium alloys can be produced using conventional metal deformation processes based on large strains at relatively-low temperatures.

The evolution of severely-strained microstructures has been the subject of extensive research focused primarily on cubic materials. Microstructure refinement during large deformation is usually associated with the formation of deformation-induced high-angle grain boundaries (HAGBs) [1–3], including (i) geometrically necessary boundaries (GNBs) separating regions that deform by different combinations of slip systems and (ii) incidental dislocation boundaries consisting of ordinary cell boundaries that form by trapping glide dislocations. The formation of GNBs with moderate-to-high misorientations requires a relatively high level of strain [2].

At ambient temperature, pure titanium (having a hexagonal crystal structure) has only four independent deformation modes comprising basal or prismatic slip along the $\boldsymbol{a}$ Burgers vector in the basal plane. Therefore, tension or compression along the $\boldsymbol{c}$ axis of titanium crystals must be accommodated by twinning and/or $\boldsymbol{c}+\boldsymbol{a}$ pyramidal

Nanocrystalline Titanium. https://doi.org/10.1016/B978-0-12-814599-9.00005-5

slip. As one of the main deformation modes, twinning in titanium typically occurs and is most readily observed during the early stages of deformation [4]. Being high-angle in nature, twin boundaries transform from well-defined coincident-site-lattice-type boundaries into arbitrary ones during subsequent interaction with lattice dislocations [5], thereby refining the microstructure considerably. Twin boundaries can have a marked influence on the evolution/refinement of microstructure during further straining, i.e., following the saturation of twinning. However, the nature of such microstructure evolution has been investigated to only a limited extent [1,6].

The challenge of producing an NC structure in commercially pure (CP) titanium via conventional deformation processes consists of two parts which are discussed in this chapter: (i) the efficiency of various deformation processes for the formation of the NC structure and (ii) the mechanisms and kinetics of microstructure evolution depending on specific deformation conditions. The mechanical properties of titanium alloys produced by various methods are discussed in Chapter 6.

5.2 PRODUCTION OF NANOCRYSTALLINE SHEETS BY FLAT ROLLING

One obvious way to increase twinning and thus refine the grain size is to decrease the deformation temperature [7]. However, the deformability of CP Ti decreases with decreasing temperatures, thus limiting the choice of working method. Flat (plane-strain) rolling imposes a largely-compressive state of stress, thus enabling deformation of CP Ti at room and even cryogenic temperatures to large strains [5,8–10]. For example, a very large strain can be attained by so-called accumulative roll bonding (ARB) at room temperature [11,12]. In this method, a metal sheet is rolled to $\sim$50% thickness reduction, then cut in two and stacked together, thereby restoring the original thickness of the sheet, and rolled again. The microstructure formed in CP Ti by ARB after a strain $\varepsilon \sim 8$ consists of elongated, heavily-deformed areas and very fine (80–100 nm) grains. Unfortunately, because of the rather high strain imposed during each pass, the influence of twinning on microstructure refinement could not be ascertained comprehensively in Ref. [12].

This section summarizes the evolution of microstructure in CP Ti during cold and cryogenic rolling to large thickness strains and quantifies the influence of deformation twinning on microstructure refinement via electron backscatter diffraction (EBSD), transmission electron microscopy (TEM), and optical microscopy. At the end of the section, an example of a large ($500 \times 250 \times 0.3\,mm^3$) NC sheet produced by flat rolling is shown.

The material for the present investigation comprised a 4-mm thick sheet of CP titanium. In the as-received condition, the microstructure of the program material consisted of equiaxed grains with an average size of 15 μm (Fig. 5.1A). The grains contained no twins and essentially no internal substructure.

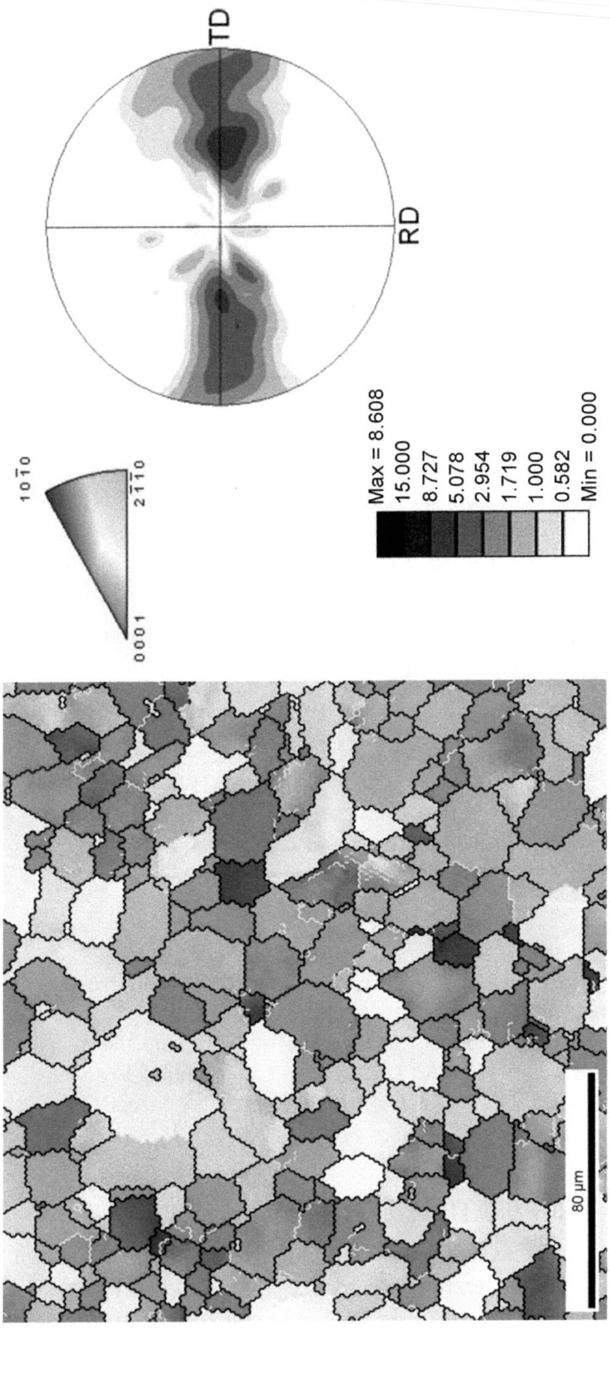

FIG. 5.1

(A) Normal direction EBSD inverse-pole-figure map taken from the rolling plane of the as-received CP Ti program material and (B) the corresponding (0002) pole figure.

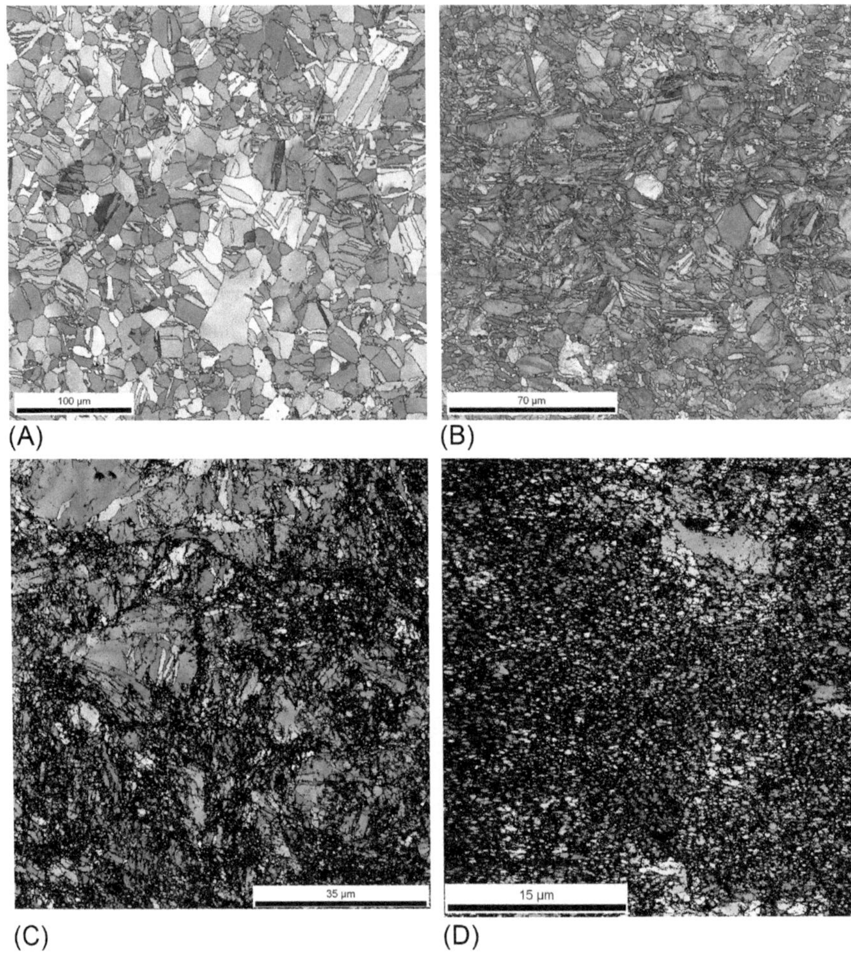

FIG. 5.2

Normal direction EBSD inverse-pole-figure maps taken on the mid-thickness rolling plane of CP Ti samples cold rolled to a thickness strain ε_{th} of (A) 0.1, (B) 0.36, (C) 0.92, or (D) 2.66. EBSD maps of highly-deformed material ($\varepsilon_{th}=0.92$ and 2.66) are shown in the as-scanned conditions for points with CI > 0.1 (C, D). The map corresponding to $\varepsilon_{th}=2.66$ (D) was obtained after annealing at 623 K for 1 h. Rolling direction is horizontal in all cases.

5.2.1 ROOM-TEMPERATURE BEHAVIOR

Normal direction (ND) inverse-pole-figure maps taken from the rolling plane of CP Ti cold rolled at room temperature to a thickness strain $\varepsilon_{th}=0.1$, 0.36, 0.92, and 2.66 are shown in Fig. 5.2; the last map was obtained from a specimen additionally annealed at 623 K for 1 h. After $\varepsilon_{th}=0.1$–0.36, a large number of twins was observed

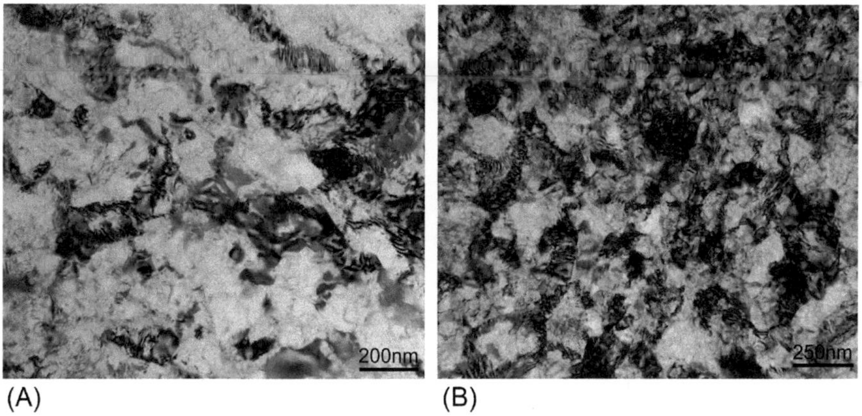

FIG. 5.3

TEM micrographs of cold rolled CP Ti: (A) $\varepsilon_{th}=0.92$ and (B) $\varepsilon_{th}=2.66$.

within the majority of the grains; however, some grains contained no twins even after a strain $\varepsilon_{th}=0.36$ (Fig. 5.2B). A pronounced difference between the microstructures after $\varepsilon_{th}=0.1$ and $\varepsilon_{th}=0.36$ was revealed by the development of substructure in the latter case. Cold rolling to a thickness strain of 0.92 increased the number of small, equiaxed grains considerably and resulted in the formation of dense substructure within large grains (Fig. 5.2C). The largest thickness reduction ($\varepsilon_{th}=2.66$) gave rise to a microstructure comprising a mixture of small, equiaxed grains and large grains containing many subboundaries (Fig. 5.2D).

TEM observations showed that cold rolling to a thickness strain $\varepsilon_{th}=0.92$ gave rise to a cellular microstructure with a high dislocation density (Fig. 5.3A). The cell boundaries were rather wide and loose. The size of the cells varied over a wide range from a hundred nanometers to a few micrometers. Separate (sub)grains with a size of 100–200 nm and thin clear boundaries were also observed in the structure. The dislocation density within such grains was lower than that in the matrix.

Cold rolling to $\varepsilon_{th}=2.66$ led to a considerable refinement of the microstructure and simultaneously to an increase in dislocation density (Fig. 5.3B). The size of cells was found to be approximately 200 nm. Small subgrains (~100 nm) were also seen in the background of the cell structure.

Further analysis of the EBSD data suggested that the initial increase in the fraction of low-angle grain boundaries (LAGBs) was associated mainly with an increase in the number of boundaries with an intermediate low-angle misorientation (Fig. 5.4A and B). Moreover, in the strain interval between 0.1 and 0.92, the fraction of those boundaries which had low-to-medium misorientation was very low. However, the misorientation bin with the fewest boundaries shifted to higher angles with thickness strain. Specifically, there were almost no boundaries with misorientations of ~10 degrees at $\varepsilon_{th}=0.1$, while boundaries with a misorientation of 20 degrees were scarce after 0.36 and 0.92 thickness strain (Fig. 5.4 C and D). Both the trend

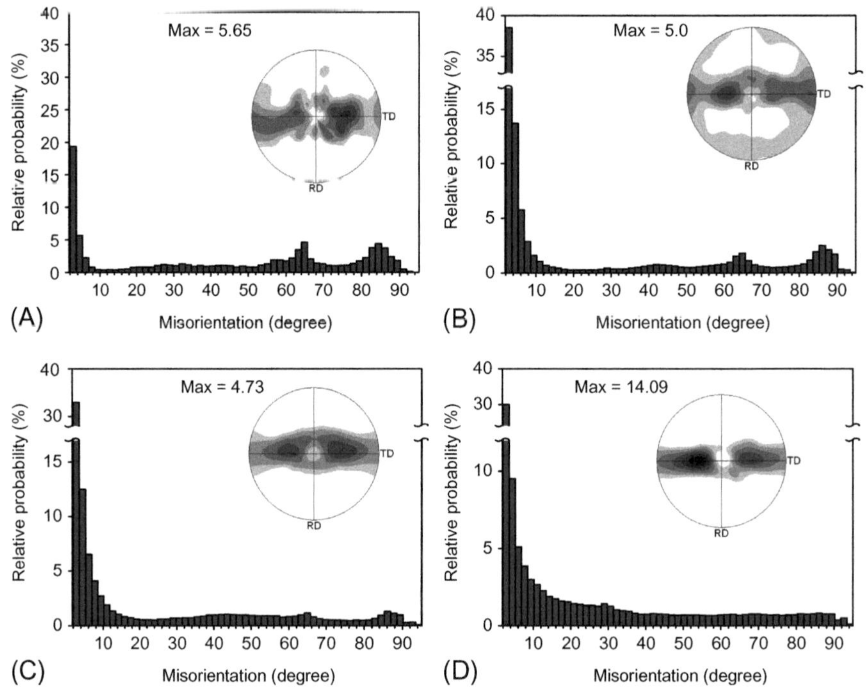

FIG. 5.4

Misorientation distributions and (0002) pole figures of CP Ti cold rolled to a true thickness strain of (A) 0.1, (B) 0.36, (C) 0.92, or (D) 2.66.

in the evolution of the misorientation distributions and the change in the fraction of LAGBs (and the corresponding behavior of the HAGBs) are typical for various metallic materials [13,14] and can be attributed to the gradual increase in cell/subgrain misorientation and the more rapid decrease in HAGB spacing than the subgrain size with thickness strain, respectively.

The high-angle boundaries exhibited two peaks corresponding to 65 and 85 degrees following ε_{th} of 0.1–0.36 (Fig. 5.4A and B). Mirroring the microstructure observations in the inverse-pole-figure maps (Fig. 5.2A and B), these peaks can be associated with the formation of 65.62 degrees $\langle 10\bar{1}0\rangle$ and 84.78 degrees $\langle 2\bar{1}\bar{1}0\rangle$ boundaries corresponding to $\{11\bar{2}2\}\langle 11\bar{2}\bar{3}\rangle$ and $\{10\bar{1}2\}\langle\bar{1}011\rangle$ twins, respectively. The formation of the same twin families was reported in Ref. [6] for CP Ti cold-rolled to ε_{th} between 0.1 and 0.5.

The evolution of the overall texture was typical for the cold rolling of CP Ti [6,8,10]. In the as-received condition, the (0002) pole figure consisted primarily of a split texture with basal poles tilted ~30 degrees from the ND toward the transverse direction (TD) (Fig. 5.1B); weaker components were located approximately 80–90 degrees from the ND toward the TD. Texture evolution during cold rolling

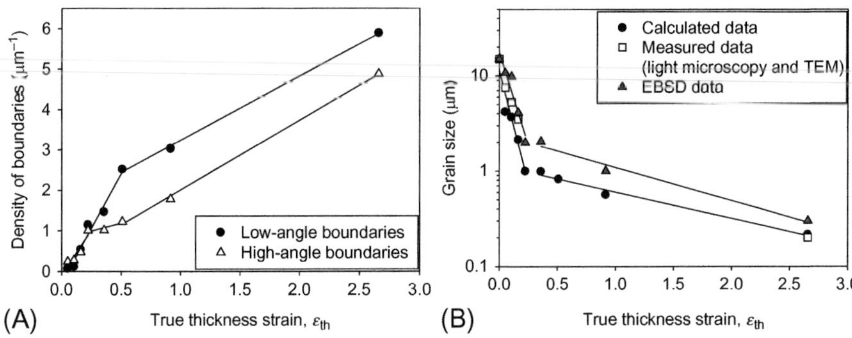

FIG. 5.5

Dependence on thickness strain of (A) the density of low- and high-angle boundaries and (B) grain size. Grain size was measured by optical microscopy and TEM after low and high strain, respectively.

to $\varepsilon_{th}=0.1$–0.36 was associated with some weakening of basal poles and the appearance of new weak peaks due to twinning (Fig. 5.4A and B). By $\varepsilon_{th}=0.92$, the influence of twinning on texture evolution was almost saturated, and the texture (Fig. 5.4C) had started to become similar to that observed in the initial condition. Likewise, pole figures after $\varepsilon_{th}=2.66$ were very similar to that observed in the initial condition (Figs. 5.1B and 5.4D). However, the texture components after heavy cold rolling were sharper and had higher intensity.

Although it is convenient to discuss microstructure evolution in terms of the HAGB fraction, it should be recognized that this characteristic can depend on the evolution of other features; e.g., the fraction of HAGBs can decrease with merely an increase in the number of LAGBs. Hence, more appropriate quantitative insight can be obtained from an analysis of EBSD data in terms of the density of LAGBs and HAGBs (i.e., boundary length per unit area) as a function of thickness strain (Fig. 5.5A). These results suggested three distinct stages in the formation of HAGBs during cold rolling while only two stages were observed for LAGBs.

The first stage in the formation of HAGBs (ε_{th} between 0 and ~0.2 in Fig. 5.5A) can be associated with marked twinning as observed in Fig. 5.2A. Twinning also influences the dislocation structure developed during deformation. The interaction between dislocations and twins in hcp metals gives rise to dislocation pile-ups at the twin–matrix interfaces and local stress concentrations [15]. Such an effect leads to Hall–Petch hardening due to a reduction in the effective slip distance by twin boundaries which act as barriers to slip [16]. Another hardening factor, proposed by Basinski et al. [17] and experimentally demonstrated by Salem et al. [18], arises from the transition of dislocations from glissile to sessile configurations within the twins, thereby contributing to the formation of low-angle dislocation walls and subboundaries (Fig. 5.4A). Because twin formation and the Basinski effect are closely interrelated, it is not surprising that the rates of increase of the densities of HAGBs and LAGBs were almost equal to each other at low levels of deformation.

It should be noted that during twinning, the factors that lead to strengthening (the Hall–Petch effect and the Basinski mechanism) may compete with texture softening due to lattice reorientation of the twinned regions, thereby resulting in a decreased rate of strain hardening [18]. In addition, because twinning accommodates a finite fraction of the imposed strain, the amount of required slip in the matrix (and hence the dislocation activity that controls microstructure refinement) is decreased relative to that necessary in the absence of twinning. Thus, microstructure refinement due to the formation of deformation-induced HAGBs may be shifted to larger strains as a result of twinning. In agreement with previous observations [6,16,18], twinning in the present work was found to saturate at a ε_{th}th $\sim$0.2.

At higher ε_{th}, i.e., $0.2 \leq \varepsilon \leq 0.51$, the rate of increase in LAGB density did not change (Fig. 5.5A). By contrast, a *second* stage of HAGB formation at a rate noticeably lower than that in the first stage was observed. During this stage, the boundaries associated with the twin–matrix interfaces evolved with continued straining into arbitrary HAGBs as a result of twin/dislocation interactions [5]. In addition, dislocation substructure evolved as a result of classical dynamic recovery during which dislocation multiplication was mitigated by dislocation annihilation and the formation of subboundaries [18]. Consequently, after the exhaustion of twinning, the rate of increase of the dislocation density with strain might be expected to be close to that in the material which deforms by dislocation glide alone but at higher initial dislocation density. Nevertheless, a necessary condition for the formation of deformation-induced HAGBs appears to be a high level of stress at boundaries between areas having different slip systems [18] and/or the activation of multiple slip systems [19]. That is to say, deformation inhomogeneity promotes the formation of HAGBs. A decrease in effective grain size, such as that due to twinning, typically results in more homogeneous plastic flow [20] and thereby should shift the formation of deformation-induced HAGBs to larger strains. In this regard, the rate of microstructure refinement with strain has been found to be lower in an aluminum alloy with an initial fine grain microstructure relative to that in the same material with a coarse starting microstructure [21]. For ε_{th} between approximately 0.2 and 0.5, therefore, the density of high-angle boundaries in CP Ti changed slowly, and the majority of microstructure evolution was associated with the formation of substructure within initial grains and twins (Figs. 5.2B and 5.4B). Thus, the second stage may be viewed as an incubation period for the accumulation of a sufficient level of dislocation density for the subsequent formation of HAGBs.

The *third*, and final, stage in microstructure evolution (Fig. 5.5A) occurred at ε_{th} above $\sim$0.5 and was most probably associated with the formation of high-angle deformation-induced boundaries [2] while the rate of formation of LAGBs decreased measurably. Assuming that the new small grains essentially originated from stain-induced subgrains, the subboundaries transformed into grain boundaries resulting in a decreased rate of LAGB formation (Fig. 5.5A).

The internal consistency of the present results was demonstrated by comparing the measured grain size as a function of imposed thickness strain with that calculated from the density of HAGBs, ρ_b, i.e., $d = 1/\rho_b$. The result (Fig. 5.5B) showed

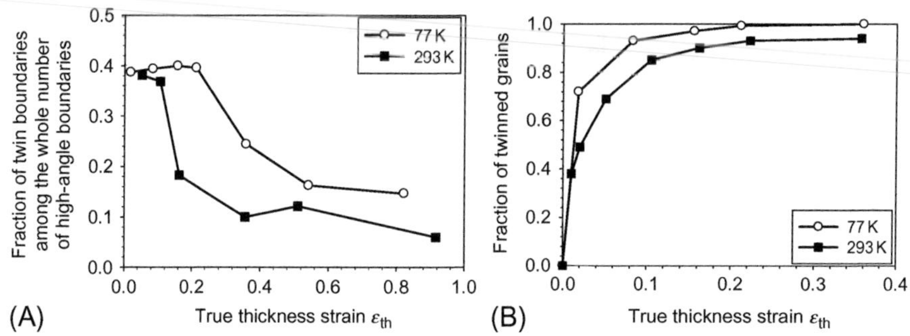

FIG. 5.6

Twin characteristics as a function of true thickness strain at room temperature (293 K) and 77 K: (A) EBSD-determined fraction of high-angle boundaries corresponding to twins and (B) fraction of twinned grains determined by optical microscopy.

reasonable agreement between the calculations and grain-size measurements based on EBSD as well as optical and TEM micrographs. The semilogarithmic plot in Fig. 5.5B also revealed distinct stages corresponding to (i) the twinning regime and (ii) the formation of deformation-induced HAGBs. Twinning resulted in rapid microstructure refinement but saturated at a grain size of only ~1–2 μm, a finding in reasonable agreement with the results of Refs. [8,22,23]. Refinement of the microstructure to grain sizes of the order of 200 nm with further deformation was noted to occur much more slowly. It should be noted that the incubation (second) period in the grain size–thickness strain plot (Fig. 5.5B) was quite small and therefore can hardly be viewed as a distinct stage.

5.2.2 CRYOGENIC VS ROOM-TEMPERATURE BEHAVIOR

A decrease in rolling temperature to 77 K resulted in a considerable increase in deformation twinning and, as a result, in more-pronounced microstructure refinement. The different extents of twinning during rolling at room temperature (RT) vs 77 K were quantified with the aid of EBSD analysis (Fig. 5.6A). Here, the aggregate fraction of twin boundaries among all of the high-angle boundaries is shown as a function of strain; the twin boundaries were binned according to the Brandon criterion (calculated as $\Delta\theta = 15\,\text{degrees}/\sqrt{\Sigma}$ relative to the perfect twin misorientation [24]). The maximum fraction of twin boundaries (~0.4) was similar for both rolling temperatures. However, the range of strains for marked twin formation at 77 K ($\varepsilon_{th} \leq 0.2$) was approximately twice as long as that at room temperature ($\varepsilon_{th} \leq 0.1$). During further deformation, the fraction of twin boundaries meeting the Brandon criterion decreased quite rapidly at room temperature; a minimum value of ~0.1 was attained at $\varepsilon_{th} = 0.2$–0.3. During cryo-rolling, on the other hand, the fraction of boundaries with twin misorientations decreased more slowly, attaining the value of ~0.16 at $\varepsilon_{th} \sim 0.5$. Another metric of the more-extensive twinning at the lower rolling

FIG. 5.7

(A) Grain/subgrain size as a function of true thickness strain ε_{th} in CP Ti rolled at 293 and 77 K and TEM micrographs of CP Ti rolled to a true thickness strain ε_{th} of 2.66 at (B) 293 K or (C) 77 K. The grain size data in (A) were obtained by means of EBSD analysis (for the interval of $\varepsilon_{th}=0$–0.92) and TEM ($\varepsilon_{th}=2.66$).

temperature was the fraction of twinned grains (Fig. 5.6B). Specifically, the fraction of grains containing at least one twin reached 100% by $\varepsilon_{th} \sim 0.2$ during cryo-rolling while the maximum fraction of twinned grains at room temperature was 0.93.

The average grain size during rolling to various strain levels ($\varepsilon_{th} \leq 0.92$) at both temperatures (Fig. 5.7A) was also obtained from EBSD measurements; for the largest strain (i.e., $\varepsilon_{th}=2.66$), the EBSD measurements were supplemented by TEM observations. From a broad perspective, the rate of microstructure refinement was measurably greater in the case of deformation at 77 K (Fig. 5.7A). Moreover, the shapes of the two curves were noticeably different. Nevertheless, the grain size decreased very quickly with strain at both temperatures for $\varepsilon_{th} \leq 0.2$, obviously due to twinning. In the interval of thickness strain between 0.2 and 0.35, the grain size during RT rolling did not change, most likely because of the saturation of twinning. During cryo-rolling, on the other hand, the grain size continued to decrease in the strain interval between 0.2 and 0.92, albeit at a lower rate than at lower strains.

Deformation above $\varepsilon_{th}=0.35$ at room temperature resulted in a gradual decrease in grain/subgrain size to 200 nm at $\varepsilon_{th}=2.66$ (Fig. 5.7B). Large deformation via cryo-rolling (to $\varepsilon_{th}=2.66$) led to a greater degree of microstructure refinement and resulted in a grain/subgrain size of ~80 nm (Fig. 5.7C).

The present results thus indicated that twinning plays a crucial role during the first stage of microstructure evolution. However, as soon as the size of stable grains/subgrains becomes smaller than the average thickness of the twins or the spacing between the twins, twinning and twin interactions per se should not be considered a mechanism of microstructure refinement. According to Refs. [8,22], the minimum grain size at which twinning can still operate at room temperature is d ~ 0.7–0.9 μm. The rate of microstructure formation and the size of subgrains/grains during further rolling depend on the strain hardening/dynamic recovery relationship at both room and cryogenic temperatures. In the case of cryo-rolling, a higher initial strain hardening rate (due to the rapid generation of twins and then due to a rapid increase in dislocation density) combined with slower dynamic recovery increases the flow stress of Ti to a very high level [8]. The high flow stress should result in a smaller grain size in comparison with room-temperature behavior. Indeed, the results of the present work indicated a decrease in grain/subgrain size from ~200 to ~80 nm when the deformation temperature was decreased from 293 to 77 K (Fig. 5.7B and C). However, the decrease in deformation temperature does not change the physics of microstructure formation per se but rather increases the kinetics of the process.

5.2.3 POSTROLLING STABILIZATION HEAT TREATMENT

To determine the parameters for stabilization heat treatment of highly-deformed CP Ti, specimens rolled at room temperature to $\varepsilon_{th}=2.66$ were annealed for 1 h at different temperatures. As shown in Fig. 5.8A, the microhardness relative to that of the

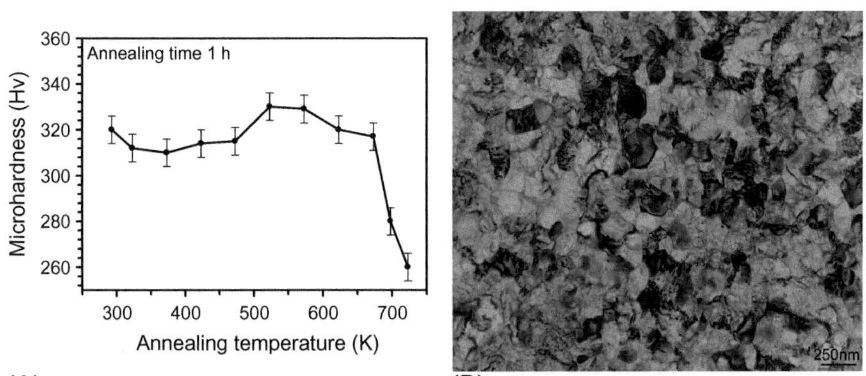

(A) (B)

FIG. 5.8

(A) Effect of annealing temperature on the microhardness of CP Ti cold rolled to a thickness strain $\varepsilon_{th}=2.66$ and (B) TEM micrograph of CP Ti rolled to a thickness strain ε_{th} of 2.66 at 293 K and then annealed at 623 K. Annealing time was 1 h in all cases.

as-rolled specimen was more-or-less unchanged until 673 K and then decreased at higher temperatures suggesting the development of recrystallization. Furthermore, TEM revealed that annealing at 623 K for 1 h resulted in a noticeable decrease in dislocation density (by a factor of 1.3), a thickening of cell boundaries, and an increase in the number of globular, almost dislocation-free subgrains (Fig. 5.8B). The size of new subgrains was ~100–200 nm. Consequently, such changes in microstructure appeared to correspond to well-developed recovery and the *beginning* of recrystallization. Therefore, the preferred stabilization anneal for highly-deformed CP Ti comprises 1 h at a temperature in the range of 573–623 K.

5.2.4 PRODUCTION-SCALE PROCESSING OF NANOCRYSTALLINE CP Ti SHEETS

The occurrence of deformation twinning in CP Ti can be used for the production of sheets with an NC structure directly by rolling without preliminary microstructure refinement in preforms. The intensity of twinning depends on temperature, deformation path, purity, and the grain size of titanium. Optimization of these parameters can result in attaining the most refined and homogeneous microstructure.

Because the extent of twinning decreases noticeably in CP Ti with a grain size $d < 15$ μm [23], on the one hand, and the homogeneity of deformation diminishes with an increase in a grain size [20] on the other, the most suitable starting microstructure should have a grain size of 15–20 μm. An increase in the fraction of impurities in CP Ti also suppresses twinning; therefore CP Ti Grade 2 is more preferable for obtaining an NC microstructure by cold rolling than, for example, Grade 4. Furthermore, strain path has a moderate effect on the extent of twinning. In comparison to the behavior for multipass unidirectional rolling, processes involving changes in rolling direction by ±90 degrees show ~5% increase in the fraction of the twinned grains. This effect is likely associated with the activation of new twin families when the rolling direction is change. The contribution of strain path change is also dependent on the initial texture of the material and should be taken into account during processing to form an NC microstructure.

Although deformation temperature is a very important processing parameter, rolling at cryogenic temperature is not practical on a commercial production scale. Rolling at elevated temperatures may be more attractive, and typically does result in a desirable grain size and high strength level. For example, the grain size after rolling to a thickness strain of 2.66 at 623 K was ~400 nm (in contrast to ~80 or ~200 nm after rolling at 77 K or room temperature, respectively).

The preferred commercial processing route for the manufacture of nanocrystalline CP Ti sheets thus consists of multipass cross (±90 degrees) rolling at room temperature to an accumulated strain of ~0.3 (at which twinning becomes saturated) and then unidirectional rolling to a total final thickness strain of $\varepsilon \sim 2.66$ (93% total reduction). The choice of deformation temperature (room temperature) is dictated primarily by a reasonable balance between desirable grain size (and corresponding mechanical properties) which form during deformation and manufacturability. The

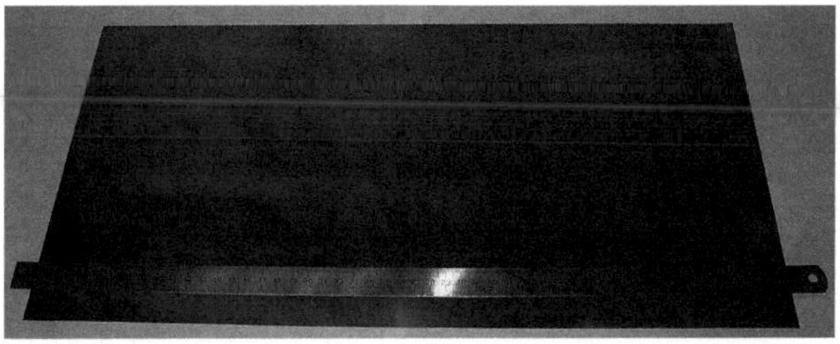

FIG. 5.9

Sheet of CP Ti measuring 500 × 250 × 0.3 mm^3 with an NC microstructure.

value of strain per pass should be ~0.05 to exploit more fully the effect of deformation twinning on microstructure refinement. After rolling, sheets should be annealed at ~573–623 K for 1 h to decrease the internal stresses. Using these parameters, a sheet measuring 500 × 250 × 0.3 mm^3 was produced (Fig. 5.9). The grain size and texture in the sheet corresponded to those developed in the laboratory specimens (Figs. 5.8B and 5.4D, respectively). An evaluation of microstructure and mechanical properties at various locations within the sheet showed a high degree of uniformity.

5.3 PRODUCTION OF BULK MILL PRODUCTS WITH NANOCRYSTALLINE STRUCTURE BY MULTIAXIAL FORGING

Multiaxial forging (MAF) is a technological process applicable for primary hot working of titanium alloy ingots/billets in the β or high α+β phase fields. Typically, MAF consists of sequential flat-die compressive deformations of a workpiece along a variety of directions [25,26]. The use of MAF for microstructure refinement in two-phase titanium alloys was first suggested by Salishchev et al. [27]. To this end, the temperature of MAF was reduced to 773 K, thereby enabling the formation of a microstructure with a grain size of ~60 nm. Multidirectional deformation of various magnitudes can achieve almost unlimited values in total strain, thereby yielding a uniform globular microstructure throughout a workpiece. On the basis of experiments discussed in Section 5.5, the minimum temperature at which MAF of CP titanium Grade 2 can be done has been determined to be ~673 K.

5.3.1 "ABC" FORGING OF CP Ti

To determine mechanical behavior and corresponding microstructure evolution for CP Ti, a sample measuring 18 × 14 × 12 mm^3 was extracted from a hot rolled bar and compressed successively along three orthogonal directions. This simplified version of MAF is also known as "abc" deformation, Fig. 5.10A [28]; a disadvantage of "abc" deformation is the considerable heterogeneity in final microstructure due to local effective strains at the center of the workpiece which are much larger than those

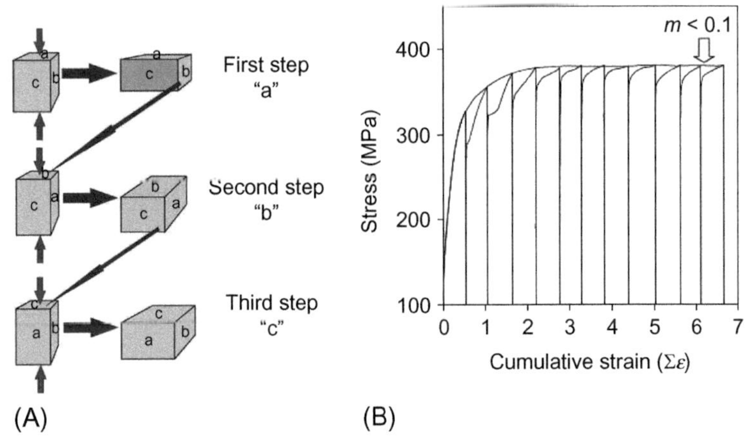

FIG. 5.10

(A) Schematic illustration of "abc" deformation and (B) cumulative $\sigma - \Sigma\varepsilon$ curve for "abc" deformation of CP Ti at 673K and strain rate of 10^{-3} s^{-1}.

in the vicinity of the surfaces which contact the dies. Prior to each rotation, a prismatic shape was restored by machining the curved faces. The initial strain rate and the true strain per deformation step were 10^{-3} s^{-1} and ~0.5, respectively.

The plastic-flow-behavior of CP Ti during "abc" deformation at 673K and a strain rate of 10^{-3} s^{-1} in terms of flow stress vs cumulative strain ($\sigma - \Sigma\varepsilon$) is shown in Fig. 5.10B. The measurements indicated that the yield stress increased noticeably during the first four increments of deformation. Further straining did not give rise to noticeable changes in yield or flow stress. Thus, near-steady flow was achieved during the final stages of the $\sigma - \Sigma\varepsilon$ curve. In addition, it was found that the yield stress at the beginning of each increment was smaller than that at the end of the previous step.

During the first compression step, localized shear bands formed a "forging cross" at the center of the workpiece (Fig. 5.11A). The shear bands crossed each other at each successive workpiece rotation/forging step thereby refining the microstructure

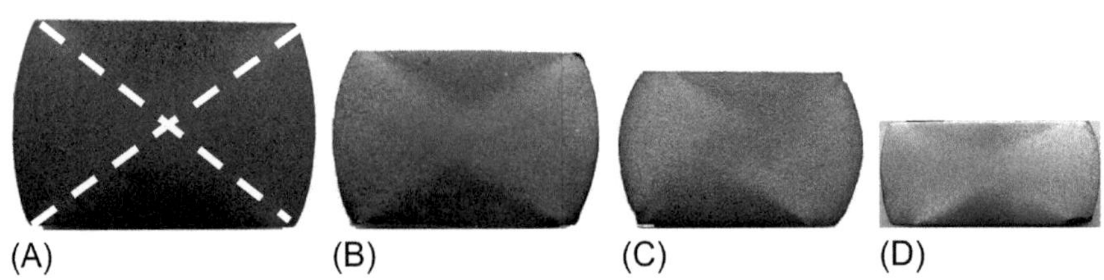

FIG. 5.11

Evolution of the macrostructure in CP Ti subjected to "abc" deformation at 673K comprising (A) one, (B) two, (C) three, or (D) seven increments of deformation.

FIG. 5.12

Microstructure of CP Ti after (A) 1, (B) 2, (C) 7, or (D) 12 increments of "abc" deformation at 673 K and $10^{-3}\,s^{-1}$ to a total strain of $\varepsilon \sim 0.5$, 1.1, 3.8, or 6.6, respectively.

at the center of the specimen. The peripheral areas which contacted the dies experienced smaller deformation in comparison to the center (Fig. 5.11B–D).

Microstructure evolution of CP Ti Grade 2 during the "abc" deformation at 673 K/$10^{-3}\,s^{-1}$ is shown in Fig. 5.12. After the first increment of deformation (Fig. 5.12A), the microstructure was heterogeneous. The initial grains were divided by narrow (100–200 nm) deformation twins belonging to the $\{10\bar{1}1\}\langle\bar{1}012\rangle$ family (having a twin/matrix misorientation of 57.42 degrees about the $\langle 2\bar{1}\bar{1}0\rangle$ axis) which is typical for this temperature [29]. Due to local crystallographic conditions, some grains did not undergo twinning. In some areas, shear bands with a high dislocation density and cell structure were also formed. Twin–twin and twin–shear band interactions fragmented the initial grains leading to a considerable refinement in microstructure and the formation of a lamellar-type microstructure along planes of maximum shear stresses. Twinning and an increase in dislocation density resulted in marked strain hardening during the first increment of "abc" deformation (Fig. 5.10B).

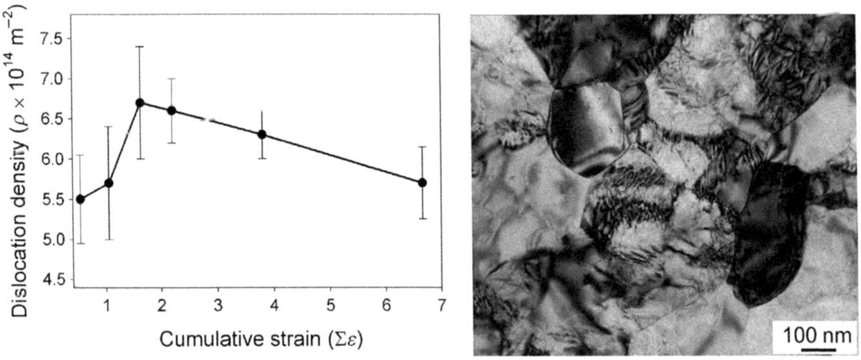

(A) (B)

FIG. 5.13

(A) Dislocation density and (B) recrystallized microstructure of CP Ti after seven increments ($\Sigma\varepsilon \sim 3.8$) of "abc" deformation at 673 K and 10^{-3} s^{-1}.

After the second increment of "abc" deformation, the microstructure at the center of the workpiece was much more homogeneous. The lamellar-like features formed during the first step became less obvious due to intersections and the shifting of lamellae when new lamellae with other spatial orientations formed (Fig. 5.12B). The dislocation density also increased during the second stage (Fig. 5.13A).

Additional changes in the loading direction during "abc" deformation increased microstructure homogeneity and decreased the grain size (Fig. 5.12C and D). Successive activity of different deformation systems also resulted in the formation of deformation-induced high-angle boundaries. Because the orientation of those boundaries was correlated with the metal-flow direction during each increment, changes of the loading direction resulted in 3-D intersections of deformation-induced boundaries and thus considerable microstructure refinement.

Another mechanism of grain refinement at 673 K was dynamic recrystallization (DRX). In particular, the transformation of the fragmented structure into (sub)grains can be ascribed to *continuous* DRX (CDRX) [30]. This process resulted in a decrease in dislocation density (Fig. 5.13A). As a result, almost dislocation-free globular grains were developed at a strain above $\Sigma\varepsilon = 2$ (Fig. 5.13B). A cumulative strain of $\Sigma\varepsilon \sim 6.6$ produced a microstructure consisting of equiaxed grains/subgrains with a mean size of ~200 nm.

The results obtained by EBSD analysis were in good agreement with TEM observations and flow-stress measurements. After the initial stages of "abc" deformation, the microstructure exhibited a lamellar-like morphology (Fig. 5.14A) due to the activation of twinning and the extensive formation of deformation-induced boundaries. Workpiece rotations between successive increments of deformation resulted in the crossing of lamellae formed during previous steps of "abc" deformation thereby giving rise to a relatively-uniform globular microstructure (e.g., Fig. 5.14D for seven increments of deformation). In fact, a relatively stable microstructure was even

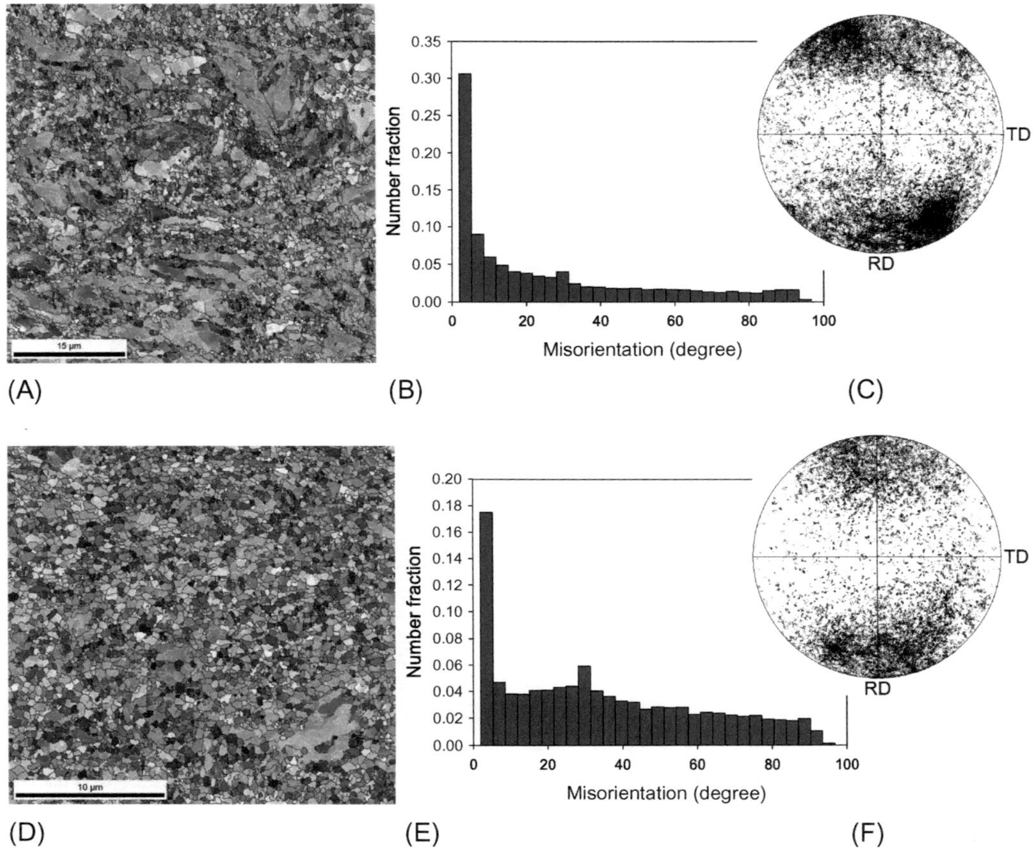

FIG. 5.14

EBSD results for CP Ti specimens after (A–C) two or (D–F) seven increments of "abc" deformation at 673 K and $10^{-3}\ s^{-1}$: (A, D) transversal-direction inverse-pole-figure maps, (B, E) misorientation distributions, and (C, F) (0002) pole figures; compression directions in (A) and (D) are vertical.

formed after four increments of compression, thereby leading to the nearly steady plastic flow shown in Fig. 5.10B. In addition, misorientation distributions from the EBSD data did not show obvious twin peaks already after the second increment of deformation (Fig. 5.14B). By contrast, an increase with strain in the fraction of boundaries with medium- to high-angle misorientation was quite obvious (Fig. 5.14B vs E).

The evolution of crystallographic texture seems to be slightly dependent on the number of increments of deformation. After each step, poles near the current compression direction (CD) were observed in (0001) pole figures (Fig. 5.14C and F). Such a behavior would appear to suggest rather considerable changes in texture and microstructure during each successive compression. As a consequence, the average misorientation gradually increased from 25 degrees (after the first step) to 35 degrees (after 12 increments of strain); a very high fraction of high-angle boundaries at the very beginning of deformation was associated with twinning.

The value of the strain rate sensitivity after the final step of "abc" deformation was $m = 0.1$, thus indicating that the steady-state flow exhibited in the cumulative $\sigma - \Sigma\varepsilon$ curve (Fig. 5.10B) was *not* related to superplasticity even though the material had developed a very-fine microstructure. In this regard, earlier results [31] have shown that superplasticity of UFG CP Ti can be mitigated by dynamic strain aging (and its associated negative rate sensitivity) inasmuch as the temperature interval for both processes ($\sim$0.3–0.4 of the melting temperature) overlap [31,32]. Therefore, it can be concluded that the presence of steady-state flow in the cumulative $\sigma - \Sigma\varepsilon$ plot indicates a balance between strain hardening and softening due to the occurrence of dynamic recovery/CDRX. Furthermore, it may also be concluded that the change in loading direction for successive increments of deformation leads to a decrease in (initial) yield stress due the dissociation of some low-angle boundaries and dislocation walls into a large number of mobile dislocations and relief of internal stresses, i.e., a Baushinger effect.

5.3.2 IMPROVED MAF OF CP Ti BILLET

To improve the homogeneity of the macrostructure and microstructure developed during MAF of CP Ti, the process must be modified to include, among other possibilities, a combination of compression (upsetting) and drawing (cogging) that result in variations in the loading direction, strain increments, etc (Fig. 5.15A). Furthermore, to attain a desirably-small grain size, it is necessary to reduce the temperature to lie within the warm/cold-working regime. In practice, the ductility of initially coarse-grained CP Ti billets may be insufficient at low temperatures to accumulate large enough strains without fracture. In such cases, microstructure refinement by MAF is achieved by initial processing at high temperatures followed by a gradual reduction in temperature as the grain size is refined and workability is increased (Fig. 5.15B). For example, it has been shown that deformation in the interval from 973 to 673 K using temperature decrements of 50–100 K at appropriate levels of cumulative strain enables the production of bulk samples measuring up to 70-mm

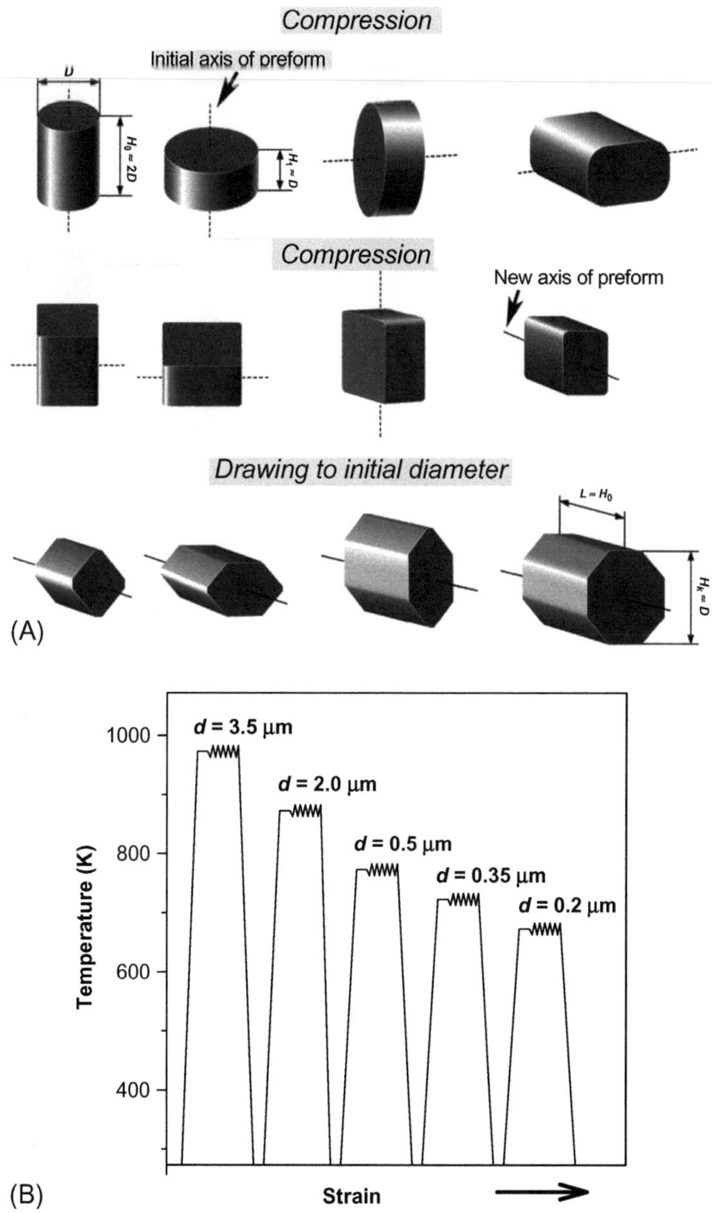

FIG. 5.15

Schematic illustration of multiaxial forging (A) and deformation path (B) for attaining an NC microstructure in CP Ti billet.

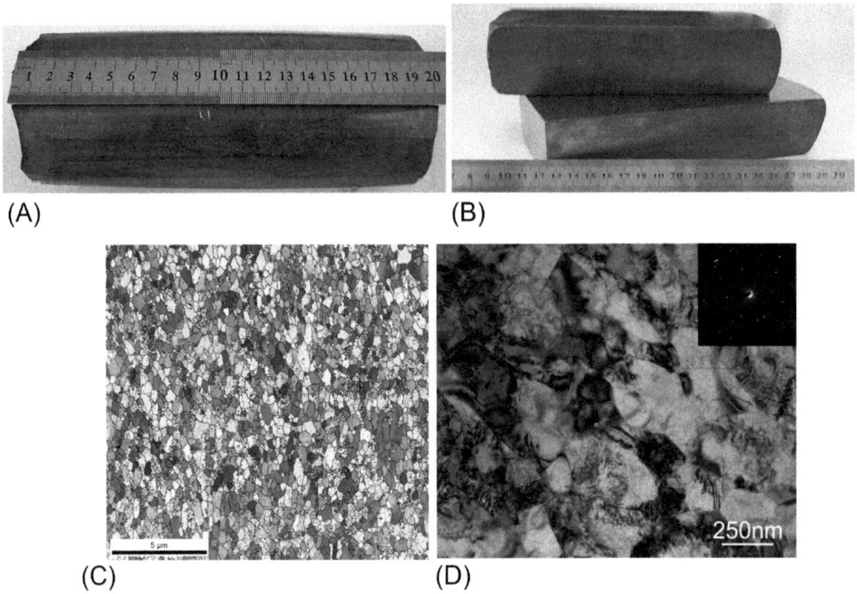

FIG. 5.16

Macrographs (A, B) and microstructure (C, D) of bulk CP Ti specimens after MAF in the temperature interval 973–673 K; (C) is an EBSD inverse-pole-figure map, and (D) is a TEM micrograph.

diameter × 120-mm length having a homogeneous macro- and microstructure with a grain size of 300–400 nm (Fig. 5.16). Following MAF, the workpiece is typically annealed for ~1 h to remove residual stresses and to stabilize mechanical properties at temperatures lower than the recrystallization temperature (573–623 K).

5.4 FORMATION OF NANOSTRUCTURE IN BARS BY COMBINED (CROSS-HELICAL AND SHAPE) ROLLING

Bars are one of the most widely-used semifinished mill products. Long bars can be produced via extrusion, swaging, and, most often, by various types of rolling (e.g., cross-helical or shape rolling). These methods are very efficient and widespread in the metalworking and machine-building industries. However, the working of bulk preforms by cross-helical rolling at room temperature is limited by rolling-mill capacity and material workability [33]. Thus, cross-helical rolling is usually performed using elevated (preheat) temperatures. Even with such an approach, radiation/conduction heat losses between passes and during rolling per se must be taken into account in process design as is necessary for any conventional (nonisothermal) metalworking operation.

The microstructure which forms during cross-helical rolling is usually different at the center and periphery of bars due to a heterogeneous distribution of strain. Because of nonisothermal hot-working conditions, the rather small strain per pass, and features of the deformation-zone geometry, deformation tends to localize in the vicinity of the bar surface [34]. During cross-helical rolling of CP Ti at 723 K, for example, it has been found that an initial microstructure comprising uniform equiaxed grains with a size of ~20 μm (Fig. 5.17A) can become noticeably heterogeneous. At the center of the bar, grains are elongated without noticeable refinement. At the periphery, on the other hand, new grains with a size of ~1 μm are formed likely due to dynamic/metadynamic recrystallization [35]. The area fraction of the recrystallized structure is 0.35 ± 0.05 [36].

More extensive deformation at the center of bars can be attained by shape rolling [36]. In addition, shape rolling can be conducted at much lower temperatures, resulting in a very fine-grained product. The combination of cross-helical and shape rolling thus enables the production of bars with a quite homogeneous UFG or NC microstructure.

During shape rolling in three stages (from 2 to 4-1, Table 5.1), microstructure evolution was associated with grain elongation along the rolling direction and the formation of low-angle boundaries which gradually evolved into high-angle ones (Fig. 5.17B and C). The latter resulted in an increase in the density of high-angle boundaries (Fig. 5.18) and corresponding microstructure refinement. The microstructure consisted of elongated grains with a transverse size of ~300 nm both in the center and at the surface of the bar (Fig. 5.17B and C). TEM analysis also revealed the formation of a partially-recrystallized microstructure elongated along the rolling direction with recrystallized grains ~300 nm in size (Fig. 5.19A).

Rolling at room temperature during the final stage (4-2, see Table 5.1) resulted in the formation of an elongated (sub)grain microstructure. The microstructure at the surface and in the center of the bar was similar. However, the fraction of globular grains at the surface was somewhat higher (Fig. 5.17D vs E). The transverse dimension of the grains was ~100–200 nm. TEM examination showed an elongated (sub) grain microstructure with a very high dislocation density; the size of the (sub)grains was ~150 nm (Fig. 5.19B).

It is worth noting that the rate of increase in the density of high-angle boundaries is noticeably higher at the bar surface in comparison with that along the axis. However, during later stages of rolling, the density of high-angle boundaries at the bar center "catches up" with the density near the surface, thereby homogenizing the microstructure across the bar (Fig. 5.18). Thus, bars with a uniform microstructure comprising 100–200 nm (sub)grains can be produced by combined (cross-helical + shape) rolling in the temperature interval 723–293 K. The formation of new small grains may be associated with *discontinuous* dynamic recrystallization (DDRX) during deformation at higher temperatures and postdynamic recrystallization during heating between the stages. At lower temperatures, by contrast, microstructure refinement is most likely controlled by the formation of deformation-induced high-angle boundaries and hence CDRX.

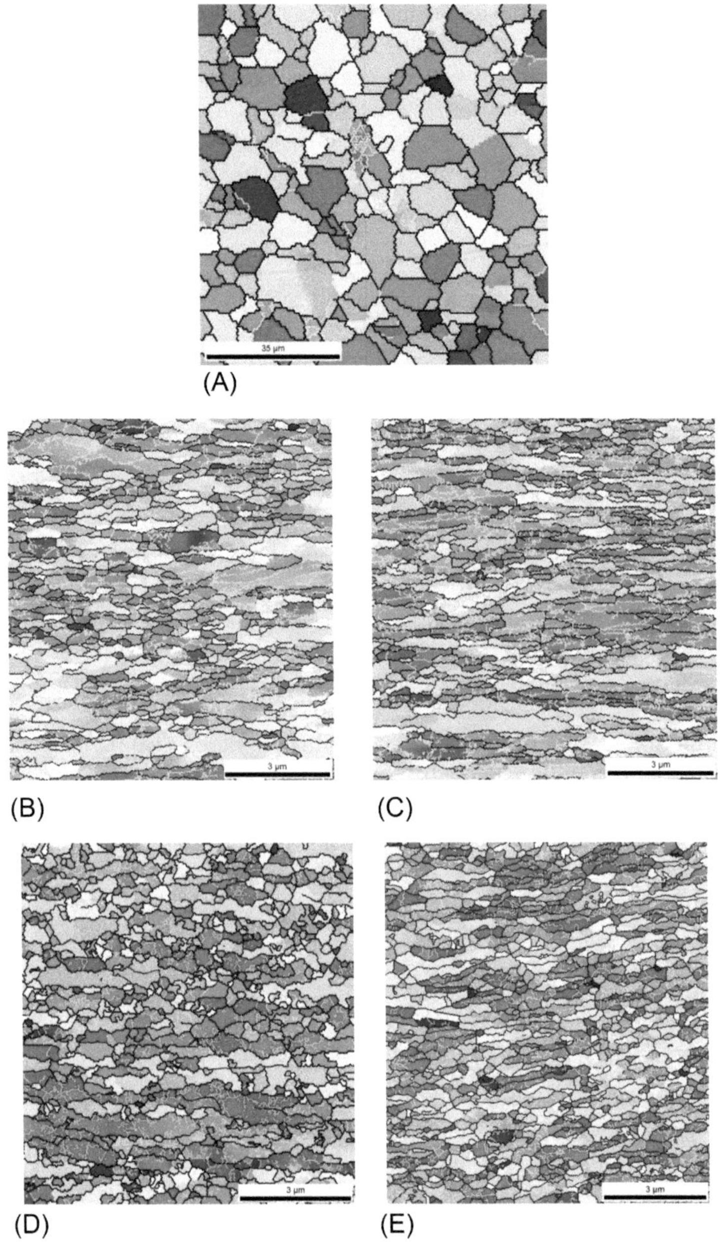

FIG. 5.17

EBSD inverse-pole-figure maps for the transverse direction of CP Ti (A) before and (B–E) after combined cross-helical + shape rolling; (B, D) along the axis and (C, E) at the surface; (B, C) stage 4-1 and (D, E) stage 4-2 (Table 5.1). The rolling direction for (B–E) is horizontal.

Table 5.1 Cross-Helical and Shape Rolling Parameters for Production of an NC Structure in CP Ti

Stage #	Type of Rolling	Temperature (K)	Accumulated Strain, $\Sigma\varepsilon$
1	Cross-helical	723	0.38
2	Shape	623	1.02
3	Shape	623	1.38
4-1	Shape	623	1.83
4-2	Shape	293	1.83

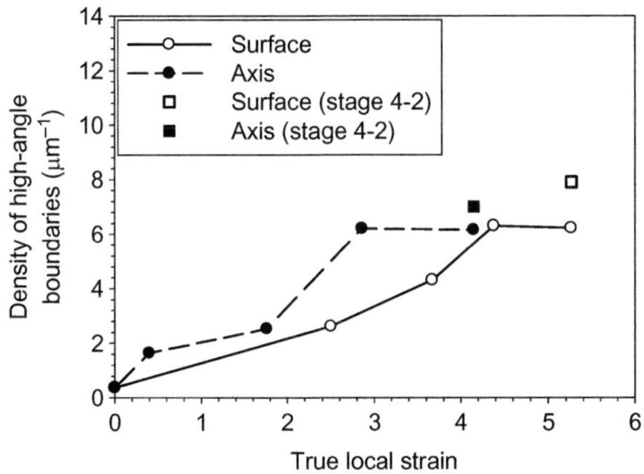

FIG. 5.18

Density of high-angle boundaries developed in CP Ti during combined (cross-helical + shape) rolling as a function of local strain.

5.5 MECHANISMS AND KINETICS OF STRUCTURE REFINEMENT DURING WARM/COLD DEFORMATION OF CP Ti

Microstructure refinement during the hot/warm plastic working of metals is typically interpreted in terms of DRX [13]. The characteristic stress –strain curves and grain structures which evolve during DRX have been studied in detail as a function of the temperature-compensated strain rate, or Zener–Hollomon parameter Z [13]:

$$Z = \dot{\varepsilon}\exp(Q/RT), \tag{5.1}$$

in which $\dot{\varepsilon}$ denotes the strain rate, R is the gas constant, T is the absolute temperature, and $Q = -R[\partial \ln\dot{\varepsilon}/\partial(1/T)]_{\sigma}$ is an apparent activation energy for DRX. Depending on Z, DRX in materials with low-to-medium stacking fault energies develops via either *continuous* (high Z, low-temperature domain) or *discontinuous* (low Z, high-temperature domain) mechanisms [13,30]. DDRX is associated with conventional nucleation and growth of new recrystallized grains [37,38]. In the case of

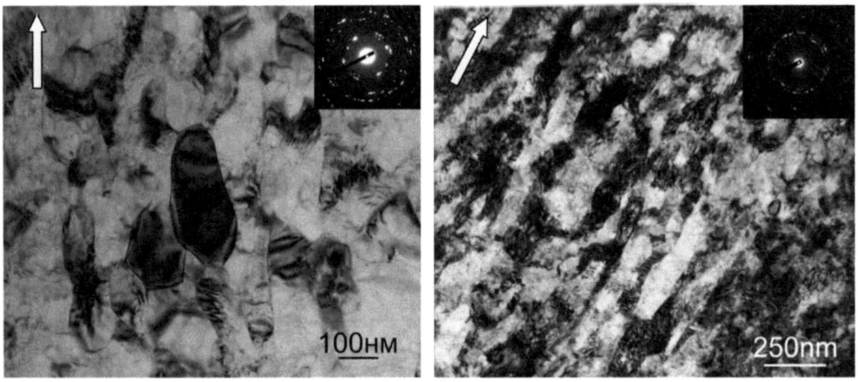

(A) (B)

FIG. 5.19

TEM images of CP Ti after the final stages of combined (cross-helical + shape) rolling at (A) 623 K (stage 4-1) or (B) 293 K (stage 4-2). Longitudinal sections: the rolling directions are shown by *arrows*.

CDRX, the new grains form due to an increase in subgrain misorientations resulting from the progressive accumulation of dislocations in subboundaries [39,40].

Under steady-state flow conditions, the size of recrystallized grains (D) can usually be expressed as a power-law function of the flow stress (σ_s):

$$\sigma_s = KD^{-N}, \tag{5.2}$$

in which K and N are constants [13,41]. In turn, Z can be related to σ_s as $Z = c_1 \sinh(c_2\sigma_s)^n$, where c_1, c_2, and n are constants. It has been shown that N in Eq. (5.2) takes on a value of ~0.7 or ~0.4 for DDRX (high-temperature domain) or CDRX (low-temperature domain), respectively [42,43]. The small value of the grain-size exponent, N, in the low-temperature domain suggests a rapid decrease in the size of the dynamically-recrystallized grains with a decrease in deformation temperature. It should be noted that the subgrain size can also be related to the flow stress through Eq. (5.2) with the subgrain size exponent equal to ~1 [13,30]. Therefore, a decrease in deformation temperature leads to a grain size approaching the size of deformation-induced subgrains. Because the grain size cannot be smaller than the subgrain size, the above relationships suggest an apparent limit for grain refinement by cold working.

The DRX grain size as a function of deformation temperature and flow stress as a function of DRX grain size for CP Ti are shown in Fig. 5.20A and B, respectively. The relationship between the flow stress and the deformation-induced subgrain size is also shown in Fig. 5.20B for reference. Three regions with different temperature dependence of the grain size are evident (Fig. 5.20A). These regions correspond to hot (923–1123 K), warm (673–923 K), and cold ($T < 673$ K) deformation conditions. An UFG microstructure (with a size <1 μm) can be developed in CP Ti at temperatures lower than ~800 K, while an NC structure (with a size $\leq$100–150 nm) forms during deformation at room or cryogenic temperatures.

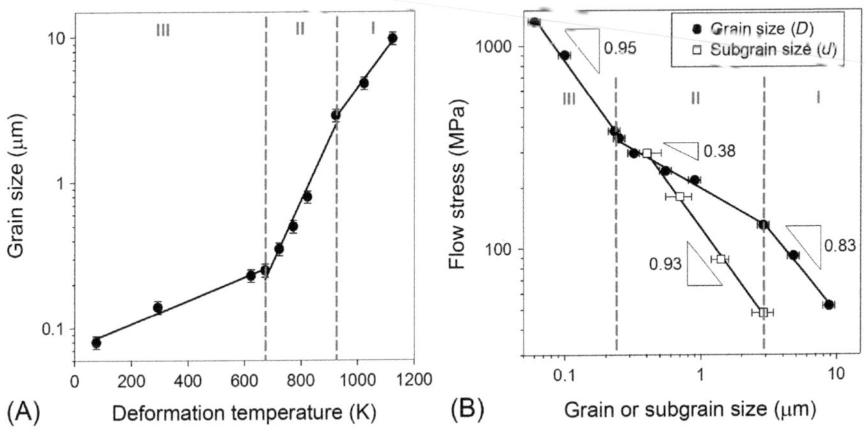

FIG. 5.20

Dependence of (A) grain size on temperature and (B) flow stress on grain or subgrain size in CP Ti. Regions I, II, and III correspond to hot, warm, and cold deformation, respectively.

The dependence of the size of the deformation-induced grains on the flow stress (Fig. 5.20B) also exhibits three distinct regimes, each with a different grain-size exponent, i.e., $N=0.83$ for the low-stress domain (hot deformation), $N=0.38$ in the moderate-stress domain (warm deformation), and $N=0.95$ in the high-stress domain (cold deformation). The change in grain-size exponent with decreasing deformation temperature can be interpreted in terms of the operative recrystallization mechanism [7,42,43]. In general, the decrease of N from values close to 1 to those lower than 0.5 is commonly related to the change in the recrystallization mechanism from discontinuous to continuous.

During DDRX under hot-working conditions, the microstructure consists of grains at various stages of the recrystallization cycle including rather large work-hardened ones [37]. Therefore, the dynamic grain size developed by DDRX is much larger than the size of deformation-induced subgrains (region I in Fig. 5.20). During CDRX under warm working conditions, the new grains develop as a result of gradual transformation of GNBs, which separate microvolumes with different combinations of slip systems, into HAGBs during continuing deformation [5]. Grain growth, which is a diffusion-controlled process, is suppressed by a decrease in temperature; only limited grain growth can take place at warm-working temperatures. Therefore, the size of DRX grains approaches the size of deformation subgrains with decreasing deformation temperature (region II in Fig. 5.20). Under cold-working conditions (region III in Fig. 5.20), the grain-size exponent of 0.95 is similar to that during DDRX in the hot-working regime. However, the grain-formation mechanisms for the two cases are quite different. Microstructure evolution during cold (and cryogenic) deformation does *not* involve grain-boundary migration over long distances characteristic of DDRX. Rather, new nanograins with a size as small as 80 nm (Fig. 5.7C) are almost free of dislocations in their interiors. It can be concluded,

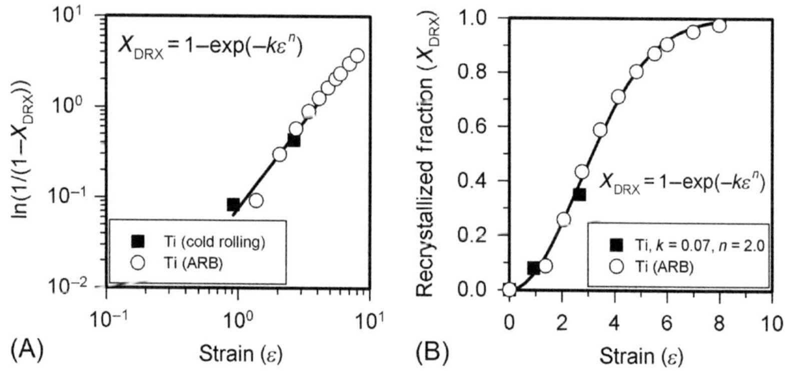

FIG. 5.21

The dependence of the recrystallized fraction on strain for dynamic recrystallization of CP Ti during cold rolling at room temperature: (A) $\ln(1/(1-X_{DRX}))$ vs ε and (B) X_{DRX} vs ε.

The data for ARB were taken from D. Terada, S. Inoue, N. Tsuji. Microstructure and mechanical properties of commercial purity titanium severely deformed by ARB process. J. Mater. Sci. 42 (2007) 1673–1681.

therefore, that these nanograins develop at large strains in place of subgrains developed during preceding deformation.

The kinetics of microstructure refinement due to nucleation-and-growth processes such as that during DDRX are often described by a Johnson–Mehl–Avrami–Kolmogorov (JMAK) equation [44] of the following form:

$$X_{DRX} = 1 - \exp\left(-k\left(\varepsilon - \varepsilon_c\right)^n\right), \tag{5.3}$$

in which X_{DRX} denotes the recrystallized fraction, ε is strain, and k, n, and ε_c are constants which depend on material and processing conditions [45,46].

Although CDRX does not involve nucleation and growth, the JMAK relation still seems to be applicable at least from a first-order phenomenological perspective. For example, a plot of $\ln(1/(1-X_{DRX}))$ as a function of strain (typically used to estimate the JMAK exponent n) is shown in Fig. 5.21A for CP Ti subjected to cold rolling at room temperature. Data for the fraction of ultrafine grains developed during ARB of CP Ti [12] are also shown in this figure. The fact that both sets of experimental results exhibit a linear relation in the log–log scale suggests that the grain refinement kinetics in titanium during cold deformation can be described by the JMAK relation (Eq. 5.3). The corresponding plot of X_{DRX} as a function of strain is shown in Fig. 5.21B. Here the data points represent the experimental data, and the smooth curve the fit of Eq. (5.3) using the value of n derived from Fig. 5.21A.

5.6 SUMMARY

1. Three stages of microstructure evolution are evident during rolling of CP Ti to a true thickness strain $\varepsilon = 2.66$ at room temperature. The first is associated with

twinning. Being one of the main modes of deformation, twinning refines the microstructure but shifts the formation of high-angle, deformation-induced boundaries and the accompanying grain refinement to larger strains. The final stage follows an intermediate, so-called incubation, strain interval. A decrease in rolling temperature to 77 K results in more extensive twinning. Rolling to a true thickness strain of 2.66 yields a microstructure with a grain size of ~80 or 200 nm at 77 or 293 K, respectively.

2. MAF of CP Ti at 573 K and 10^{-3} s^{-1} to a cumulative strain of $\Sigma\varepsilon = 6.6$ results in the formation of a microstructure with a grain size of ~0.2 μm. During MAF, new small grains develop as a result of the formation and crossing of deformation-induced, high-angle boundaries created by slip and twinning. During later/final increments of deformation, a steady-state flow stress is observed. This observation is not indicative of superplastic flow, but rather the occurrence of dynamic recovery and/or CDRX.
3. Combined (cross-helical + shape) rolling in the temperature interval 723–293 K can be used to produce bars with a uniform, elongated microstructure with a (sub) grain size of 100–200 nm.
4. The relationship, $\sigma \propto D^{-N}$, between the flow stress (σ) and the size (D) of deformation-induced grains has been shown to apply for CP Ti in the temperature interval 77–1123 K. The specific behavior is broken into three discrete regimes, each with a different grain-size exponent (N). A grain-size exponent of $N = 0.83$ is obtained under conditions of hot deformation at 923–1123 K, in which new grains form as a result of DDRX. The development of continuous dynamic recrystallization under conditions of warm deformation at 673–873 K gives rise to a grain-size exponent of $N = 0.38$. Last, a power law relation with a grain-size exponent of $N = 0.95$ is found for cold deformation at 77–623 K, which is very similar to the stress–subgrain size relationship over the entire temperature range.
5. The kinetics of grain refinement during large-strain cold-to-warm working conditions can be expressed by a Johnson–Mehl–Avrami–Kolmogorov-type equation similar to that used to describe DDRX during hot working of many metallic materials. Grain subdivision in titanium during cold working is assisted by deformation twinning thereby increasing the rate of grain refinement.
6. Production of bulk specimens by flat (plane-strain) rolling, MAF, or combined (cross-helical and shape) rolling at low temperatures has been demonstrated.

ACKNOWLEDGMENTS

The authors are grateful to the personnel of the Joint Research Center, Technology and Materials, Belgorod State University, for their assistance with instrumental analysis.

REFERENCES

[1] V.V. Rybin, Large Plastic Strains and Fracture of Metals, Metallurgy, Moscow, 1986.
[2] D.A. Hughes, N. Hansen, Microstructure and strength of nickel at large strains, Acta Mater. 48 (2000) 2985–3004.
[3] N. Hansen, D.J. Jensen, Development of microstructure in FCC metals during cold work, Philos. Trans. R. Soc. Lond. A 357 (1999) 1447–1469.
[4] J.W. Christian, S. Mahajan, Deformation twinning, Prog. Mater. Sci. 39 (1995) 1–157.
[5] G. Salishchev, S. Mironov, S. Zherebtsov, A. Belyakov, Changes in misorientations of grain boundaries in titanium during deformation, Mater. Charact. 61 (2010) 732–739.
[6] Y.B. Chun, S.H. Yu, S.L. Semiatin, S.K. Hwang, Effect of deformation twinning on microstructure and texture evolution during cold rolling of CP-titanium, Mater. Sci. Eng. A 398 (2005) 209–219.
[7] A. Belyakov, S. Zherebtsov, G. Salishchev, Three-stage relationship between flow stress and dynamic grain size in titanium in a wide temperature interval, Mater. Sci. Eng. A 628 (2015) 104–109.
[8] N. Bozzolo, N. Dewobroto, H.R. Wenk, F. Wagner, Microstructure and microtexture of highly cold-rolled commercially pure titanium, J. Mater. Sci. 42 (2007) 2405–2416.
[9] S.V. Zherebtsov, G.S. Dyakonov, A.A. Salem, V.I. Sokolenko, G.A. Salishchev, S. L. Semiatin, Formation of nanostructures in commercial-purity titanium via cryo-rolling, Acta Mater. 61 (2013) 1167–1178.
[10] G.S. Dyakonov, S. Mironov, S.V. Zherebtsov, S.P. Malysheva, G.A. Salishchev, A. A. Salem, S.L. Semiatin, Grain-structure development in heavily cold-rolled alpha-titanium, Mater. Sci. Eng. A 607 (2014) 145–154.
[11] Y. Saito, H. Utsunomiya, N. Tsuji, T. Sakai, Novel ultra-high straining process for bulk materials-development of the accumulative roll-bonding (ARB) process, Acta Mater. 47 (1999) 579–583.
[12] D. Terada, S. Inoue, N. Tsuji, Microstructure and mechanical properties of commercial purity titanium severely deformed by ARB process, J. Mater. Sci. 42 (2007) 1673–1681.
[13] F. Humphreys, M. Hatherly, Recrystallization and Related Annealing Phenomena, Elsevier, Oxford, 2004.
[14] G. Langford, M. Cohen, Microstructural analysis by high-voltage electron diffraction of severely drawn iron wires, Metall. Trans. A 6 (1975) 901–910.
[15] M.H. Yoo, Slip, twinning, and fracture in hexagonal close-packed metals, Metall. Trans. A. 12 (1981) 409–418.
[16] A.A. Salem, S.R. Kalidindi, R.D. Doherty, Strain hardening of titanium: role of deformation twinning, Acta Mater. 51 (2003) 4225–4237.
[17] Z.S. Basinski, M.S. Szczerba, M. Niewczas, J.D. Embury, S.J. Basinski, The transformation of slip dislocations during twinning of copper-aluminum alloy crystals, Rev. Metal. 94 (1997) 1037–1043.
[18] A.A. Salem, S.R. Kalidindi, R.D. Doherty, S.L. Semiatin, Strain hardening due to deformation twinning in α-titanium: mechanisms, Metall. Mater. Trans. A 37 (2006) 259–268.
[19] A.A. Salem, Z. Horita, T.G. Langdon, T.R. McNelley, S.L. Semiatin, Strain-path effects on the evolution of microstructure and texture during the severe-plastic deformation of aluminum, Metall. Mater. Trans. A 37 (2006) 2879–2891.
[20] M.A. Meyers, K.K. Chawla, Mechanical Behavior of Materials, Cambridge University Press, New York, 2009.

[21] H. Jazaeri, F.J. Humphreys, The transition from discontinuous to continuous recrystallization in some aluminium alloys: I the deformed state, Acta Mater. 52 (2004) 3239–3250.

[22] Q. Yu, Z.-W. Shan, J. Li, X. Huang, L. Xiao, J. Sun, E. Ma, Strong crystal size effect on deformation twinning, Nature 463 (2010) 335–338.

[23] S.V. Zherebtsov, G.S. Dyakonov, G.A. Salishchev, A.A. Salem, S.L. Semiatin, The influence of grain size on twinning and microstructure refinement during cold rolling of commercial-purity titanium, Metall. Mat. Trans. A 47 (2016) 5101–5113.

[24] D.G. Brandon, The structure of high-angle grain boundaries, Acta Metall. 14 (1966) 1479–1484.

[25] S.V. Zherebtsov, G.A. Salishchev, R.M. Galeyev, O.R. Valiakhmetov, S.Y. Mironov, S. L. Semiatin, Production of submicrocrystalline structure in large-scale Ti-6Al-4V billet by warm severe deformation processing, Scr. Mater. 51 (2004) 1147–1151.

[26] S. Zherebtsov, E. Kudryavtsev, S. Kostjuchenko, S. Malysheva, G. Salishchev, Strength and ductility-related properties of ultrafine grained two-phase titanium alloy produced by warm multiaxial forging, Mater. Sci. Eng. A 536 (2012) 190–196.

[27] G.A. Salishchev, O.R. Valiakhmetov, R.M. Galeyev, Formation of submicrocrystalline structure in the titanium alloy VT8 and its influence on mechanical properties, J. Mater. Sci. 28 (1993) 2898–2902.

[28] G.A. Salishchev, S.V. Zherebtsov, O.R. Valiakhmetov, R.M. Galeyev, S.Y. Mironov, Development of submicrocrystalline titanium alloys using "abc" isothermal forging, Mater. Sci. Forum 447–448 (2004) 459–464.

[29] U. Zwicker, Titan und Titanlegierungen, Springer, Berlin Heidelberg New York, 1974.

[30] T. Sakai, A. Belyakov, R. Kaibyshev, H. Miura, J.J. Jonas, Dynamic and post-dynamic recrystallization under hot, cold, and severe plastic deformation conditions, Prog. Mater. Sci. 60 (2014) 130–207.

[31] S.P. Malysheva, G.A. Salishchev, R.M. Galeev, V.N. Danilenko, M.M. Myshlyaev, A. A. Popov, Changes in the structure and mechanical properties of submicrocrystalline titanium during deformation in a temperature range of (0.15-0.45)T_m, Phys. Met. Metallogr. 95 (2003) 390–397.

[32] H. Conrad, Effect of interstitial solutes on the strength and ductility of titanium, Prog. Mater. Sci. 26 (1981) 123–403.

[33] V.I. Betekhtin, Y.R. Kolobov, V. Sklenicka, A.G. Kadomtsev, M.V. Narykova, J. Dvorak, E.V. Golosov, B.K. Kardashev, I.N. Kuz'menko, Effect of a defect structure on the static and long-term strength of submicrocrystalline VT1-0 titanium fabricated by plastic deformation during screw and lengthwise rolling, Tech. Phys. 60 (2015) 66–71.

[34] S.V. Parshin, Metal deformability during formation of profiled pipes from a sheet under compression in a gauge, Russ. J. of Non-Ferrous Met. 49 (2008) 31–33.

[35] Z. Zeng, Y. Zhang, S. Jonsson, Deformation behaviour of commercially pure titanium during simple hot compression, Mater. Des. 30 (2009) 3105–3111.

[36] N.V. Lopatin, Microstructure evolution in pure titanium during warm deformation by combined rolling processes, Mater. Sci. Eng. A 556 (2012) 704–715.

[37] M.J. Luton, C.M. Sellars, Dynamic recrystallization in nickel and nickel-iron alloys during high temperature deformation, Acta Metall. 17 (1969) 1033–1043.

[38] T. Sakai, J.J. Jonas, Dynamic recrystallization: mechanical and microstructural considerations, Acta Metall. 32 (1984) 189–209.

[39] N.D. Bakhteyeva, V.I. Levit, Influence of temperature and rate of deformation on dynamic recrystallization of single crystals of a Ni-based alloy, Phys. Met. Metallogr. 55 (1983) 124–130.

[40] S. Gourdet, F. Montheillet, A model of continuous dynamic recrystallization, Acta Mater. 51 (2003) 2685–2699.
[41] F. Montheillet, J.J. Jonas, G.L. Trigg (Ed.), Encyclopedia of Applied Physics, vol. 16, VCH Publishers, 1996, pp. 205–225.
[42] A. Belyakov, T. Sakai, H. Miura, R. Kaibyshev, Grain refinement under multiple warm deformation in 304 type austenitic stainless steel, ISIJ Int. 39 (1999) 592–599.
[43] A.M. Wusatowska-Sarnek, H. Miura, T. Sakai, Nucleation and microtexture development under dynamic recrystallization of copper, Mater. Sci. Eng. A 323 (2002) 177–186.
[44] W. Roberts, Strength of Metals and Alloys (ICSMA-7), Ed. by H.J. McQueen et al. Pergamon Press, Oxford, UK, 1986, p. 1859.
[45] A. Belyakov, S. Zherebtsov, M. Tikhonova, G. Salishchev, Kinetics of grain refinement in metallic materials during large strain deformation, Mater. Phys. Mech. 24 (3) (2015) 224–231.
[46] A. Morozova, E. Borodin, V. Bratov, S. Zherebtsov, A. Belyakov, R. Kaibyshev, Grain refinement kinetics in a low alloyed Cu-Cr-Zr alloy subjected to large strain deformation, Materials 10 (2017) 1394.

SECTION 2

Properties of nanocrystalline titanium determining its applications

CHAPTER

Advanced mechanical properties

6

Sergey Zherebtsov*, Irina P. Semenova†, Halina Garbacz‡, Maciej Motyka§

*Laboratory of Bulk Nanostructured Materials, Belgorod State University, Belgorod, Russia**

Institute of Physics of Advanced Materials, Ufa State Aviation Technical University, Ufa, Russia†

Faculty of Materials Science and Engineering, Warsaw University of Technology, Warsaw, Poland‡

Faculty of Mechanical Engineering and Aeronautics, Rzeszow University of Technology, Rzeszow, Poland§

6.1 INTRODUCTION

Ever since Hall and Petch introduced their well-known relationship between the lower yield point of low-carbon steels and grain size [1,2], a great deal of effort has been devoted to increasing the strength of metallic materials via microstructure refinement. This is particularly important with respect to nanostructured material in which the level of strength can be comparable to that of theoretical strength [3]. The Hall–Petch equation has the form:

$$\sigma_y = \sigma_0 + kd^{-\frac{1}{2}} \tag{6.1}$$

where σ_y is the yield stress, σ_0 is a frictional stress required to move dislocations, k is the Hall–Petch slope, and d is the grain size. This relation is quite universal and applicable to metallic materials with different crystalline lattice and types of structure [4]. As (micro) hardness in metallic materials correlates with strength, the Hall–Petch relationship is often used in the form

$$H_v = H_0 + k_h D^{-\frac{1}{2}} \tag{6.2}$$

with the similar physical meaning of the variables (e.g., see Ref. [5]). The fatigue limit also tends to increase with decreasing grain size [6–8]. An increase in the strength-related properties can be very important for titanium to expand the field of its application. However, considerable microstructure refinement usually results in a decrease in ductility-related properties (such as tensile elongation, impact toughness, crack propagation resistance) [4,9] because of a low ability of such materials to strain hardening, the fast loss of plastic flow stability and, as a result, more intensive strain localization. At elevated temperatures microstructure refinement usually results in some improvement of superplastic properties [10,11], however, in

Nanocrystalline Titanium. https://doi.org/10.1016/B978-0-12-814599-9.00006-7

commercially pure (CP) Ti with the ultrafine-grained (UFG) or nanocrystalline (NC) structure this phenomena can be depressed [12]. In this chapter some the most important mechanical properties, including Young's modulus, strength, (micro) hardness, ductility, fatigue limit, crack growth resistance, impact toughness, and superplasticity of NC CP Ti will be reviewed in more detail.

6.2 YOUNG'S MODULUS

Young's modulus is a measure of the interatomic bonds force and depends only slightly on the microstructure morphology of materials. Indeed, in microcrystalline (MC) materials the degree of plastic deformation and the size of grains have no significant influence on the value of the Young's modulus; however, one of the characteristic features of the nanostructured materials is lower values of their elastic constants in comparison with those in MC materials [13–17]. A decrease in the values of elastic modulus is usually ascribed to the substantially increased volumetric share of grain boundaries and triple junctions [18]. One of the reasons for a decrease in the Young's modulus is considered to be the elastic deformation of the grain boundaries [19].

Several geometric models have been proposed, designed for assessing the volumetric share of atoms located within the region of grain boundaries as a function of the average grain diameter [20–23]. The volumetric share of the grain boundaries can be calculated from the Palumbo relationship [23]:

$$f_G = 1 - \left(\frac{d - \Delta}{d}\right)^3 \qquad (6.3)$$

where f_G is the volumetric share of the grain boundaries, d is the grain diameter, and Δ is the thickness of the grain boundary (usually taken to be 1 nm). Using Eq. (6.3) the authors of Ref. [23] have shown that the volumetric share of the atoms within such grain boundary (1 nm width) rapidly increases when the grain diameter becomes smaller than 40 nm. In the materials with a grain diameter of 5 nm, the relative volumetric share of atoms in grain boundaries reaches 50%. Within this range of grain sizes, the effect of the grain boundary surface area may be significant.

The effect of the volumetric share of the grain boundaries on the elastic properties of nanomaterials can be traced by considering the behavior of Ni–P produced using the electrodeposition method. With the grain sizes equal to 4.1 and 28.9 nm, the volumetric proportion of the grain boundaries was 60% and 10%, respectively. In the sample with the grain size of 28.9 nm, the Young modulus was 205 GPa, i.e., it was comparable with the literature data reported for a coarse-grained (CG) counterparts (207 GPa [17]). In samples with a grain size of 4.1 nm, the elastic modulus was lower (178 GPa).

In NC titanium, the volumetric share of grain boundaries is an order of magnitude lower than that in Ni–P, where the decrease in the Young's modulus is substantial. Hence this may be not the only factor which influence on the elastic modulus value. High density of dislocations observed in NC titanium may be another factor which can promote the effect of microplasticity. This effect consists in reversible plastic deformation of isolated subgrains or reversible displacements of some segments of anchored dislocations [24]. The reversible plastic deformation can decrease the measured values of the elastic modulus. In the literature, this phenomenon is often called as "relaxation effect" [25,26]. The occurrence of the reversible deformations in NC metals was detected in the experiments with NC aluminum, copper, and gold [24,27].

The value of the Young's modulus can also be associated with the technology of manufacturing of nanomaterials which may be responsible for the formation of microstructural defects such as porosity and microcracks. Many early results on the Young's modulus measurements in NC materials appeared to be quite contradictory depending on the NC structure production method. In those NC materials which were produced by inert gas condensation (IGC) the elastic modulus was significantly lower in comparison with the same material produced by other methods [28]. At same time a material with a grain size of 7 nm produced by electrodeposition had the Young's modulus comparable with that of a CG counterpart. The basic reason of this substantial decrease in the elastic modulus was considered to be the discontinuity of the microstructure of the IGC-produced nanomaterials associated with their porosity [29–31]. Results reported by Shen et al. [32] and later by Zhou et al. [17] confirmed that the value of the Young modulus only slightly depends on grain size in nonporous NC Ni–P alloys prepared by electrodeposition. A decreased value of the Young's modulus was observed in materials with a grain size below 20 nm only.

When dealing with NC metals produced by various severe plastic deformation (SPD) techniques, the factor of internal defects can be neglected because the density of these materials is very high. SPD-processed materials are characterized by deformation texture and increased residual stresses. The real value of the Young's modulus of NC Ti may therefore be even lower. Nevertheless, the results obtained can be utilized for a qualitative analysis of the effect of a grain size on the Young's modulus of Ti. It can be inferred from this analysis that the elastic modulus in titanium is already decreased at grain sizes of about 90 nm. It is worth noting that the ab-initio calculations performed by Dobosz et al. [33] have shown that the Young's modulus in the region of grain boundaries may be lower by about 30% than that in the region of grain interiors.

Another factor which can influence on the value of the Young's modulus is the method of measurement. The nanoindentation method is most often used for assessing the properties of elastic nanomaterials [15,21,30,31]. In small volumes of NC samples, the Young's modulus is rarely determined using the static tensile test or the ultrasonic method. When measuring by the nanoindentation method, it is of great importance to prepare the surface of the sample properly. Mechanical grinding and polishing induce stresses in the near-surface zone of the sample resulting in its strengthening. In order to avoid this effect, the samples are subjected to electrolytic polishing. The roughness (R_a) of the MC and NC samples prepared by this way appeared to be similar (44.1 and 37.1 nm, respectively).

The measurements of the Young's modulus by the cyclic uniaxial tensile test were conducted at various strain rates (approximately 10^{-4}–10^{-3} s^{-1}). The range of the stress applied to the samples during the tensile test was selected individually for each material examined depending on the yield stress (10–250 MPa for MC Ti, and 10–800 MPa for NC Ti). Each sample was subjected to a loading cycle during which the stress increased from its minimum to maximum value (which is always lower than the yield stress) and then in the reverse direction from the maximum to the minimum. This sequence constituted one loading cycle. During each test conducted at given parameters, this cycle was repeated five times. The calculations were performed using only the data obtained when the stress was increasing.

The values of the Young's modulus obtained using different testing methods are given in Table 6.1 for MC and NC Ti; the latter was obtained by hydrostatic extrusion (HE). One can see that grain refinement to the NC regime resulted in a decrease in the Young's modulus of Ti [34].

It should however be noted that the Young's modulus of a polycrystalline sample measured by the nanoindentation method is a little overestimated in comparison with the literature data [35,36] due to certain limitations of this method as the deformation is not strictly elastic. It is estimated that the Young's modulus measured by the nanoindentation method becomes more overvalued when the compressive stress increases [37–39].

In the case of uniaxial tensile test, the results obtained for MC conditions are in agreement with the literature data. Moreover, it was found that an increase in a strain rate results in an increase in the Young's modulus in both materials. A dependence of the Young's modulus on a strain rate was also experimentally shown by Sakai et al. [40] with NC gold. The sensitivity of NC materials to a change in a strain rate depends on the deformation mechanism, the material grain size, and crystalline structure. In NC metals a decrease in the stress sensitivity for a strain rate is the main factor responsible for a strong tendency to strain localization within the shear bands.

Table 6.1 Measured Values of the Young's Modulus

Investigation Method	Condition of Ti	Young's Modulus Value (GPa)	ΔE
Nanoindentation	MC	126.0	4.01
	NC	110.0	2.10
Tension (slower)	MC	104.4	1.98
	NC	91.2	0.89
Tension (faster)	MC	108.1	1.74
	NC	92.9	2.28

6.3 STRENGTH AND DUCTILITY

Strength and ductility are the main characteristics of a material determining almost all other mechanical properties. For example, hardness, high-cycle fatigue (HCF) limit, and wear resistance are accepted to be depended on strength while fracture toughness, low-cycle fatigue (LCF), and impact toughness can rather be considered as the ductility-dependent properties. As the grain size has a strong impact on both strength and ductility, NC materials were expected to have some "unusual" properties [41,42].

The main requirement for structural material is associated with a reasonable balance between high strength and acceptable ductility. Fatigue resistance and the toughness are the most important "secondary" properties, which are directly governed by strength and ductility. According to the Hall–Petch Eq. (6.1) the maximum strength can be attained by microstructure refinement to the NC regime. For conventional alloys a reduction in grain size to several micrometers usually leads to an increase in ductility. Then ductility increasing can also be expected as the grain size is reduced to nanoscale. However, in the UFG and NC range, strength and ductility are mutually exclusive properties, which means that extremely strong materials usually have unacceptably low ductility. Therefore, simultaneous substantial improvement of both strength and ductility is a very challenging task.

A considerable (in some cases many times) increase in strength in comparison with "usual" CG conditions of the same material was reported starting from the early studies of NC materials [43,44]. Some experimental data on microhardness, strength and ductility of NC CP Ti (Grades 1–4) produced by various methods are collected in Table 6.2.

One of the earliest results showing increased strength of NC CP Ti with a grain size of ~100 nm produced by HPT to a true logarithmic strain of 7 at room temperature was described in Ref. [46]. Because of very small specimens, strength characteristics were determined using a miniaturized bend test technique. Tensile strength and ductility of CP Ti with the same grain size produced by HPT was evaluated in Ref. [5]. This later experiment showed very high strength with the yield stress of the order of ~1200 MPa and reasonable ductility. Numerous experiments conducted in the last few decades have shown that room temperature strength of the NC CP Ti produced by various methods is ~30%–100% higher than that of the CG counterparts. It is worth noting that in the context of grain refinement and strengthening of bulk titanium, one of the most effective method is HE [56–58]. HE at ambient temperature led to an impressive increase in the tensile strength (by 150% on average) due to reduction in a grain size to 50–100 nm. However, in comparison to other SPD methods, HE resulted in a considerable loss in ductility.

In fact the level of strength that can be attained in NC CP Ti is comparable to many titanium alloys (Table 6.2, [63,64]). Microhardness of CP Ti also shows considerable growth with decreasing grain size (Table 6.2). It is worth noting that a very high level of strength can be attained in CP Ti by various combinations of SPD

Table 6.2 Mechanical Properties of NC CP Ti Produced by Various Methods

Grade	Processing	Grain Size (μm)	HV (MPa)	YS (MPa)	UTS (MPa)	Elongation (%)	Reference
1	Unprocessed	23	1590	275	407	35.0	[45]
1	HPT 5 GPa 8 turns	0.1	2800	1050[a]	1380[a]	n/a	[46]
1	HPT 5 GPa 8 turns	0.12	2255	790	950	14.0	[5]
1	ECAP 8 Bc 400°C	0.3	n/a	650	810	15.0	[47]
1	ECAP 8 Bc RT	0.2	2640	710	790	19.0	[48]
2	Unprocessed	40	2270	354	487	30.0	[45]
2	HPT 3 GPa 10 turns	0.13	2990	~800	940	23.0	[49]
2	ARB 8 cycles	0.08–0.1	n/a	631	864	6.0	[50]
2	ECAP 10 Bc 400°C	0.45	2765	582	645	20.6	[45]
2	ECAP 10 Bc 400°C, 77% CR	0.19	3600	736	945	14.5	[45]
2	ECAP 12 E 300°C	0.25	n/a	564	771	22.8	[51]
2	ECAP 6 Bc 300°C	n/a	2440	615	790	32.0	[52]
2	ECAP 6 Bc 300°C, 70% CR	0.15–0.2	2840	785	910	34.0	[52]
2	ECAP 6 Bc 300°C, 70% LTR	0.15–0.2	2735	915	1030	32.0	[52]
2	MAF 450°C	0.4	n/a	760	810	20.0	[53]
2	MAF 450°C CR 95%	0.1	n/a	1210	1265	9.0	[54]
2	HE RT $\varepsilon=1.8$	0.1–0.5	n/a	985	1080	14.0	[55]
2	HE RT $\varepsilon=3.21$	0.073	2609	860	1053	8.1	[56]
2	HE RT $\varepsilon=3.77$	0.055	2678	1040	1141	5.6	[57]
2	HE RT $\varepsilon=4.60$	0.062	2736	1045	1190	6.9	[56]
2	HE RT $\varepsilon=5.47$	0.047	2745	1245	1320	7.5	[58]
2	HE RS	0.09	2840	1050	1130	13	[59]
2	CR −196°C $\varepsilon=2.66$	0.08	2910	870	1180	3.8	[60]
4	Unprocessed	35	n/a	630	740	25.0	[45]
4	HPT 6GPa 10 turns	0.105	4460	1400	1600	7.0	[61]
4	ECAP 8 Bc 400°C	n/a	3290	750	815	19.0	[45]
4	ECAP 8 Bc 400°C, 83% CR	n/a	4250	1006	1135	10.7	[45]
4	ECAP 4 Bc 450°C, TMP (swaging + drawing) 300°C	0.15	n/a	1150	1240	11.0	[62]

ARB, *accumulative roll bonding;* CR, *cold rolling;* ECAP, *equal-channel angular pressing;* HE, *hydrostatic extrusion;* HPT, *high-pressure torsion;* LTR, *rolling at subzero temperature (−100°C);* MAF, *multiaxial forging;* RS, *rotary swaging;* RT, *room temperature;* TMP, *thermomechanical processing.*

[a] *Determined using a miniaturized bend test technique.*

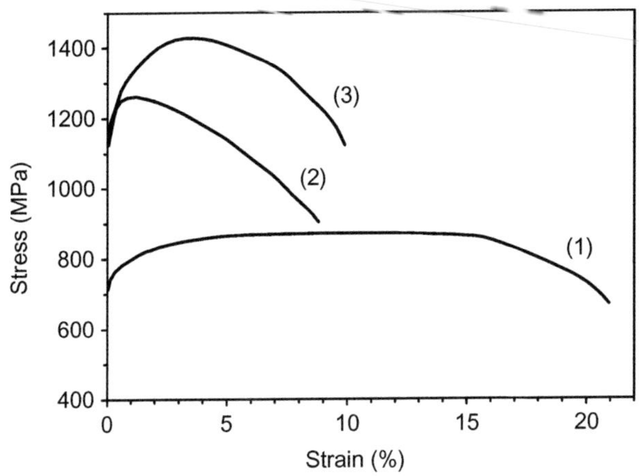

FIG. 6.1

Engineering tensile curves for NC Ti Grade 4 samples produced by ECAP and TMP before (2) and after annealing at 350°C for 6h (3) in comparison with a CG condition (1) [65].

methods, such as ECAP followed by swaging and drawing or multiaxial forging and subsequent cold rolling. In these cases strong dislocation (substructure) strengthening and possibly texture strengthening contribute into strength of the specimens in addition to the Hall–Petch strengthening.

Meanwhile the decrease in grain size to the NC regime results in a considerable decrease in ductility (Table 6.2). This decrease can attain 70% in some cases; however, typical loss in ductility is 20%–30%. It should be noted that both the total elongation and the uniform deformation decrease with decreasing grain size. A typical example of the engineering stress–strain curves for NC CP Ti Grade 4 with a grain size of 150nm produced by ECAP and TMP (warm rolling and drawing with the total accumulated strain of 80%) and for a CG condition of the same CP Ti are shown in Fig. 6.1 [65]. One can see that the NC condition shows much smaller total elongation as well as deformation before necking.

At the same time, after the strain localization (the uniform strain of the NC condition was about 0.5%), when the plastic deformation has become concentrated in the neck, the specimen shows additional ~8.5% elongation. As the volume involved into the plastic deformation after the strain localization is considerably smaller than the nominal size of the gauge, the total elongation does not reflect ductility of the NC condition properly. In some cases ductility of UFG or NC materials should be discussed in terms of the area reduction which can be almost a twice as high as that for a "usual" MC condition [66]. Hence, under some "soft" deformation schemes (rolling, compression) the NC condition can manifest quite a high ductility although ductility under tensile stresses is low.

Such behavior is a common feature of the UFG and NC materials irrespective of the method by which the structure is produced. Early necking and reduced tensile elongation, relative to their MC counterparts, has been observed in UFG and NC materials produced by SPD, consolidation of nanosize particles or electrodeposition [4,9,42,43,47]. The plastic instability in the NC materials which manifests itself either as shear bands or by "early" necking is usually attributed to the lack of an effective hardening mechanism [4,9,42]. Indeed, the number of dislocations that become stored into the microstructure during the tensile test is no longer significant to make a difference over the already-high existing dislocation density (especially if the NC material was produced using SPD). As a result, the flow stress does not increase remarkably during the tensile test.

Among some suggested ways to improve ductility of NC materials [42], only one approach is applicable for a single-phase low-alloyed CP Ti. The most obvious method to increase ductility without considerable loss in strength in case of NC Ti is associated with low-temperature annealing (controlling recrystallization). An increase in ductility is associated with a decrease in dislocation density and thereby with an increase in the ability of the material to strain hardening. An example of a positive effect of such treatment on both the total elongation and uniform elongation can be seen in Fig. 6.1 (curve (3)) [65]. However, it should be noted that strength of the NC material can also increase after annealing. The increase in both strength and ductility was ascribed to changes in the grain boundary state due to structure recovery and the formation of impurity segregation at the boundaries. It should be noted that a similar effect of short annealing on mechanical properties was reported earlier for NC CP Ti produced by HPT [67]. Therefore, annealing is able to increase the ductility-related properties of NC Ti, however, this improvement is supposed to be quite limited due to the corresponding grain growth and decrease in strength.

One of the promising approaches to increase ductility of NC material is associated with nanotwinning [68], which produces a "dynamic Hall–Petch effect" and decreases the propensity of the material to plastic instability. This approach is possible for materials with a pronounced proclivity for twinning; however, in titanium the minimum grain size at which twinning is still active was found to be ~0.9 for room and ~0.7 μm for cryogenic (−196°C) temperatures [60].

Another method which can be potentially used for CP Ti is associated with formation of a bimodal structure with micrometer-sized grains embedded into the NC or UFG matrix [69]. In a somewhat simplistic fashion it was claimed that the nanoscale fraction of the grain population was responsible for high strength, while the coarse grains contributed to ductility by promoting the strain hardening necessary to stabilize the tensile deformation against strain localizations. A positive effect of the bimodal structure on tensile ductility was found in CP Ti with a so-called harmonic structure produced by powder metallurgy methods [70]. However, in the case of controlling recrystallization of bulk NC specimens produced by SPD, an increase in ductility in bimodal structures is usually associated with the corresponding decrease in strength [42] that limits the application of this approach considerably.

6.4 FATIGUE DURABILITY

Cyclic properties of NC Ti under monotonic loading is very important along with the strength and ductility for prospective engineering applications of these materials. As the fatigue resistance is in general correlated to strength of the material under monotonic loading, the fatigue properties of CP Ti were also found to be improved due to the grain refinement [7,8,71–73]. This empirical dependence is clearly seen in Fig. 6.2; the results show the HCF limit—tensile strength relation for NC CP Ti Grade 2 and Grade 4 produced by ECAP [8,47,74,75] or multiaxial forging [53]. Although all the experimental points follow the common trend pretty well, this relation is not a firm law and can vary depending upon the chemical composition, the grain refinement process used, and the loading conditions. That is why the increase in fatigue resistance in NC metals is not always as large as one might expect, considering the increase in strength [76]. This is particularly important in case of the LCF: regardless of the SPD technique used to produce the NC condition, the LCF endurance in NC metals can be even lower than that for the CG counterparts [4,77].

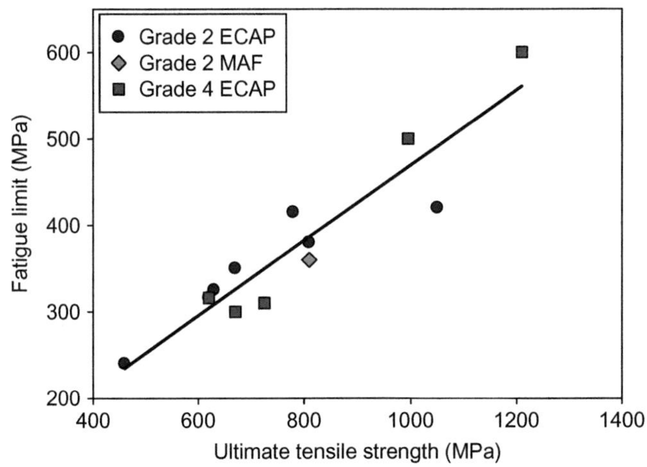

FIG. 6.2

Relation between the tensile strength and the fatigue limit for NC CP Ti Grade 2 and Grade 4 produced by ECAP or MAF.

Data from Y. Estrin, A. Vinogradov, Fatigue behaviour of light alloys with ultrafine grain structure produced by severe plastic deformation: an overview. Int. J. Fatigue 32 (2010) 898–907; A.Y. Vinogradov, V.V. Stolyarov, S. Hashimoto, R.Z. Valiev. Cyclic behavior of ultrafine-grain titanium produced by severe plastic deformation. Mater. Sci. Eng. A 318 (2001) 163–173; S. Zherebtsov, G. Salishchev. Production, properties and application of ultrafine-grained titanium alloys. Mater. Sci. Forum 838–839 (2016) 294–301; V.V. Stolyarov, I.V. Alexandrov, Yu.R. Kolobov, M. Zhu, T. Zhu, T. Lowe. in: X.R. Wu, Z.G. Wang (Eds.) Fatigue'99, vol. 3. Higher Education Press, PR China, 1999, 1345–1351; W.J. Kim, C.Y. Hyun, H.K. Kim. Fatigue strength of ultrafine-grained pure Ti after severe plastic deformation. Scripta Mater. 54 (2006) 1745–1750.

The LCF and HCF regimes are usually distinguished on the basis of the applied strain amplitude [76,78]. The fatigue life during HCF is associated with the resistance of the material to crack initiation, while LCF is controlled by crack propagation which depends on the material's tolerance to defects. Microstructure refinement and the corresponding strengthening increases the resistance to crack initiation, thereby increasing the HCF endurance. Meanwhile the inferior LCF performance of NC CP Ti (as well as of all NC and UFG metals in general [4]) can be attributed to limited ductility, which promotes early crack initiation and greater availability of grain boundaries in orientations favorable for crack propagation in a material with a finer grain structure [8]. Therefore, the strategy of increasing the fatigue limit via microstructure refinement is applicable mainly for the HCF regime.

Several examples of the application of this strategy for improvement of fatigue performance combining the effects of structure refinement and strain hardening with respect to CP titanium are shown in Fig. 6.3. The S–N curves demonstrate considerably improved the HCF properties upon the reduction of the grain size in comparison with a CG (d=9 μm) condition of titanium [79]. It should be noted however that the LCF fatigue resistance of the material also does not degrade after eight ECAP pressings using Route Bc [47]. A significant improvement of HCF properties is achieved already after the first ECAP passes [8]; an increase in strain during SPD resulted in additional increase in fatigue limit in accordance with the strengthening strategy. For instance, using a combination of equal-channel angular pressing and subsequent thermal and mechanical treatment, it is possible to increase the fatigue endurance limit of commercial-purity titanium by a factor of 1.5 [72,80].

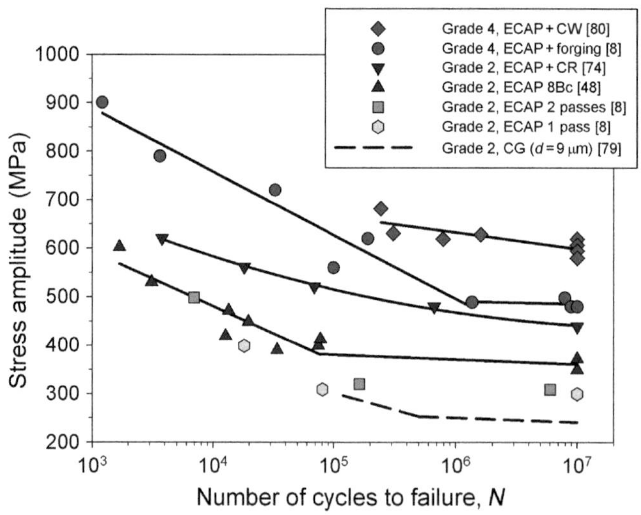

FIG. 6.3

S-N fatigue curves for CP titanium with different grain size obtained by ECAP processing (*CW*, cold working) [8].

A considerable increase in fatigue resistance of NC CP Ti was attained via other than ECAP methods of SPD, for example, multiaxial forging [53] or hydrostatic extrusion [81]. The NC titanium obtained by hydrostatic extrusion ($d \approx 50\,\text{nm}$) shows an increase in fatigue strength by ~75% (from 160 to 280 MPa) in comparison to the reference CG Ti with a grain size of 20 μm. The extrusion-induced grain refinement of titanium was found to result in a smoother fracture surface than that of the CG material. Multiaxial forging resulted in microstructure and properties similar to those that are obtained by ECAP (Fig. 6.2).

6.5 FRACTURE TOUGHNESS

Quite limited data are available in the literature on crack growth behavior in NC CP titanium. However, noticeable deterioration in the fatigue crack growth resistance can be clearly seen based on the published results. Fatigue crack growth response of a MC condition and ECAP-processed NC CP Ti in terms of the variation of da/dN as a function of ΔK at a load ratio of $R=0.1$ is shown in Fig. 6.4A [7,8]. Grain refinement from the fine-grained to the NC regime leads to a reduction in ΔK_{th} by a factor of 2.5. The rate of fatigue crack propagation is more than an order of magnitude higher in the NC Ti, over a wide range of applied loading, reflecting the trend captured for other NC materials [4,7,8].

The most recent results obtained for NC CP Ti Grade 4 produced by ECAP were obtained in Ref. [82]. The fatigue crack growth curves for the CG ($d=16$ μm) or NC Ti ($d=250\,\text{nm}$) are shown in Fig. 6.4B (results for two specimens of each condition are shown). It is seen that crack growth rates for the NC Ti specimens shifts to higher values while the threshold stress intensity range ΔK for crack growth becomes lower, thereby indeed confirming a decrease in the fatigue crack growth resistance with decreasing the grain size.

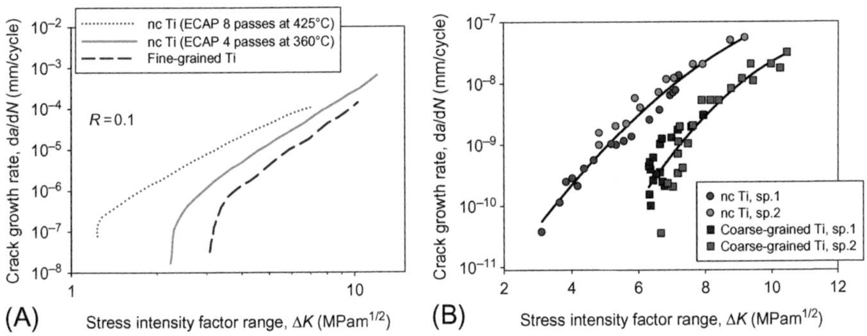

FIG. 6.4

Fatigue crack growth rate as a function of the stress intensity factor range ΔK in Grade 2 (A) [7,8] or Grade (4) (B) [82] CP Ti.

These results are in agreement with data on fatigue tests of notched specimens. It was shown in Ref. [72] that the notch sensitivity increases with a decrease in a grain size and the fatigue endurance limit of notched specimens of an NC Ti produced by ECAP and subsequent thermomechanical treatment is lower that of the CG counterparts [72].

The Charpy impact test of CP Ti Grade 2 also shows some decrease in the impact energy (from 0.88 to 0.76 MJ/m^2) with decreasing the grain size from 55 μm to 650 nm [83]. However, it worth noting that ECAP at room temperature using a die with a channel angle of 135 degrees in the route Bc results in a considerable increase in the impact energy (from 0.33 to 0.76 MJ/m^2 after one and three passes of ECAP, respectively) thereby suggesting that grain refinement to the NC regime is not necessarily associated with degradation in fracture strength characteristics.

The higher both rate of crack growth or notch sensitivity of the NC materials in comparison with the CG counterparts can be associated with the shrinkage of plastic zone in front of the crack tip [84,85]. The radius of the plastic zone can be found as seen in Ref. [85]: $r_y = (1/2\pi)(K/\sigma_y)^2$, where K denotes the stress intensity factor in the crack tip and σ_y is the yield stress of the material. Small grain size and a very high density of dislocations [60,65] result in increased yield strength of NC materials while the fracture toughness trends to decrease with microstructure refinement [4].

Structurally, the size of the plastic zone in front of the crack tip is related to the grain size. When grains are equal to or smaller than the plastic zone, the spreading of the plastic deformation is associated with the passing strain through grain boundaries. Obviously, the presence of a number of boundaries restricts deformation volume in the vicinity of the crack tip, thereby decreasing crack propagation energy [85]. In addition, the path of the crack propagation in the case of UFG materials should be much smoother in comparison to CG counterparts [4,42].

6.6 SUPERPLASTICITY

The ability to produce large and complex samples in one step, with a good surface finish, high accuracy, and low deformation force are the main advantages of superplastic (SP) deformation. Microstructure refinement to the nanorange leads to a significant reduction in the temperature for SP deformation. The decrease in SP deformation temperature associated with microstructure refinement can be predicted from the Bird–Mukherjee–Dorn equation [86]:

$$\dot{\varepsilon} = A\frac{DGb}{kT}\left(\frac{b}{d}\right)^p\left(\frac{\sigma}{G}\right)^n, \tag{6.4}$$

in which $\dot{\varepsilon}$ is the strain rate, kT is Boltzmann's constant times the absolute temperature, D is the dominant temperature-dependent diffusivity, G is the temperature-dependent shear modulus, b is Burgers vector, d is the grain size, n is the stress exponent ($n = 1/m$, where m is the strain-rate-sensitivity exponent); and p and A are constants. The low-temperature SP properties of two-phase titanium alloys with an UFG or NC microstructure developed via large deformation has been discussed in a number of earlier works [10,11,87–89].

This benefit has attracted great interest due to the possibility to conduct superplastic forming (SPF) and/or SP-assisted diffusion bonding at temperatures much lower than those typically used for commercial production (e.g., see Refs. [90,91]). For example, successful diffusion bonding of Ti–6Al–4V with a grain size 200–400 nm at the very low temperature of 600°C was reported in Ref. [92]. In addition, lower SPF temperatures can result in: (i) the retention of high service-temperature strength which is intrinsic to UFG or NC metals; (ii) reduced surface oxidation on SPF parts; (iii) the possibility of using less expensive tool materials; and (iv) decreased flow stress that considerably reduces tool wear during SPF.

However, studies on the SPF of CP titanium even at elevated temperatures are quite limited because of poor superplastic characteristics in comparison with two-phase titanium alloys [93]. Zhu et al. [94] have shown that due to cavitation and rapid grain growth, CP titanium does not show good superplasticity at 600–800°C. The maximum elongation-to-failure value was 188% at 600°C and an initial strain rate of 0.001/s. Recrystallization and intensive grain growth was observed in Ref. [95] during deformation of two grades of ECAP CP Ti (Grade 2 and Grade 4) in the temperature interval 600–900°C.

An effort to evaluate the SP characteristics of NC CP Ti Grade 2 alloy was made in Ref. [12] during deformation in a temperature range of (0.15–0.45) T_m (where T_m is the melting temperature). Formation of a NC microstructure with a grain size 100 nm via warm multiaxial forging was accompanied by dynamic strain aging (DSA) [96]. The development of DSA in a temperature range of 300–425°C resulted in anomalous changes in mechanical properties: decreased plasticity, increased flow stress, and increased rate of strain hardening. Some signs of SP deformation was found at 450°C and a strain rate of $1 \times 10^{-4} s^{-1}$ when the total elongation δ and the strain rate sensitivity m were found to be $\delta = 110\%$ and $m = 0.3$ [12]. Analysis of the mechanical properties and activation energies suggest quite a narrow temperature-strain rate interval of superplasticity in the NC Ti. The lower temperature boundary is limited by the development of DSA while the upper boundary is limited by grain growth.

Therefore, it can be concluded that SP properties of NC CP Ti are rather low, and can considerably limit the use of SP-based processes for working this material.

6.7 SUMMARY

Based on the analysis of experimental data for the Young's modulus, the strength- and ductility-related properties as well as superplasticity of NC CP Ti the following summary can be noted.

The Young's modulus of NC titanium produced by a SPD method is ~10% lower than that of MC counterparts; however, the effect of microstructure refinement to the NC regime on the Young's modulus of Ti is a multi-aspect problem and depends on many factors, both structural and those associated with the investigation method. Grain refinement to the NC interval by SPD of CP Ti increases strength by 30%–100%. A very high level of strength can be attained in CP Ti by combinations of various methods of SPD. Hardness and HCF limit of CP Ti also increased considerably with microstructure refinement to the NC regime. However, considerable microstructure refinement results also in a decrease in ductility by 20%–70% and ductility-related properties (tensile elongation, impact toughness, crack propagation resistance) because of the low ability of NC Ti to strain hardening, the fast loss of plastic flow stability, and more intensive strain localization. Superplastic characteristics of the NC CP Ti was found to be quite low because of the development of DSA in the same temperature interval.

REFERENCES

[1] E.O. Hall, The deformation and ageing of mild steel: discussion of result, Proc. Phys. Soc. B64 (1951) 747–753. London.

[2] N.J. Petch, The cleavage strength of polycrystals, J. Iron Steel Inst. 174 (1953) 25–28.

[3] T.R. Smith, R.W. Armstrong, P.M. Hazzledine, R.A. Masumura, C.S. Pande, Pile-up based hall-petch considerations at ultra-fine grain sizes, Mater. Res. Soc. Symp. Proc. 362 (1995) 31–37.

[4] M.A. Meyers, A. Mishra, D.J. Benson, Mechanical properties of nanocrystalline materials, Prog. Mater. Sci. 51 (2006) 427–556.

[5] A.V. Sergueeva, V.V. Stolyarov, R.Z. Valiev, A.K. Mukherjee, Advanced mechanical properties of pure titanium with ultrafine grained structure, Scr. Mater. 45 (2001) 747–752.

[6] R.W. Armsrong, The influence of polycrystal grain size on several mechanical properties of materials, Met. Trans. 1A (1970) 1169–1176.

[7] T. Hanlon, E.D. Tabachnikova, S. Suresh, Fatigue behavior of nanocrystaline metals and alloys, Int. J. Fatigue 27 (2005) 1147–1158.

[8] Y. Estrin, A. Vinogradov, Fatigue behaviour of light alloys with ultrafine grain structure produced by severe plastic deformation: an overview, Int. J. Fatigue 32 (2010) 898–907.

[9] K.S. Kumar, H. Van Swygenhoven, S. Suresh, Mechanical behavior of nanocrystalline metals and alloys, Acta Mater. 51 (2003) 5743–5774.

[10] Y.G. Ko, C.S. Lee, D.H. Shin, S.L. Semiatin, Low-temperature superplasticity of ultra-fine-grained Ti-6Al-4V processed by equal-channel angular pressing, Metall. Mater. Trans. A 37A (2006) 381–391.

[11] A.V. Sergueeva, V.V. Stolyarov, R.Z. Valiev, A.K. Mukherjee, Superplastic behavior of ultrafine-grained Ti-6Al-4V alloy, Mater. Sci. Eng. A 323 (2002) 318–325.

[12] S.P. Malysheva, G.A. Salishchev, R.M. Galeev, V.N. Danilenko, M.M. Myshlyaev, A. A. Popov, Changes in the structure and mechanical properties of submicrocrystalline titanium during deformation in a temperature range of (0.15–0.45)Tm, Phys. Met. Metallogr. 95 (2003) 390–397.

[13] H. Gleiter, Nanostructured materials: basic concepts and microstructure, Acta Mater. 48 (2000) 1–29.

[14] X. Qing, G. Xingming, The scale effect on the yield strength of nanocrystalline materials, Int. J. Solids Struct. 43 (2006) 7793–7799.

[15] G.E. Fougere, L. Riester, M. Ferber, J.R. Weertman, R.W. Siegel, Young's modulus of nanocrystalline Fe measured by nanoindentation, Mater. Sci. Eng. A 204 (1995) 1–6.
[16] J.R. Weertman, Hall-Petch strengthening in nanocrystalline metals, Mater. Sci. Eng. A 166 (1993) 161–167.
[17] Y. Zhou, U. Erb, K.T. Aust, G. Palumbo, The effects of triple junctions and grain boundaries on hardness and Young's modulus in nanostructured Ni-P, Scr. Mater. 48 (2003) 825–830.
[18] C. Koch, Bulk behavior of nanostructured materials, in: R.W. Siegel, E. Hu, M.C. Roco (Eds.), Nanostructure Science and Technology—A Worldwide Study, WTEC Loyola College in Maryland, 2000, p. 93.
[19] H. Huang, F. Spaepen, Tensile testing of free-standing Cu, Ag and Al thin films and Ag/Cu multilayers, Acta Mater. 48 (2000) 3261–3269.
[20] S.R. Phillpot, D. Wolf, H. Gleiter, A structural model for grain boundaries in nanocrystalline materials, Scr. Metall. Mater. 33 (1995) 1245–1251.
[21] T. Mütschele, R. Kircheim, Hydrogen as a probe for the average thickness of a grain boundary, Scr. Metall. Mater. 21 (1987) 1101–1104.
[22] A. Caro, H. Van Swygenhoven, Grain boundary and triple junction enthalpies in nanocrystalline metals, Phys. Rev. B 63 (2001) 1–5. 134101.
[23] G. Palumbo, S.J. Thorpe, K.T. Aust, On the contribution of triple junctions to the structure and properties of nanocrystalline materials, Scr. Metall. Mater. 24 (1990) 1347–1350.
[24] J.C. Gibeling, W.D. Nix, Observations of anelastic backflow following stress reductions during creep of pure metals, Acta Metall. 29 (1981) 1769–1784.
[25] E. Bonetti, E.G. Campari, L. Del Bianco, L. Pasquini, E. Sampaolesi, Mechanical behavior of nanocrystalline iron and nickel in the quasi-static and low frequency anelastic regime, Nanostruct. Mater. 11 (1999) 709–720.
[26] Z. Hung, L.Y. Gu, J.R. Weertman, Temperature dependence of hardness of nanocrystalleve copper in low-temperature range, Scr. Mater. 37 (1997) 1071–1075.
[27] J. Rajagopalan, J. Han, M.T. Saif, Plastic deformation recovery in freestanding nanocrystalline aluminum and gold thin films, Science 315 (2007) 1831–1834.
[28] G.W. Nieman, J.R. Weertman, R.W. Siegel, Mechanical behavior of nanocrystalline Cu and Pd, J. Mater. Res. 6 (1991) 1012–1027.
[29] V.D. Krstic, U. Erb, G. Palumbo, Effect of porosity on Young's modulus of nanocrystalline materials, Scr. Metall. Mater. 29 (1993) 1501–1504.
[30] R. Zugic, B. Szpunar, V.D. Krstic, U. Erb, Effect of porosity on the elastic response of brittle materials: an embedded-atom method approach, Philos. Mag. A. 75 (1997) 1041–1055.
[31] P.G. Sanders, J.A. Eastman, J.R. Weertman, Elastic and tensile behavior of nanocrystalline copper and palladium, Acta Mater. 45 (1997) 4019–4025.
[32] T.D. Shen, C.C. Koch, T.Y. Tsui, G.M. Pharr, On the elastic moduli of nanocrystalline Fe, Cu, Ni, and Cu-Ni alloys prepared by mechanical milling/alloying, J. Mater. Res. 10 (1995) 2892–2896.
[33] R. Dobosz, M. Muzyk, Z. Pakiela, M. Lewandowska, K.J. Kurzydlowski, Simulations of the elastic properties of nanomaterials using multiscale modelling methods, Mech. Mater. 67 (2013) 74–78.
[34] K. Topolski, T. Brynk, H. Garbacz, Elastic modulus of nanocrystalline titanium evaluated by cyclic tensile method, Arch. Civ. Mech. Eng. 16 (2016) 927–934.
[35] ASTM, n.d. Grade 2 Titanium, UNS R50400 [www.matweb.com].

[36] ASTM, n.d. Grade 5 Titanium, UNS R56400 [www.matweb.com].

[37] S. Kamiya, H. Nagasawa, K. Yamanobe, H. Hanyu, M. Saka, A comparative study of the mechanical properties of PVD coating evaluated by new techniques and conventional methods, Thin Solid Films 469–470 (2004) 248–253.

[38] W.C. Oliver, G.M. Pharr, An improved technique for determining hardness and elastic modulus using load and displacement sensing indentation experiments, J. Mater. Res. 7 (1992) 1564–1583.

[39] R. Polini, P. D'Antonio, S. Lo Castro, V.F. Traversa, Cutting performance and indentation behavior of diamond films on co-cemented tungsten carbide, Surf. Coat. Technol. 123 (2000) 78–83.

[40] S. Sakai, H. Tanimoto, H. Mizubayashi, Mechanical behavior of high-density nanocrystalline gold prepared by gas deposition method, Acta Mater. 47 (1999) 211–217.

[41] R.Z. Valiev, I.V. Alexandrov, Y.T. Zhu, T.C. Lowe, Paradox of strength and ductility in metals processed by severe plastic deformation, J. Mater. Res. 17 (2002) 5–8.

[42] Y. Estrin, A. Vinogradov, Extreme grain refinement by severe plastic deformation: a wealth of challenging science, Acta Mater. 61 (2013) 782–817.

[43] R.Z. Valiev, R.K. Islamgaliev, I.V. Alexandrov, Bulk nanostructured materials from severe plastic deformation, Prog. Mater. Sci. 45 (2000) 103–189.

[44] R.Z. Valiev, A.V. Korznikov, R.R. Mulyukov, Structure and properties of ultrafine-grained materials produced by severe plastic deformation, Mater. Sci. Eng. A 168 (1993) 141–148.

[45] V.V. Stolyarov, L. Zeipper, B. Mingler, M. Zehetbauer, Influence of post-deformation on CP-Ti processed by equal channel angular pressing, Mater. Sci. Eng. A 476 (2008) 98–105.

[46] A.A. Popov, I.Y. Pyshmintsev, S.L. Demakov, A.G. Illarionov, T.C. Lowe, A. V. Sergeyeva, R.Z. Valiev, Structural and mechanical properties of nanocrystalline titanium processed by severe plastic deformation, Scr. Mater. 37 (1997) 1089–1094.

[47] A.Y. Vinogradov, V.V. Stolyarov, S. Hashimoto, R.Z. Valiev, Cyclic behavior of ultrafine-grain titanium produced by severe plastic deformation, Mater. Sci. Eng. A 318 (2001) 163–173.

[48] X. Zhao, X. Yang, X. Liu, X. Wang, T.G. Langdon, The processing of pure titanium through multiple passes of ECAP at room temperature, Mater. Sci. Eng. A 527 (2010) 6335–6339.

[49] C.T. Wang, N.M.G. Gee, R.J.K. Wood, T.G. Langdon, Processing of an ultrafine-grained titanium by high-pressure torsion: an evaluation of the wear properties with and without a TiN coating, J. Mech. Behav. Biomed. Mater. 17 (2013) 166–175.

[50] D. Terada, S. Inoue, N. Tsuji, Microstructure and mechanical properties of commercial purity titanium severely deformed by ARB process, J. Mater. Sci. 42 (2007) 1673–1681.

[51] G. Purcek, O. Saray, O. Kul, I. Karaman, G.G. Yapici, M. Haouaoui, H.J. Maier, Mechanical and wear properties of ultrafine-grained pure Ti produced by multi-pass equal-channel angular extrusion, Mater. Sci. Eng. A 517 (2009) 97–104.

[52] V.L. Sordi, M. Ferrante, M. Kawasaki, T.G. Langdon, Microstructure and tensile strength of grade 2 titanium processed by equal-channel angular pressing and by rolling, J. Mater. Sci. 47 (2012) 7870–7876.

[53] S. Zherebtsov, G. Salishchev, Production, properties and application of ultrafine-grained titanium alloys, Mater. Sci. Forum 838–839 (2016) 294–301.

[54] G.A. Salishchev, R.M. Galeev, S.P. Malysheva, S.V. Zherebtsov, S.Y. Mironov, O. R. Valiakhmetov, E.I. Ivanisenko, Formation of submicrocrystalline structure in titanium and titanium alloys and their mechanical properties, Metal Sci. Heat Treat. 48 (2006) 63–69.
[55] S. Zherebtsov, W. Lojkowski, A. Mazur, G. Salishchev, Structure and properties of hydrostatically extruded commercially pure titanium, Mater. Sci. Eng. A 527 (2010) 5596–5603.
[56] K. Topolski, W. Pachla, H. Garbacz, Progress in hydrostatic extrusion of titanium, J. Mater. Sci. 48 (2013) 4543–4548.
[57] K. Topolski, H. Garbacz, P. Wiecinski, W. Pachla, K.J. Kurzydlowski, Mechanical properties of titanium processed by hydrostatic extrusion, Arch. Metall. Mater. 57 (2012) 863–867.
[58] W. Pachla, M. Kulczyk, M. Sus-Ryszkowska, A. Mazur, K.J. Kurzydlowski, Nanocrystalline titanium produced by hydrostatic extrusion, J. Mater. Process. Technol. 205 (2008) 173–182.
[59] W. Pachla, M. Kulczyk, S. Przybysz, J. Skiba, K. Wojciechowski, M. Przybysz, K. Topolski, A. Sobolewski, M. Charkiewicz, Effect of severe plastic deformation realized by hydrostatic extrusion and rotary swaging on the properties of CP Ti grade 2, J. Mater. Process. Technol. 221 (2015) 255–268.
[60] S.V. Zherebtsov, G.S. Dyakonov, A.A. Salem, S.P. Malysheva, G.A. Salishchev, S. L. Semiatin, Evolution of grain and subgrain structure during cold rolling of commercial-purity titanium, Mater. Sci. Eng. A 528 (2011) 3472–3479.
[61] R.K. Islamgaliev, V.U. Kazyhanov, L.O. Shestakova, A.V. Sharafutdinov, R.Z. Valiev, Microstructure and mechanical properties of titanium (Grade 4) processed by high-pressure torsion, Mater. Sci. Eng. A 493 (2008) 190–194.
[62] I.P. Semenova, R.Z. Valiev, E.B. Yakushina, G.H. Salimgareeva, T.C. Lowe, Strength and fatigue properties enhancement in ultrafine-grained Ti produced by severe plastic deformation, J. Mater. Sci. 43 (2008) 7354–7359.
[63] C. Leyens, M. Peters, Titanium and Titanium Alloys: Fundamentals and Applications, Wiley-VCH Verlag GmbH, Weinheim, 2003.
[64] G. Lutjering, J.C. Williams, Titanium, Springer-Verlag, Berlin/Heidelberg, 2007.
[65] G.S. Semenova, G. Da Costa, W. Lefebvre, R. Valiev, Enhanced strength and ductility of ultrafine-grained Ti processed by severe plastic deformation, Adv. Eng. Mater. 12 (2010) 803–807.
[66] S. Zherebtsov, E. Kudryavtsev, S. Kostjuchenko, S. Malysheva, G. Salishchev, Strength and ductility-related properties of ultrafine grained two-phase titanium alloy produced by warm multiaxial forging, Mater. Sci. Eng. A 536 (2012) 190–196.
[67] R.Z. Valiev, A.V. Sergueeva, A.K. Mukherjee, The effect of annealing on tensile deformation behavior of nanostructured SPD titanium, Scr. Mater. 49 (2003) 669–674.
[68] Y.M. Wang, E. Ma, Three strategies to achieve uniform tensile deformation in a nanostructured metal, Acta Mater. 52 (2004) 1699–1709.
[69] Y. Wang, M. Chen, F. Zhou, E. Ma, High tensile ductility in a nanostructured metal, Nature 419 (2002) 912–915.
[70] T. Sekiguchi, K. Ono, H. Fujiwara, K. Ameyama, New microstructure design for commercially pure titanium with outstanding mechanical properties by mechanical milling and hot roll sintering, Mater. Trans. 51 (2010) 39–45.

[71] P. Cavaliere, Fatigue properties and crack behavior of ultra-fine and nanocrystalline pure metals, Int. J. Fatigue 31 (2009) 1476–1489.
[72] I.P. Semenova, G.K. Salimgareeva, V.V. Latysh, T. Lowe, R.Z. Valiev, Enhanced fatigue strength of commercially pure Ti processed by severe plastic deformation, Mater. Sci. Eng. A 503 (2009) 92–95.
[73] A. Vinogradov, Fatigue limit and crack growth in ultra-fine grain metals produced by severe plastic deformation, J. Mater. Sci. 42 (2007) 1797–1808.
[74] V.V. Stolyarov, I.V. Alexandrov, Y.R. Kolobov, M. Zhu, T. Zhu, T. Lowe, in: X.R. Wu, Z.G. Wang (Eds.), Fatigue'99, 3 Higher Education Press, PR China, 1999, pp. 1345–1351.
[75] W.J. Kim, C.Y. Hyun, H.K. Kim, Fatigue strength of ultrafine-grained pure Ti after severe plastic deformation, Scr. Mater. 54 (2006) 1745–1750.
[76] Y.A. Vinogradov, S.R. Agnew., Encyclopedia of Nanoscience and Nanotechnology, Marcel-Dekker, New York, 2003, pp. 2269–2288.
[77] H. Mughrabi, H.W. Hoppel, M. Kautz, Fatigue and microstructure of ultrafine-grained metals produced by severe plastic deformation, Scr. Mater. 51 (2004) 807–812.
[78] S. Suresh, Fatigue of Materials, Cambridge University Press, 1991.
[79] N.G. Turner, W.T. Roberts, Fatigue behavior of titanium, Trans. Metall. Soc. AIME 242 (1968) 1223–1230.
[80] I.P. Semenova, A.V. Polyakov, G.I. Raab, T.C. Lowe, R.Z. Valiev, Enhanced fatigue properties of ultrafine-grained Ti rods processed by ECAP-Conform, J. Mater. Sci. 47 (2012) 7777–7781.
[81] H. Garbacz, Z. Pakiela, K.J. Kurzydlowsli, Fatigue properties of nanocrystalline titanium, Rev. Adv. Mater. Sci. 25 (2010) 256–260.
[82] S. Fintova, M. Arzaghi, I. Kubena, L. Kunz, C. Sarrazin-Baudoux, Fatigue crack propagation in UFG Ti grade 4 processed by severe plastic deformation, Int. J. Fatigue 98 (2017) 187–194.
[83] R. Naseri, M. Kadkho Dayan, M. Shariati, Static mechanical properties and ductility of biomedical ultrafine-grained commercially pure titanium produced by ECAP process, Trans. Nonferrous Metals Soc. China 27 (2017) 1964–1975.
[84] M.A. Meyers, K.K. Chawla, Mechanical Behavior of Materials, Cambridge University Press, New York, 2009.
[85] M.K. Rabinovich, M.V. Markushev, Influence of fine-grained structure and superplastic deformation on the strength of aluminium alloys, J. Mater. Sci. 31 (1996) 4997–5001.
[86] J.E. Bird, A.K. Mukherjee, J.E. Dorn, Experimental correlations between high-temperature creep behavior and structure, in: D.G. Brandon, A. Rosen (Eds.), Quantitative Relation Between Microstructure and Properties, Israel Universities Press, Jerusalem, Israel, 1969, pp. 255–342.
[87] S.L. Semiatin, P.N. Fagin, J.F. Betten, A.P. Zane, A.K. Ghosh, G.A. Sargent, Plastic flow and microstructure evolution during low-temperature superplasticity of ultrafine Ti-6Al-4V sheet material, Metall. Mater. Trans. A 41 (2010) 499–512.
[88] S.V. Zherebtsov, E.A. Kudryavtsev, G.A. Salishchev, B.B. Straumal, S.L. Semiatin, Microstructure evolution and mechanical behavior of ultrafine Ti-6Al-4V during low-temperature superplastic deformation, Acta Mater. 121 (2016) 152–163.
[89] H. Matsumoto, K. Yoshida, S.-H. Lee, Y. Ono, A. Chiba, Ti–6Al–4V alloy with an ultrafine-grained microstructure exhibiting low-temperature-high-strain-rate superplasticity, Mater. Lett. 98 (2013) 209–212.
[90] O.A. Kaibyshev, Superplasticity of Alloys, Intermetallics, and Ceramics, Springer-Verlag, Berlin, 1992.

[91] T.G. Nieh, J. Wadsworth, O.D. Sherby, Superplasticity in Metals and Ceramics, Cambridge University Press, Cambridge, 1997.

[92] R.Y. Lutfullin, A.A. Kruglov, R.V. Safiullin, M.K. Mukhametrakhimov, O.A. Rudenko, Processing properties of nano- and submicro-crystalline Ti–6Al–4V titanium alloy, Mater. Sci. Eng. A 503 (2009) 52–54.

[93] K. Ghosh, C.H. Hamilton, Superplastic forming and diffusion bonding of titanium alloys, Def. Sci. J. 36 (1986) 153–177.

[94] X.J. Zhu, M.J. Tan, W. Zhou, Enhanced superplasticity in commercially pure titanium alloy, Scr. Mater. 52 (2005) 651–655.

[95] S.V. Sajadifar, G.G. Yapici, Effect of purity levels on the high-temperature deformation characteristics of severely deformed titanium, Metal. Mater. Trans. A 48 (2017) 999–1012.

[96] H. Conrad, Effect of interstitial solutes on the strength and ductility of titanium, Prog. Mater. Sci. 26 (1981) 123–403.

CHAPTER

7 Strengthening mechanisms and super-strength of severely deformed titanium

Grigory S. Dyakonov*, Sergey Mironov†, Irina P. Semenova*, Ruslan Z. Valiev*,‡

*Institute of Physics of Advanced Materials, Ufa State Aviation Technical University, Ufa, Russia**

Department of Materials Processing, Graduate School of Engineering, Tohoku University, Sendai, Japan†

Laboratory for Mechanics of Bulk Nanostructured Materials, Saint Petersburg State University, Saint Petersburg, Russia‡

7.1 INTRODUCTION

It is well known that severe plastic deformation (SPD) leads to a significant transformation of the microstructure in titanium. To date, several SPD techniques have been developed for ultrafine-grained (UFG) structure formation in metals and alloys [1]. The following SPD techniques are used most often for UFG structure formation: high-pressure torsion (HPT), equal-channel angular pressing (ECAP), hydrostatic extrusion (HE), and multiaxial forging (MAF). Each of these techniques has its own distinctive features. For instance, the use of the HPT technique enables accumulating a very large strain in a sample and producing a structure with an extremely small grain size ($d \sim 120$ nm) and high strength values of $\sigma_B = 950$ MPa [2]. However, samples after HPT processing have small dimensions, which is a very significant drawback of this technique. Processing of titanium by hydrostatic extrusion leads to the formation of a structure with a grain/subgrain size of about 55 nm (assessed using transmission electron microscopy [TEM]) and an ultimate tensile strength (UTS) circa 1050 MPa [3]. However, after hydrostatic extrusion the microstructure often has a sharp crystallographic and metallographic texture, which has an effect on the anisotropy of mechanical properties. The MAF technique is based on stage-by-stage structure refinement in the process of repeated operations of uniaxial compression, and is distinguished by its labor intensity. As a result of the use of the MAF for the processing of titanium, bulk billets with a mean grain size of about 200 nm and a UTS of 810 MPa were produced [4]. Another efficient technique for UFG structure formation is ECAP. Using this technique, samples with a mean grain size of about 0.28 μm and a UTS of 710 MPa were produced [5]. Extraordinary properties were produced when a modification of the ECAP process, ECAP-Conform (ECAP-C), was used [6]. The use of this technique ensured the formation of a

Nanocrystalline Titanium. https://doi.org/10.1016/B978-0-12-814599-9.00007-9

structure with a mean grain size of about 0.2 μm (assessed via TEM) and high strength properties of $\sigma_{0.2} = 1050$ MPa and $\sigma_B = 1130$ MPa. The ECAP-C technique is different from other SPD techniques with its high productivity and the capability to produce billets with a length of up to 3 m with high strength characteristics. The data visibly demonstrate that the use of SPD processing techniques opens up possibilities for the creation of ultrastrength materials.

The practical interest in production of nanocrystalline (NC) or UFG titanium via SPD is primarily driven by a concomitant strengthening effect. Initially, this effect was simply attributed to the drastic grain refinement. However, the remarkable complexity of deformation-induced microstructures established recently has conclusively demonstrated that the microstructure–strength relationship cannot be readily explained in terms of the Hall–Petch relationship. This requires a more fundamental understanding of the mechanisms of hardening of UFG materials obtained by intense plastic deformation and several research attempts have been undertaken recently in this area. Considering the importance of this problem, this chapter provides a brief review of the current understanding of the microstructure–strength relationship in severely deformed titanium.

7.2 STRENGTHENING EFFECT FROM SEVERE PLASTIC DEFORMATION

When discussing strengthening mechanisms for materials, it is first useful to highlight the most common trends in strengthening behavior of titanium under SPD. These are briefly summarized in this section.

One of the most important observations is a gradual saturation of the strengthening effect with accumulated strain, as shown in Fig. 7.1A–C. Specifically, the deformed material typically experiences relatively rapid hardening at low strains but then the strengthening efficiency significantly decreases (Fig. 7.2A). Such behavior has been observed during HPT (e.g., see Ref. [7]), ECAP (e.g., see Refs. [8–10]), accumulated roll bonding (e.g., see Ref. [11]) and MAF [12]. Moreover, it has also been reported for more conventional techniques including cyclic channel-die compression [13], hydrostatic extrusion [14–16], and cold rolling [17,18], as shown in Figs. 7.1D–F and 7.2B. Therefore, the saturation of the strengthening effect appears to be a common trend, and, if so, it may reflect a very specific character of operating strengthening mechanisms seen in Fig. 7.1.

As expected, the strengthening efficiency has been found to be sensitive to deformation temperature. Specifically, processing at 500°C has been shown to impart only a moderate hardening effect [19] whereas deformation at 590°C resulted in material softening [20]. On the other hand, a reduction of the processing temperature below ~450°C typically gives rise to significant strengthening [7,11,13–15,21–27]. Considering that recrystallization temperature of titanium is well accepted to be 500–550°C, the strength of such low-temperature deformed material is expected to be essentially contributed by substructure, i.e., dislocations and low-angle

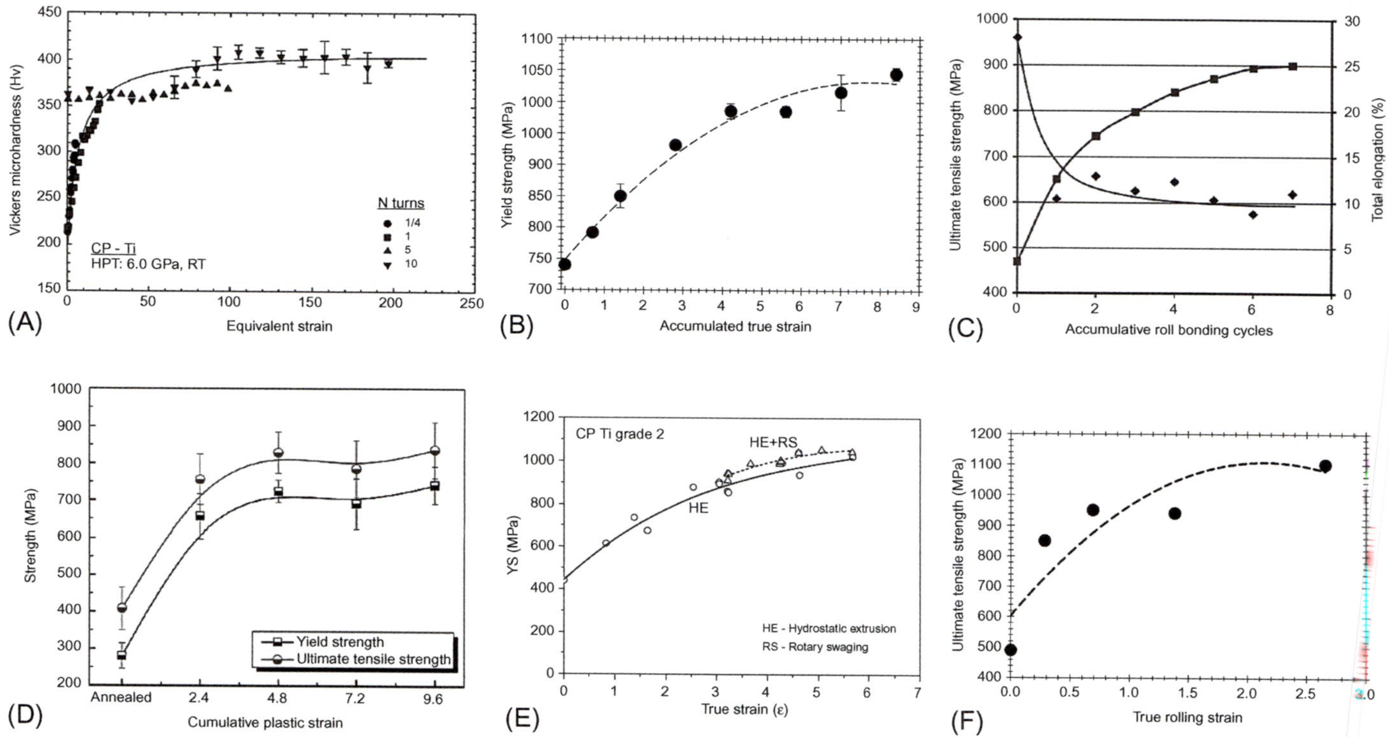

FIG. 7.1

Effect of accumulated strain on strength characteristics of titanium during (A) high-pressure torsion, (B) equal-channel angular pressing, (C) accumulated roll bonding, (D) cyclic channel-die compression, (E) hydrostatic extrusion, and (F) cold rolling.

Panel (A): After Shirooyeh [7]. Panel (B): After G.S. Dyakonov, S. Mironov, I.P. Semenova, R.Z. Valiev, S.L. Semiatin, Microstructurer evolution and strengthening mechanisms in commercial-purity titanium subjected to equal-channel angular pressing, Mater. Sci. Eng. A *701 (2017) 289–301, https://doi.org/10.1016/j.msea.2017.06.079. Panel (C): After J.L. Milner, F. Abu-Farha, C. Bunget, T. Kurfess, V.H. Hammond, Grain refinement and mechanical properties of CP-Ti processed by warm accumulated roll bonding,* Mater. Sci. Eng. A *561 (2013), 109–117, https://doi.org/10.1016/j.msea.2012.10.081. Panel (D): After S.S. Satheesh Kumar, K. Priyasudha, M. Sudhakara Rao, T. Raghu, Deformation homogeneity, mechanical behavior and strain hardening characteristics of titanium severely plastically deformed by cyclic channel die compression method,* Mater. Des. *101 (2016), 117–129, https://doi.org/10.1016/j.matdes.2016.04.004. Panel (E): After W. Pachla, M. Kulczyk, S. Przybysz, J. Skiba, K. Wojciechowski, M. Przybysz, K. Topolski, A. Sobolevwski, M. Charkiewicz, Effect of severe plastic deformation realized by hydrostatic extrusion and rotary swaging on the properties of CP Ti grade 2,* J. Mater. Process. Technol. *221 (2015), 255–268, https://doi.org/10.1016/j.jmatprotec.2015.02.027. Panel (F): After S. Malysheva, G. Salishchev, S. Mironov, S. Zherebtsov, Production of nanostructure in titanium by cold rolling,* Mater. Sci. Forum *584–586 (2008), 759–764, https://doi.org/10.4028/www.scientific.net/MSF.584-586.759.*

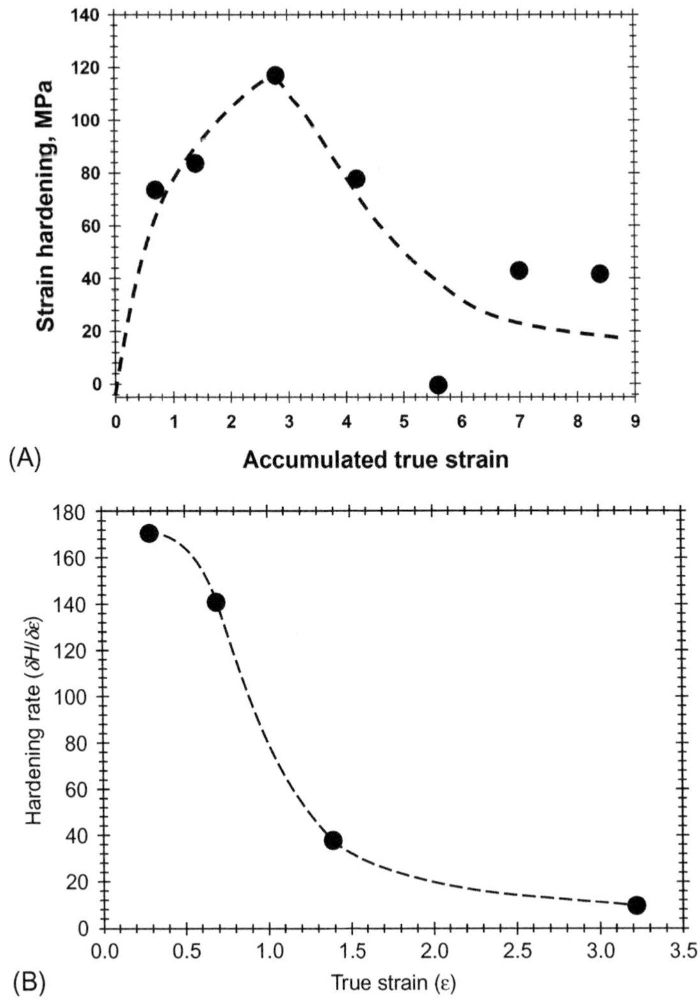

FIG. 7.2

Effect of accumulated strain on hardening rate during (A) equal-channel angular pressing and (B) cold rolling.

Panel (A): After G.S. Dyakonov, S. Mironov, I.P. Semenova, R.Z. Valiev, S.L. Semiatin, Microstructurer evolution and strengthening mechanisms in commercial-purity titanium subjected to equal-channel angular pressing, Mater. Sci. Eng. A 701 (2017) 289–301, https://doi.org/10.1016/j.msea.2017.06.079. Panel (B): After G.S. Dyakonov, S. Mironov, S.V. Zherebtsov, S.P. Malysheva, G.A. Salishchev, A.A. Salem, S.L. Semiatin, Grain-structure development in heavily cold-rolled alpha-titanium, Mater. Sci. Eng. A *607 (2014), 145–154, https://doi.org/10.1016/j.msea.2014.03.141.*

boundaries (LABs). In this processed condition, the material usually shows a strength level comparable with that of Ti–6Al–4V alloy (UTS $\sim$ 880 MPa).

It is also worth noting that the severely deformed material often exhibits substantial anisotropy of strength characteristics which is usually attributed to crystallographic texture [9,21,25,28]. This kind of anisotropy of properties is apparently related to the favorable and unfavorable orientations of crystallites, with respect to the acting external stresses.

7.3 MICROSTRUCTURE EVOLUTION DURING SEVERE PLASTIC DEFORMATION

The strengthening effect induced by SPD is governed by its underlying microstructural evolution. Thus, it is worth giving detailed consideration to this process.

It is important to emphasize that HPT of titanium may result in $\alpha \rightarrow \omega$ martensitic transformation [7,29–35] and therefore microstructural changes in this case are considered separately.

7.3.1 THE $\alpha \rightarrow \omega$ MARTENSITIC TRANSFORMATION DURING HIGH-PRESSURE TORSION

The martensitic transformation is commonly believed to be induced by relatively high applied pressure and does not develop below its critical limit, as shown in Fig. 7.3A. However, due to the sensitivity of the transformation to impurity content as well as to the pressure conditions (hydrostaticity and applied time), the reported threshold stresses significantly scatter in magnitude. It should be noted that microstructural measurements in the material processed by ECAP revealed only negligible fraction of the ω phase [25,36] and therefore the martensitic transformation seems to be primarily attributable to the HPT only.

In the context of this work, of particular importance is a pronounced grain refinement (Fig. 7.3B) as well as abrupt material hardening (Fig. 7.1A) associated with the martensitic transformation.

The transformation is believed to be governed by a shear (i.e., diffusionless) mechanism which results in the $(0001)_\alpha||(11-20)_\omega$, $[11-20]_\alpha||[0001]_\omega$ orientation relationship between the phases [33]. Details of this process are largely unclear; however, it is only suggested that nucleation of the martensitic phase is promoted by mechanical twinning [32].

It is believed that the presence of oxygen in titanium suppresses the formation of the ω phase because this element increases both the energy of the martensitic phase and an energy barrier for the transformation [29].

The martensitic phase has relatively low thermal stability. Specifically, an annealing of the deformed material at 100–200°C may induce the reverse $\omega \rightarrow \alpha$ phase transformation [29,31–34], which results in relatively coarse recrystallized α grains [34]. Moreover, the activation energy of this transformation was found to be close to that for self-diffusion in α titanium [34]. Therefore, the reverse transformation is believed to be governed by diffusion-controlled mechanisms [33,34].

For adiabatic heating induced during high-pressure torsion, the martensitic transformation was found to be sensitive to a strain rate [33]. Specifically, an increase in the strain rate (and the concomitant adiabatic heating) may suppress the $\alpha \rightarrow \omega$ phase transformation [33].

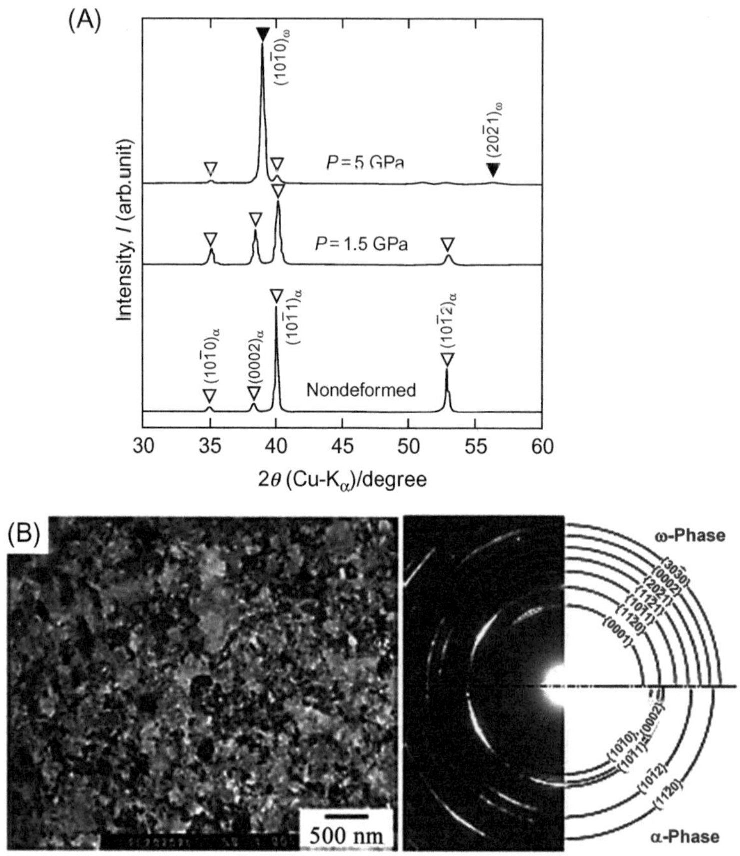

FIG. 7.3

The development of martensitic ω phase during high-pressure torsion: (A) X-ray measurements and (B) transmission-electron-microscopy observations.

Panel (A): After Y. Todaka, J. Sasaki, T. Moto, M. Umemoto, Bulk submicrocrystalline ω-Ti produced by high-pressure torsion straining, Scr. Mater. *59 (2008), 615–618, https://doi.org/10.1016/j.scriptamat.2008.05.015. Panel (B): After N. Adachi, Y. Todaka, K. Irie, M. Umemoto, Phase transformation kinetics of ω−phase in pure Ti formed by high-pressure torsion,* J. Mater. Sci. *51 (2016), 2608–2615, https://doi.org/10.1007/s10853-015-9574-z.*

7.3.2 MICROSTRUCTURAL CHANGES IN THE α PHASE

It is importantly to realize that microstructural changes are very sensitive to deformation temperature as well as to the chosen processing technique. Thus, despite the extensive microstructural studies over the last two decades, the accumulated data are still limited and common trends of the microstructural evolution are difficult to derive. In this regard, this section is primarily focused on the most well-studied case, i.e., the microstructural changes induced by ECAP conducted below recrystallization temperature.

The microstructural evolution during warm ECAP has been found to be a relatively complex process which may include several stages [10]. At relatively low accumulated strains (equivalent to 1 or 2 ECAP passes), the microstructural changes are markedly influenced by mechanical twinning [10,22,23,37–41], as exemplified in Fig. 7.4A and B. The $\{10\bar{1}1\}$ twinning is often reported to predominate [22,23,37–41] but other twinning modes may also operate [10]. An activation of a particular twinning mode is thought to depend on deformation temperature, material purity, and initial texture, as well as on the employed ECAP route. The twins usually have lenticular morphology, often extend across entire grains, and sometimes are grouped in series (Fig. 7.4A). Remarkably, only a limited number of twin variants are usually found in the grain interior (typically one or two) thus perhaps indicating a variant selection. The twinning provides relatively rapid grain refinement (Fig. 7.5A) but the evolved microstructures

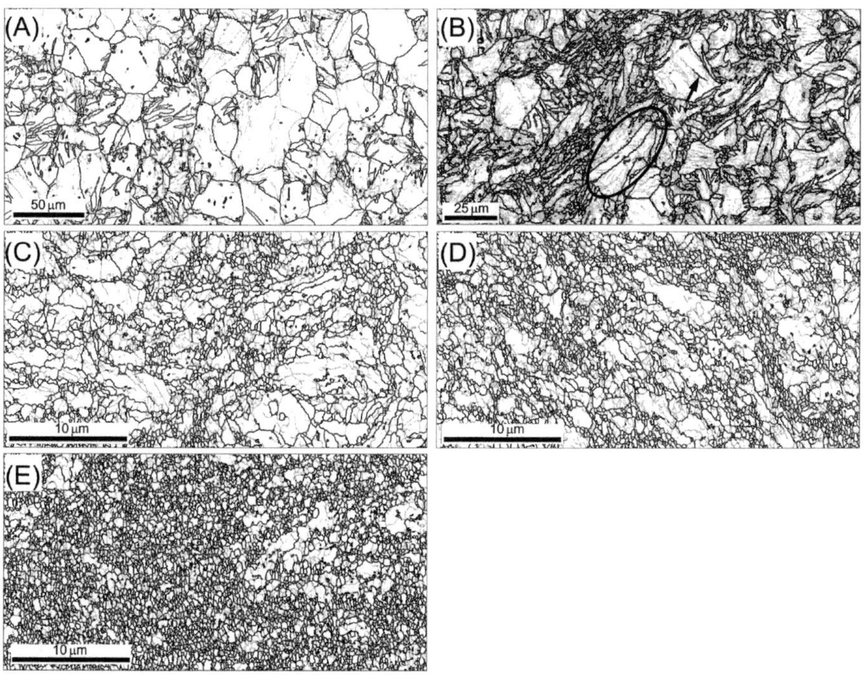

FIG. 7.4

EBSD grain boundary maps showing microstructures developed after (A) 1 ECAP pass ($\varepsilon \sim 0.7$), (B) 2 ECAP passes ($\varepsilon \sim 1.4$), (C) 4 ECAP passes ($\varepsilon \sim 2.8$), (D) 6 ECAP passes ($\varepsilon \sim 4.2$), and (E) 12 ECAP passes ($\varepsilon \sim 8.4$). In the maps, LABs, HABs, and {1012} twin boundaries are depicted by *red* (*gray* in the print version), *black*, and *blue* (*dark gray* in the print version) *lines*, respectively. Note the difference in the scales. See Section 7.3.2 for details.

After G.S. Dyakonov, S. Mironov, I.P. Semenova, R.Z. Valiev, S.L. Semiatin, Microstructurer evolution and strengthening mechanisms in commercial-purity titanium subjected to equal-channel angular pressing, Mater. Sci. Eng. A 701 (2017) 289–301, https://doi.org/10.1016/j.msea.2017.06.079.

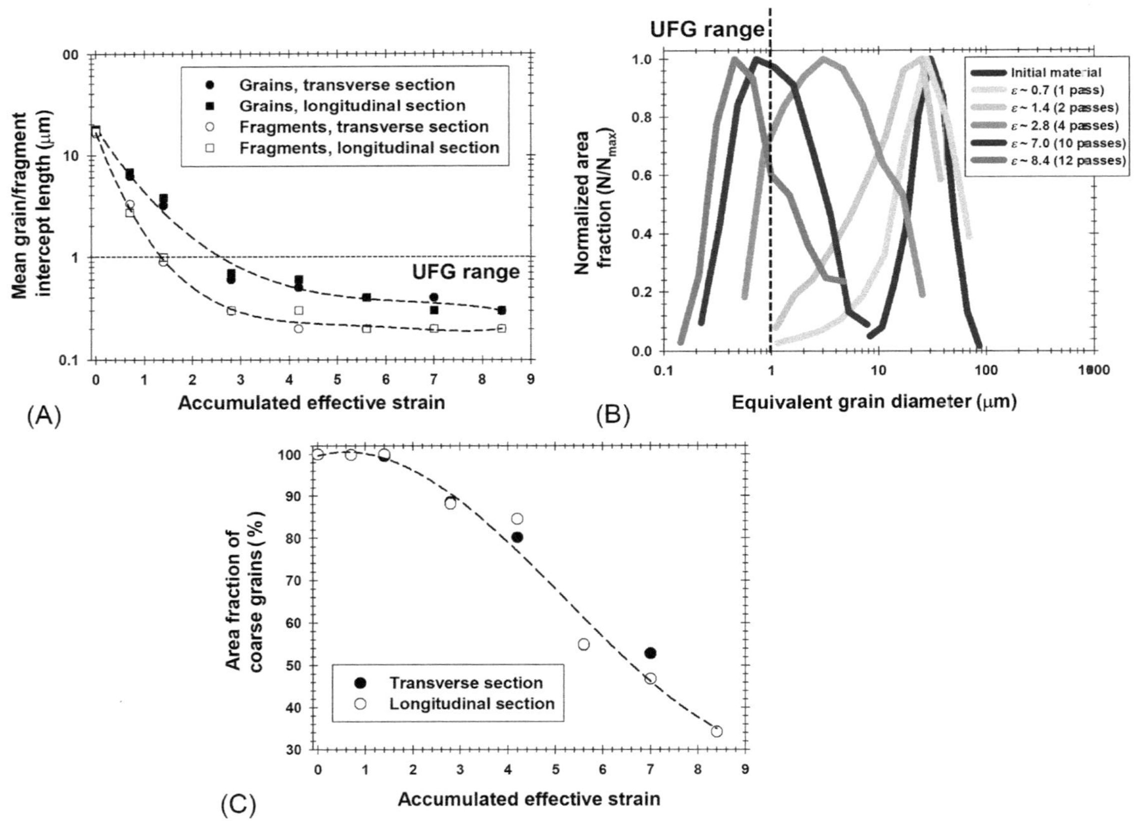

FIG. 7.5

Effect of accumulated ECAP strain on (A) mean grain/fragment intercept length, (B) grain size distribution, and (C) fraction of retained coarse grains. *Note*: UFG range indicates ultrafine-grained range of grain sizes.

After G.S. Dyakonov, S. Mironov, I.P. Semenova, R.Z. Valiev, S.L. Semiatin, Microstructurer evolution and strengthening mechanisms in commercial-purity titanium subjected to equal-channel angular pressing, Mater. Sci. Eng. A 701 (2017) 289–301, https://doi.org/10.1016/j.msea.2017.06.079.

are inhomogeneous (Figs. 7.4A and 7.5B) and thus the volume fraction of the UFG structure may be relatively low (Fig. 7.5C).

In addition to the twins, deformation-induced LABs also extensively develop in grain interior (Fig. 7.4A and B) thus rapidly decreasing a fraction of high-angle boundaries (HABs) (Fig. 7.6A). Importantly, the LABs tend to preferentially cluster near the original grain boundaries or within twins, as shown in Fig. 7.4B. In some cases, the deformation-induced boundaries accumulate misorientation above 15 degrees thus transforming into HABs (an example is shown with an arrow in Fig. 7.4B). This observation is typically interpreted as an evidence of continuous recrystallization [22,39,42]. Despite the extensive subboundary development, dislocation density is reported to remain relatively low at this stage [8,37,38], as illustrated in Figs. 7.7A and 7.8.

An increase of the accumulated strain is believed to gradually suppress the twinning due to the concomitant grain refinement, as shown in Fig. 7.4C–E. Remarkably, misorientations across the twin boundaries are typically observed to deviate from the ideal twin-matrix orientation relationship thus transforming them into random HABs; an example of the transformed twin is circled in Fig. 7.4B. This phenomenon is usually ascribed to diverging crystallographic rotations of the twin and surrounding matrix during ongoing strain [43,44].

At this stage, dislocation density is reported to increase rapidly, reaching $\sim 10^{15}\ m^{-2}$ [8]. This enhances the formation of deformation-induced boundaries (Fig. 7.6C). Importantly, the latter process preferentially occurs along the original grain boundaries, thus giving rise to a distinctly bimodal structure [6,9,10,22–25,36,39,42,45], as illustrated in Fig. 7.4C. It is worth noting that the development of the bimodal structure was also observed during MAF [12], as well as rolling [18,45], and thus it appears to be a more or less common characteristic of severely deformed titanium. Unlike cubic metals, no clear cell structure is usually found within grains [8,10] and dislocation distribution is fairly inhomogeneous between the fine-grained domains and remnants of original coarse grains, as shown in Fig. 7.7B and C.

The further increase in the accumulated strain results in the increase of the fraction of the fine grains which gradually consume the coarse-grained (CG) remnants (Figs. 7.4D and E, and 7.5C) [10]. The mean grain size changes relatively slowly at this stage (Fig. 7.5A) but grain size distribution gradually shifts towards UFG range (Fig. 7.5B) [10]. Fraction of HABs as well as mean misorientation of LABs gradually increases (Fig. 7.6A and B, respectively) [10] whereas density of free dislocations tends to saturate at the level of $\sim 10^{15}\ m^{-2}$ (Fig. 7.8) [8]. At higher density, dislocations cores are believed to overlap thus transforming the dislocations into grain boundaries. Thus, the revealed saturation of the dislocation density at this critical level appears to be reasonable.

After true accumulated strain of ~4–5, the total grain boundary area in the material also reaches a saturation limit and further microstructural evolution is dictated by a transformation of LABs into high-angle ones [10], as shown in Fig. 7.6C. In contrast to the saturation of the dislocation density mentioned earlier, the rationale for

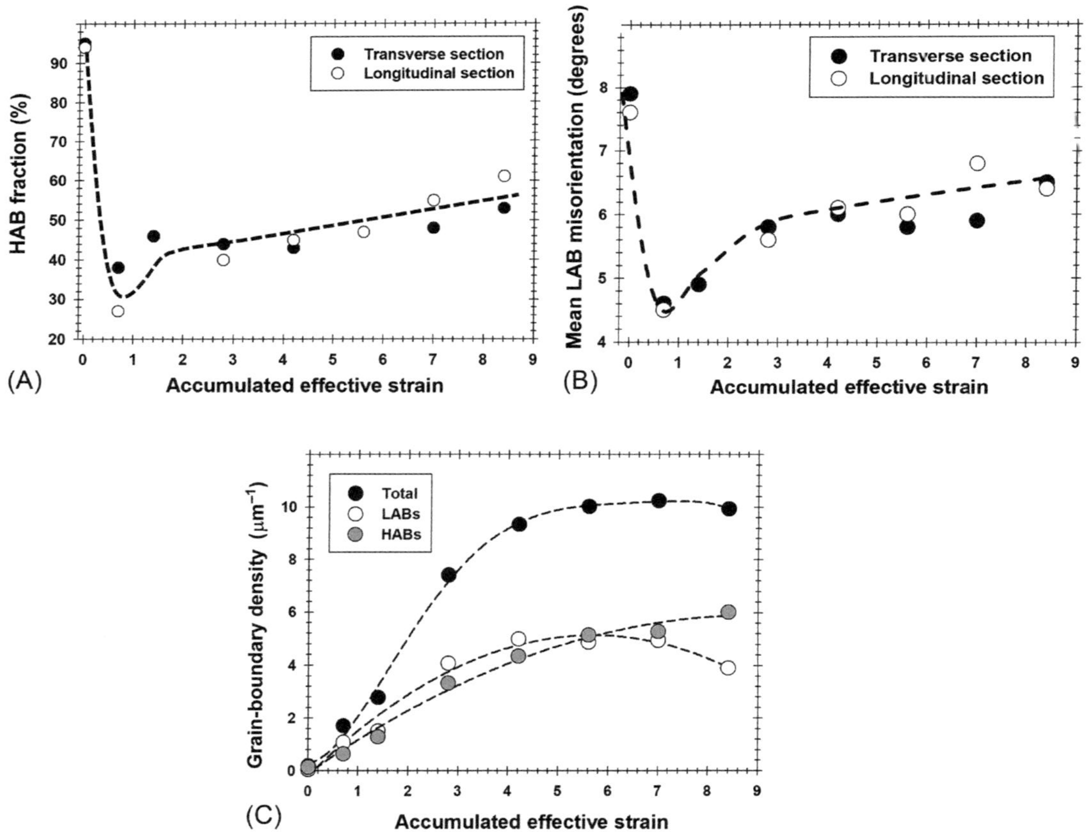

FIG. 7.6

Effect of accumulated ECAP strain on (A) fraction of HABs, (B) mean misorientation of LABs, and (C) grain boundary density.

After G.S. Dyakonov, S. Mironov, I.P. Semenova, R.Z. Valiev, S.L. Semiatin, Microstructurer evolution and strengthening mechanisms in commercial-purity titanium subjected to equal-channel angular pressing, Mater. Sci. Eng. A 701 (2017) 289–301, https://doi.org/10.1016/j.msea.2017.05.079.

FIG. 7.7

Typical transmission-electron-microscopy images of microstructure after (A) relatively low accumulated ECAP strain and after relatively high accumulated ECAP strain: (B) fine-grained region and (c) Coarse-grained region.

Panel (A): After D.H. Shin, I. Kim, J. Kim, Y.T. Zhu, Shear strain accommodation during severe plastic deformation of titanium using equal channel angular pressing, Mater. Sci. Eng. A *334 (2002), 239–245, https://doi.org/10.1016/S0921-5093(01)01813-5. Panels (B) and (C): After G.S. Dyakonov, S. Mironov, I.P. Semenova, R.Z. Valiev, S.L. Semiatin, Microstructurer evolution and strengthening mechanisms in commercial-purity titanium subjected to equal-channel angular pressing, Mater. Sci. Eng. A 701 (2017) 289–301, https://doi.org/10.1016/j.msea.2017.06.079.*

the revealed saturation of the total grain boundary area is totally unclear. Nevertheless, both these effects are in line with the observed saturation of material strength (Figs. 7.1 and 7.2). This relationship is discussed in the following section.

7.4 STRENGTHENING MECHANISMS

On the basis of detailed microstructural characterization, the various contributions to the strength of commercial-purity titanium may be estimated and interpreted within the context of yield strength measurements. Considering the limited character of the microstructural data mentioned in the previous section, the analysis of the strengthening mechanisms is also restricted to warm ECAP only.

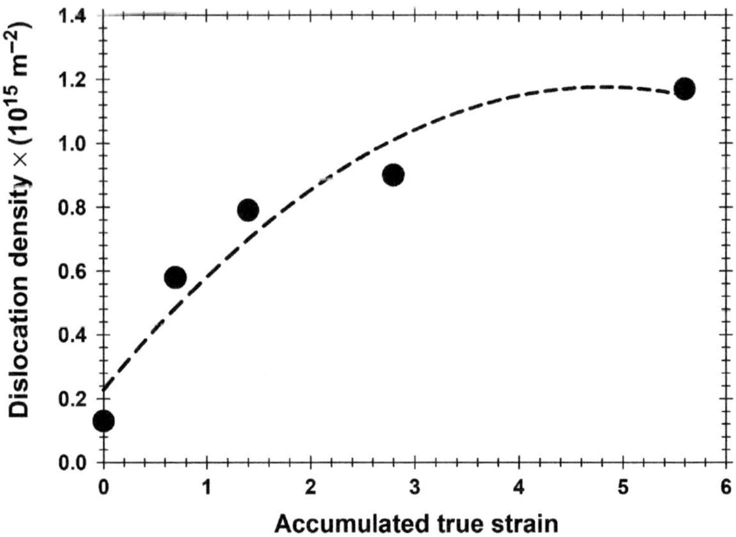

FIG. 7.8

Effect of accumulated ECAP strain on dislocation density.

After D.V. Gunderov, A.V. Polyakov, I.P. Semenova, G.I. Raab, A.A. Churakova, E.I. Gimaltdinova, I. Sabirov, J. Segurado, V.D. Sitdikov, I.V. Alexandrov, N.A. Enikeev, R.Z. Valiev, Evolution of microstructure and mechanical properties of commercial pure Ti during ECAP-conform processing and drawing, Mater. Sci. Eng. A *562 (2013), 128–136, https://doi.org/10.1016/j.msea.2012.11.007.*

Assuming that the strengthening mechanisms act independently and thus are additive in nature, the yield strength can be expressed as the following:

$$\sigma = \sigma_0 + \sigma_s + \sigma_{gb} \tag{7.1}$$

in which, σ_0, σ_s, and σ_{gb} denote the threshold stress, substructure strengthening, and grain boundary strengthening, respectively. This relationship is often used to calculate the theoretical strength of materials, but for NC or UFG structures in a number of cases it is necessary to take into account additional factors (e.g., nanoparticles, segregations, nanotwins, etc.) that can have an effect on the level of strength.

7.4.1 SOLID SOLUTION STRENGTHENING

The threshold stress can be expressed as the sum of the friction (Peierls–Nabarro) stress (σ_{P-N}) and the solid solution strengthening ($\Delta\sigma_{ss}$):

$$\sigma_0 = \sigma_{P-N} + \Delta\sigma_{ss} \tag{7.2}$$

The friction stress is reported to be negligible low, i.e., $\sigma_{P-N} \approx 0.52$ MPa [20] and thus the threshold stress is dominated by the solid solution strengthening. In turn, the solid solution strengthening in commercial-purity titanium is well accepted to be governed by interstitial solutes such as oxygen, nitrogen, and carbon, whose content is usually converted into an oxygen equivalent [46]. Thus:

$$\sigma_0 \approx \Delta\sigma_{ss} = 0.02Gc^{1/2} \tag{7.3}$$

where G is the shear modulus of titanium and c is the oxygen equivalent in atomic percent, which is evaluated as [46]:

$$c = \mathrm{O} + 20\mathrm{N} + 0.75\mathrm{C} \tag{7.4}$$

In titanium Grade 4, the friction stress is reported to be 460 MPa [46].

7.4.2 SUBSTRUCTURE STRENGTHENING

The substructure strengthening is governed by free dislocations as well as LABs:

$$\sigma_s = \sigma_\rho + \sigma_{LAB} \tag{7.5}$$

where σ_ρ and σ_{LAB} represent strengthening effect of the dislocations and LABs, respectively.

The dislocation strengthening is usually described by the classic Taylor relationship:

$$\sigma_\rho = M\alpha Gb\rho^{0.5} \tag{7.6}$$

Here, M denotes the Taylor factor, α is a constant, b is the Burgers vector, and ρ denotes the density of free dislocations.

Due to the operation of different slip modes with essentially different threshold stresses, an estimation of the Taylor factor in titanium is not straightforward. Assuming a prevalence of prism slip, M is sometimes taken to be 2.5 [10,36,41]. Considering a possibility of additional activation of basal and pyramidal slips, the employed Taylor factor may be 3.06 [9,42].

The constant α characterizes dislocation interactions and is believed to vary from 0.1 to 0.5 [46]. For severely deformed titanium, its magnitude is usually taken to be 0.15 [6,10,36] or 0.2 [9,20,42].

Suggesting a dominant role of $\langle a \rangle$–dislocations in plastic strain, the Burgers vector is usually taken to be 0.295 nm [6,10,36]. However, assuming an activation of $\langle c + a \rangle$ slip, magnitudes of 0.34 nm [9] or 0.347 nm [20,42] are also used.

It is also worth noting that dislocation density in severely deformed titanium may be as high as 10^{15} m^{-2} (Fig. 7.8) and thus it is difficult to measure directly. Therefore, the dislocation density is sometimes derived from X-ray data [8].

The strengthening effect of the LABs may be expressed as [47]:

$$\sigma_{LAB} = M\alpha G\sqrt{\frac{3b\theta_{LAB}(1-f)}{d}} \tag{7.7}$$

where θ_{LAB} is mean misorientation of the LABs (in radian), f is the fraction of HABs, and d is mean subgrain size.

It should be noted that grain boundary characteristics of severely deformed titanium are typically studied by using the electrobackscatter diffraction (EBSD) approach. Due to limited angular resolution of this technique, however, it cannot reliably detect misorientations below 2 degrees, and thus the accurate measurements of θ_{LAB}, f, and d are difficult.

For Grade 4 Ti, the variation of substructural strengthening with increasing ECAP-C strain was analyzed (Fig. 7.9A) [10]. The effect of ECAP strain on the predicted dislocation strengthening (Fig. 7.9A) increased significantly during the first several ECAP passes but then tended to saturate at an effective strain of ~2.8. The peak dislocation strengthening was predicted to be ~140 MPa.

7.4.3 GRAIN BOUNDARY STRENGTHENING

The strength increment associated with grain boundaries is usually described in terms of the well-known Hall–Petch relationship. However, considering a development of bimodal strain structure during severe plastic deformation of titanium (Fig. 7.4B–E), it seems reasonable to formulate this strengthening effect as a sum of the contributions from the ultrafine-grains (σ_{UFG}) and the CG remnants (σ_{CG}) [36]:

$$\sigma_{gb} = \sigma_{UFG} + \sigma_{CG} = k_{HP}\left[f_{UFG}d_{UFG}^{-1/2} + (1 - f_{UFG})d_{CG}^{-1/2}\right] \tag{7.8}$$

Here, k_{HP} is the Hall–Petch constant, f_{UFG} is volume fraction of the UFG structure, whereas d_{UFG} and d_{CG} are mean grain sizes of the UFG and CG constituents, respectively.

The consequent crystallographic texture as well as suppression of mechanical twinning are believed to reduce the efficiency of grain boundary strengthening [45,48,49]. Thus, the Hall–Petch coefficient in severely deformed titanium is taken to be as low as 0.36 MPa × $m^{1/2}$ [20], 0.35 MPa × $m^{1/2}$ [42], 0.30 MPa × $m^{1/2}$ [6,10,36], or even 0.16 MPa × $m^{1/2}$ [9].

The effect of ECAP-C strain on the predicted strengthening effect of LABs and HABs is summarized in Fig. 7.9A [10]. LAB strengthening was calculated to increase rapidly with strain but saturated after an effective strain of ~4.2. This behavior was in line with the evolution of LAB density (Fig. 7.5B). The total grain boundary density at high ECAP strains (ε >4.2) reached a saturation level (Fig. 7.6C). Accordingly, the LAB strengthening in this range was governed by a balance between a reduction of the LAB area due to LAB-to-HAB transformation (Fig. 7.6C) and a gradual increase of the mean LAB misorientation (Fig. 7.6B). The peak LAB strengthening was estimated to be ~160 MPa (Fig. 7.9A).

Based on the above calculations, it can be found that substructure strengthening significantly contributes to the strength of the severely deformed titanium [9,10,20,36,42], as shown in Fig. 7.9A. Another important issue is a clear saturation of the strengthening mechanisms with strain (Fig. 7.9A), which is directly linked with the saturation of dislocation density (Fig. 7.8) and total grain boundary area (Fig. 7.6C).

Despite the uncertainty of many microstructural parameters mentioned earlier, the predicted strength usually demonstrates a good agreement with measured yield

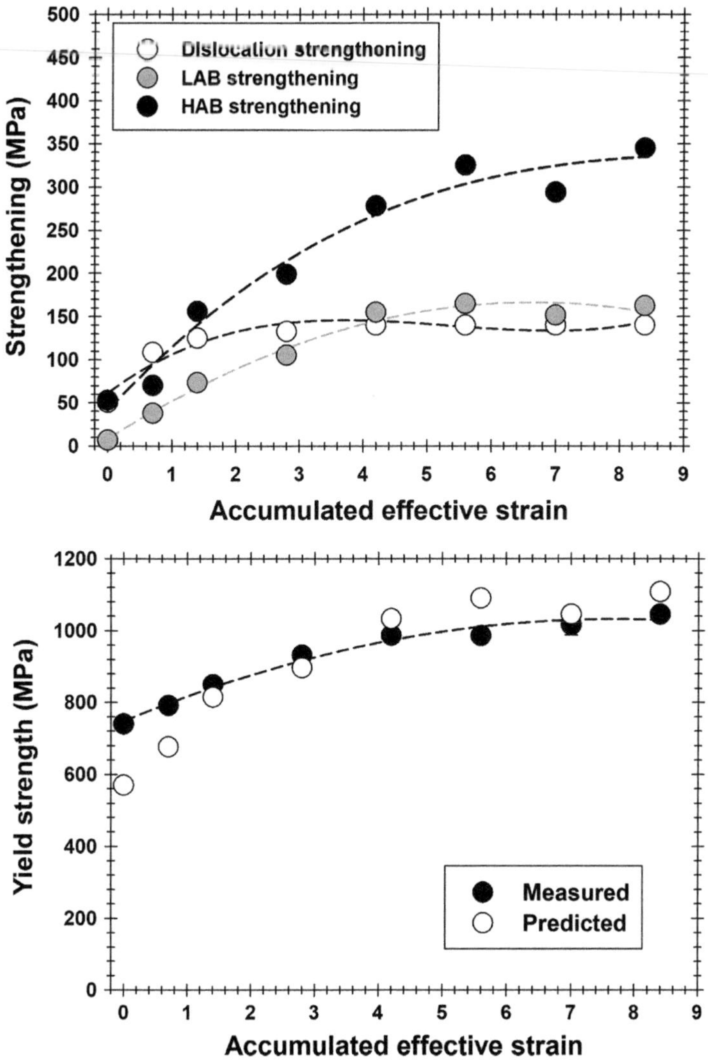

FIG. 7.9

(A) Effect of ECAP strain on the contribution of different strengthening mechanisms and (B) comparison of measured and predicted yield strengths.

After G.S. Dyakonov, S. Mironov, I.P. Semenova, R.Z. Valiev, S.L. Semiatin, Microstructurer evolution and strengthening mechanisms in commercial-purity titanium subjected to equal-channel angular pressing, Mater. Sci. Eng. A 701 (2017) 289–301, https://doi.org/10.1016/j.msea.2017.06.079.

stress [9,10,20,36,42], as exemplified in Fig. 7.9B. This perhaps shows the reliability of the calculation approach described earlier. Nevertheless, there is a possibility that additional mechanisms may also contribute to the material strength. Among them, the so-called Basinski effect [50,51] is thought to play a notable role at relatively low accumulated strains. This phenomenon is associated with transformation of glissile dislocations into sessile ones due to the reorientation of crystal structure during twinning. To accommodate a subsequent strain, new dislocations should be generated within the twins, thus increasing the local dislocation density and promoting an additional strengthening effect.

7.5 EFFECT OF CHANGE OF STRAIN PATH AND APPROACH TO SUPER-STRENGTH

In practice, SPD techniques are often combined with each other or with conventional processing methods (e.g., rolling, drawing, etc.) and typically this enables an improvement of strengthening efficiency [8,12,19,21,23,24,27,28,52]. Usually, the final processing step is conducted at a lower temperature and therefore the additional strengthening effect may be readily attributed to the increase of dislocation density and/or more pronounced grain refinement.

In some cases, however, an additional strengthening is also observed after the changing of strain path at the same temperature [52], as exemplified in Fig. 7.10. This phenomenon appears to open up a possibility of breaking through the strength saturation and achieving super-strength. At present, however, its mechanism is totally unknown. Considering the significant potential of this effect, it obviously warrants thorough study.

Another approach for achieving a high strength is based on the creation of structures with a high density of twin boundaries. It is possible to stimulate the

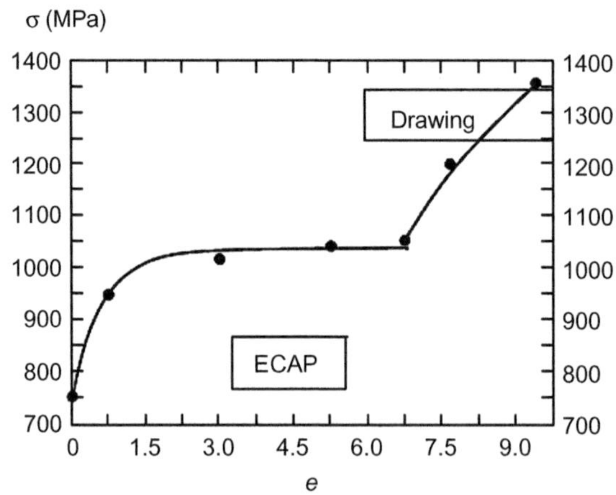

FIG. 7.10

Effect of change of strain path on material strengthening.

After I.P. Semenova, G.I. Raab, R.Z. Valiev, Nanostructured titanium alloys: new developments and application prospects, Nanotechnol. Russ. *9 (2014), 311–324, https://doi.org/10.1134/S199507801403015X.*

development of twinning in titanium due to the realization of plastic deformation at low temperatures, as was demonstrated in the cryogenic rolling of Grade 2 Ti [46]. The cryogenic deformation of titanium to a strain of $\varepsilon = 2.6$ resulted in strength enhancement up to 1180 MPa, in contrast to the initial strength of 428 MPa.

Thus, a potentially wide variety of regimes for the severe plastic deformation of titanium, as well as the possibility combine techniques, makes it possible to create specific structures that can possess super-strength.

7.6 SUMMARY

In this chapter, the current understanding of microstructure–strength relationship in severely deformed titanium was reviewed. The main conclusions are as follows.

Severe plastic deformation is a very effective tool for enhancement of strength characteristics of commercial-purity titanium to a level comparable with Ti–6Al–4V alloy.

The strengthening effect is fundamentally linked with underlying microstructural changes. Due to a specific character of microstructural processes, severe plastic deformation of titanium gives rise to heterogeneous UFG structure, which contains significant fraction of LABs and high dislocation density. As a result, the induced strengthening effect is essentially contributed by the substructure and thus cannot been simply explained in terms of the Hall–Petch relationship.

An important characteristic of severely deformed material is a saturation of dislocation density as well as total area of deformation-induced boundaries with accumulated strain. This leads to saturation of the strengthening effect.

From limited experimental data, it seems that a change of strain path may recommence the microstructural evolution and thus impart an additional strengthening to the material. However, mechanism of this phenomenon is not clear and warrants further study.

ACKNOWLEDGMENTS

This work was supported by RFBR No. 16-02-00094, the Ministry of Education and Science of the Russian Federation in the field of scientific research No. 16.7268.2017 , and Saint Petersburg State University in the framework of Call 3 project (id 26130576).

REFERENCES

[1] R.Z. Valiev, A.P. Zhilyaev, T.G. Langdon, Bulk Nanostructured Materials: Fundamentals and Applications, John Wiley & Sons., 2013.

[2] A.V. Sergueeva, V.V. Stolyarov, R.Z. Valiev, A.K. Mukherjee, Advanced mechanical properties of pure titanium with ultrafine grained structure, Scr. Mater. 45 (2001) 747–752.

[3] M. Lewandowska, K.J. Kurzydlowski, Recent development in grain refinement by hydrostatic extrusion. J. Mater. Sci. 43 (2008) 7299–7306, https://doi.org/10.1007/s10853-008-2810-z.
[4] G.A. Salishchev, R.M. Galeyev, S.P. Malysheva, M.M. Myshlyaev, Structure and density of submicrocrystalline titanium produced by severe plastic deformation. Nanostruct. Mater. 11 (1999) 407–414, https://doi.org/10.1016/S0965-9773(99)00195-6.
[5] V.V. Stolyarov, Y.T. Zhu, I.V. Alexandrov, T.C. Lowe, R.Z. Valiev, Influence of ECAP routes on the microstructure and properties of pure Ti. Mater. Sci. Eng. A 299 (2001) 59–67, https://doi.org/10.1016/S0921-5093(00)01411-8.
[6] G.S. Dyakonov, S. Mironov, I.P. Semenova, R.Z. Valiev, S.L. Semiatin, The microstructure-strength relationship in severely-deformed commercial-purity titanium, Mater. Lett. 192 (2017) 165–168. https://doi.org/10.1016/j.matlet.2016.12.032.
[7] M. Shirooyeh, J. Xu, T.G. Langdon, Microhardness evolution and mechanical characteristics of commercial purity titanium processed by high-pressure torsion, Mater. Sci. Eng. A 614 (2014) 223–231. https://doi.org/10.1016/j.msea.2014.07.030.
[8] D.V. Gunderov, A.V. Polyakov, I.P. Semenova, G.I. Raab, A.A. Churakova, E.I. Gimaltdinova, I. Sabirov, J. Segurado, V.D. Sitdikov, I.V. Alexandrov, N.A. Enikeev, R.Z. Valiev, Evolution of microstructure and mechanical properties of commercial pure Ti during ECAP-conform processing and drawing, Mater. Sci. Eng. A 562 (2013) 128–136. https://doi.org/10.1016/j.msea.2012.11.007.
[9] P. Rodriguez-Cavillo, J.M. Cabrera, Microstructure and mechanical properties of a commercial pure Ti processed by warm equal channel angular pressing, Mater. Sci. Eng. A 625 (2015) 311–320. https://doi.org/10.1016/j.msea.2014.11.082.
[10] G.S. Dyakonov, S. Mironov, I.P. Semenova, R.Z. Valiev, S.L. Semiatin, Microstructurer evolution and strengthening mechanisms in commercial-purity titanium subjected to equal-channel angular pressing, Mater. Sci. Eng. A 701 (2017) 289–301. https://doi.org/10.1016/j.msea.2017.06.079.
[11] J.L. Milner, F. Abu-Farha, C. Bunget, T. Kurfess, V.H. Hammond, Grain refinement and mechanical properties of CP-Ti processed by warm accumulated roll bonding, Mater. Sci. Eng. A 561 (2013) 109–117. https://doi.org/10.1016/j.msea.2012.10.081.
[12] Z. Bezi, G. Krallics, M. El-tahawy, P. Pekker, J. Gubicza, Processing of ultrafine-grained titanium with high strength and good ductility by a combination of multiple forging and rolling, Mater. Sci. Eng. A 688 (2017) 210–217. https://doi.org/10.1016/j.msea.2017.01.112.
[13] S.S. Satheesh Kumar, K. Priyasudha, M. Sudhakara Rao, T. Raghu, Deformation homogeneity, mechanical behavior and strain hardening characteristics of titanium severely plastically deformed by cyclic channel die compression method, Mater. Des. 101 (2016) 117–129. https://doi.org/10.1016/j.matdes.2016.04.004.
[14] W. Pachla, M. Kulczyk, M. Sus-Ryszkowska, A. Mazur, K.J. Kurzydlowski, Nanocrystalline titanium produced by hydrostatic extrusion, J. Mater. Process. Technol. 205 (2008) 173–182. https://doi.org/10.1016/j.matprotec.2007.11.103.
[15] K. Topolski, W. Pachla, H. Garbacz, Progress in hydrostatic extrusion of titanium, J. Mater. Sci. 48 (2013) 4543–4548. https://doi.org/10.1007/s10853-012-7086-7.
[16] W. Pachla, M. Kulczyk, S. Przybysz, J. Skiba, K. Wojciechowski, M. Przybysz, K. Topolski, A. Sobolevwski, M. Charkiewicz, Effect of severe plastic deformation realized by hydrostatic extrusion and rotary swaging on the properties of CP Ti grade 2, J. Mater. Process. Technol. 221 (2015) 255–268. https://doi.org/10.1016/j.jmatprotec.2015.02.027.

[17] S. Malysheva, G. Salishchev, S. Mironov, S. Zherebtsov, Production of nanostructure in titanium by cold rolling. Mater. Sci. Forum 584–586 (2008) 759–764, https://doi.org/10.4028/www.scientific.net/MSF.584-586.759.

[18] G.S. Dyakonov, S. Mironov, S.V. Zherebtsov, S.P. Malysheva, G.A. Salishchev, A.A. Salem, S.L. Semiatin, Grain-structure development in heavily cold-rolled alpha-titanium, Mater. Sci. Eng. A 607 (2014) 145–154. https://doi.org/10.1016/j.msea.2014.03.141.

[19] V.V. Stolyarov, Y.T. Zhu, T.C. Lowe, R.K. Islamgaliev, R.Z. Valiev, A two step SPD processing of ultrafine-grained titanium, Nanostruct. Mater. 11 (1999) 947–954. https://doi.org/10.1016/S0965-9773(99)00384-0.

[20] P. Luo, D.T. McDonald, W. Xu, S. Palanisamy, M.S. Dargusch, K. Xia, A modified Hall-Petch relationship in ultrafine-grained titanium recycled from chips by equal channel angular pressing, Scr. Mater. 66 (2012) 785–788. https://doi.org/10.1016/j.scriptamat.2012.02.008.

[21] G.G. Yapici, I. Karaman, H.J. Maier, Mechanical flow anisotropy in severely deformed pure titanium, Mater. Sci. Eng. A 434 (2006) 294–302. https://doi.org/10.1016/j.msea.2006.06.082.

[22] M.J. Qarni, G. Sivaswamy, A. Rosochowski, S. Boczkal, Effect of incremental equal channel angular pressing (I-ECAP) on the microstructural characteristics and mechanical behavior of commercial purity titanium, Mater. Des. 122 (2017) 385–402. https://doi.org/10.1016/j.matdes.2017.03.015.

[23] Z. Fan, H. Jiang, X. Sun, J. Song, X. Zhang, C. Xie, Microstructures and mechanical deformation behaviors of ultrafine-grained commercial pure (Grade 3) Ti processed by two-step severe plastic deformation, Mater. Sci. Eng. A 527 (2009) 45–51. https://doi.org/10.1016/j.msea.2009.07.030.

[24] V.L. Sordi, M. Ferrante, M. Kawasaki, T.G. Langdon, Microstructure and tensile strength of grade 2 titanium processed by equal-channel angular pressing and by rolling, J. Mater. Sci. 47 (2012) 7870–7876. https://doi.org/10.1007/s10853-012-6593-x.

[25] A. Jager, V. Gartnerova, K. Tesar, Microstructure and anisotropy of the mechanical properties in commercially pure titanium after equal channel angular pressing with back pressure at room temperature, Mater. Sci. Eng. A 644 (2015) 114–120. https://doi.org/10.1016/j.msea.2015.07.038.

[26] V.V. Stolyarov, Y.T. Zhu, I.V. Alexandrov, T.C. Lowe, R.Z. Valiev, Grain refinement and properties of pure Ti processed by warm ECAP and cold rolling, Mater. Sci. Eng. A 343 (2003) 45–50. https://doi.org/10.1016/S0921-5093(02)00366-0.

[27] V.V. Stolyarov, L. Zeipper, B. Mingler, M. Zehetbauer, Influence of post-deformation on CP-Ti processed by equal channel angular pressing, Mater. Sci. Eng. A 476 (2008) 98–105. https://doi.org/10.1016.jmsea.2007.04.069.

[28] I. Sabirov, M.T. Perez-Prado, J.M. Molina-Aldareguia, I.P. Semenova, G.K. Salimgareeva, R.Z. Valiev, Anisotropy of mechanical properties in high-strength ultra-fine-grained pure Ti processed via a complex severe plastic deformation route, Scr. Mater. 64 (2011) 69–72. https://doi.org/10.1016/j.scriptamat.2010.09.006.

[29] Y. Todaka, J. Sasaki, T. Moto, M. Umemoto, Bulk submicrocrystalline ω-Ti produced by high-pressure torsion straining, Scr. Mater. 59 (2008) 615–618. https://doi.org/10.1016/j.scriptamat.2008.05.015.

[30] M. Tane, Y. Okuda, Y. Todaka, H. Ogi, A. Nagakubo, Elastic properties of single-crystalline ω phase in titanium, Acta Mater. 61 (2013) 7543–7554. https://doi.org/10.1016/j.actamat.2013.08.036.

[31] K. Edalati, T. Daio, M. Arita, S. Lee, Z. Horita, A. Togo, I. Tanaka, High-pressure torsion of titanium at cryogenic and room temperatures: grain size effect on allotropic phase transformations, Acta Mater. 68 (2014) 207–213. https://doi.org/10.1016/j.actamat.2014.01.037.
[32] V.P. Pilyugin, Y.V. Khlebnikova, L.Y. Egorova, T.R. Suaridze, N.N. Resnina, A.M Patselov, Structural transformations in single-crystal titanium under high-pressure cold and cryogenic deformation, Phys. Metal. Metallogr. 116 (2015) 1203–1212. https://doi.org/10.1134/S0031918X15120066.
[33] H. Shahmir, T.G. Langdon, Characteristics of the allotropic phase transformation in titanium processed by high-pressure torsion using different rotation speeds, Mater. Sci. Eng. A 667 (2016) 293–299. https://doi.org/10.1016/j.msea.2016.05.001.
[34] N. Adachi, Y. Todaka, K. Irie, M. Umemoto, Phase transformation kinetics of ω – phase in pure Ti formed by high-pressure torsion, J. Mater. Sci. 51 (2016) 2608–2615. https://doi.org/10.1007/s10853-015-9574-z.
[35] R.B. Figueiredo, T.G. Langdon, Development of an ω-phase in Grade 2 titanium processed by HPT at high hydrostatic pressure, Mater. Res. 19 (2016) 1144–1148. https://doi.org/10.1590/1980-5373-MR-2016-0403.
[36] G.S. Dyakonov, E. Zemtsova, S. Mironov, I.P. Semenova, R.Z. Valiev, S.L. Semiatin, An EBSD investigation of ultrafine-grain titanium for biomedical applications, Mater. Sci. Eng. A 648 (2015) 305–310. https://doi.org/10.1016/j.msea.2015.09.080.
[37] D.H. Shin, I. Kim, J. Kim, Y.S. Kim, S.L. Semiatin, Microstructure development during equal-channel angular pressing of titanium. Acta Mater. 51 (2003) 983–996, https://doi.org/10.1016/S1359-6454(02)00501-3.
[38] D.H. Shin, I. Kim, J. Kim, Y.T. Zhu, Shear strain accommodation during severe plastic deformation of titanium using equal channel angular pressing. Mater. Sci. Eng. A 334 (2002) 239–245, https://doi.org/10.1016/S0921-5093(01)01813-5.
[39] Y.J. Chen, Y.J. Li, J.C. Walmsley, S. Dumoulin, S.S. Gireesh, S. Armada, P.C. Skaret, H.J. Roven, Quantitative analysis of grain refinement in titanium during equal channel angular pressing. Scr. Mater. 64 (2011) 904–907, https://doi.org/10.1016/j.scriptamat.2011.01.030.
[40] I. Kim, W.-S. Jeong, J. Kim, K.-T. Park, D.H. Shin, Deformation structures of pure Ti produced by equal channel angular pressing. Scr. Mater. 45 (2001) 575–581, https://doi.org/10.1016/S1359-6462(01)01065-X.
[41] S. Suwas, B. Beausir, L.S. Toth, J.-J. Fundenberger, G. Gottstein, Texture evolution in commercially pure titanium after warm equal channel angular extrusion. Acta Mater. 59 (2011) 1121–1133, https://doi.org/10.1016/j.actamat.2010.10.045.
[42] P. Luo, D.T. McDonald, S.M. Zhu, S. Palanisamy, M.S. Dargusch, K. Xia, Analysis of microstructure and strengthening in pure titanium recycled from machining chips by equal channel angular pressing using electron backscatter diffraction, Mater. Sci. Eng. A 538 (2012) 252–258. https://doi.org/10.1016/j.msea.2012.01.039.
[43] E.V. Nesterova, V.V. Rybin, Mechanical twinning and fragmentation of commercial purity titanium during large plastic deformation, Phys. Met. Metall. 59 (1985) 169–179.
[44] G. Salishchev, S. Mironov, S. Zherebtsov, A. Belyakov, Changes in misorientations of grain boundaries in titanium during deformation. Mater. Charact. 61 (2010) 732–739, https://doi.org/10.1016/j.matchar.2010.04.005.
[45] S.V. Zherebtsov, G.S. Dyakonov, A.A. Salem, V.I. Sokolenko, G.A. Salishchev, S.L. Semiatin, Formation of nanostructures in commercial-purity titanium via cryorolling, Acta Mater. 61 (2013) 1167–1178. https://doi.org/10.1016/j.actamat.2012.10.026.

[46] H. Conrad, Effect of interstitial solutes on the strength and ductility of titanium. Prog. Mater. Sci. 26 (1981) 123–403, https://doi.org/10.1016/0079-6425(81)90001-3.

[47] N. Hansen, Hall-Petch relation and boundary strengthening. Scr. Mater. 51 (2004) 801–806, https://doi.org/10.1016/j.scriptamat.2004.06.002.

[48] Y. Guo, T.B. Britton, A.J. Wilkinson, Slip band-grain boundary interactions in commercial-purity titanium, Acta Mater. 76 (2014) 1–12. https://doi.org/10.1016/j.actamat.2014.05.015.

[49] N. Stanford, U. Carlson, M.R. Barnett, Deformation twinning and the Hall–Petch relation in commercial purity Ti. Metall. Mater. Trans. A 39A (2008) 934–944, https://doi.org/10.1007/s11661-007-9442-9.

[50] Z.S. Basinski, M.S. Szczerba, M. Niewczas, J.D. Embury, S.J. Basinski, The transformation of slip dislocations during twinning of copper-aluminum alloy crystal. Rev. Met. Paris 94 (1997) 1037–1044, https://doi.org/10.1051/metal/199794091037.

[51] A.A. Salem, S.R. Kalidindi, R.D. Doherty, S.L. Semiatin, Strain hardening due to deformation twinning in α-titanium: mechanisms. Metall. Mater. Trans. 37A (2006) 259–268, https://doi.org/10.1007/s11661-006-0171-2.

[52] I.P. Semenova, G.I. Raab, R.Z. Valiev, Nanostructured titanium alloys: new developments and application prospects, Nanotechnol. Russ. 9 (2014) 311–324. https://doi.org/10.1134/S199507801403015X.

CHAPTER

Corrosion resistance of nanocrystalline titanium

8

Halina Garbacz*, Andrzej Królikowski†

Faculty of Materials Science and Engineering, Warsaw University of Technology, Warsaw, Poland Faculty of Chemistry, Warsaw University of Technology, Warsaw, Poland†*

8.1 INTRODUCTION

It is well known that, despite its low standard potential value (−1.6 V), titanium displays excellent corrosion resistance in most aggressive environments [1]. This is the result of its high reactivity with oxygen, which leads to the spontaneous formation of an adherent and stable oxide layer on the titanium surface in air or aqueous environments. The oxide layer can be amorphous or crystalline, depending on the rate of formation [2]. The properties of the oxide layer depend mainly on its stoichiometry and on the density of defects. The process of oxide layer growth and breakdown can be described using the point defect model (PDM), which is based on the movement of point defects under the steady state in an electric field. Oxide growth occurs by means of the transport of oxygen ions or oxide ion vacancies through the film. It is assumed that solid-state diffusion is responsible for the mass migration through the passive layer. The PDM makes it possible to perform a quantitative analysis of the concentration of point defects and their transport properties, and consequently, to estimate their diffusivity in the oxide film. In the case of titanium, the mobility of oxygen ions is the primary mechanism of ion transport. The dominant defects in a titanium oxide layer are oxygen vacancies and titanium interstitials, which means that titanium passive layer can be classified as an n-type semiconductor. According to the model, the energy necessary for ion transport increases with the thickness of the passive layer, which can limit further oxide growth [3–5]. Passive layer stability is crucial for corrosion protection, and is determined by the corrosion potential, pH, temperature, and composition of the corrosion solution. It is possible to make a theoretical prediction of the reactions and reaction products in specific conditions by analysing the Pourbaix (pH-potential diagram) (Fig. 8.1). It is evident that an oxide layer in the form of TiO_2, Ti_2O_3, or TiO is stable across a wide spectrum of pH and potential values [7]. This confirms that, although it is an active metal, titanium shows high corrosion resistance in many corrosive environments. Nevertheless, the experimental data indicate that, in the presence of halide ions (Cl^-, F^-, Br^-, I^-) the

Nanocrystalline Titanium. https://doi.org/10.1016/B978-0-12-814599-9.00008-0

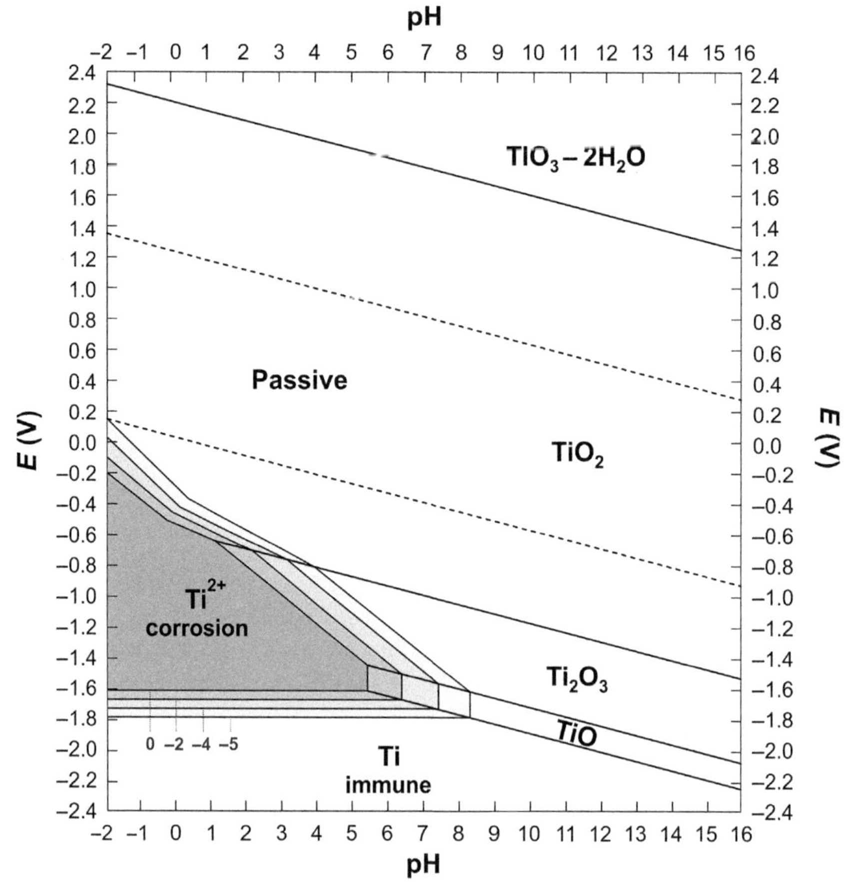

FIG. 8.1

Pourbaix diagram for titanium showing thermodynamic predictions of reactions and reaction products in water at equilibrium [6].

titanium oxide layer can be locally destroyed. According to the PDM, the process of the formation of local corrosion damages includes the following stages: the dissolution of aggressive ions from the corrosion environment into oxygen vacancies, the formation of cation vacancies at the interface between the oxide and the solution, the diffusion of cation vacancies and their annihilation by metal oxidation, and the condensation of non-annihilated vacancies to the form of voids leading to a local detachment of the passive layer and pitting corrosion [5].

According to the literature, titanium corrosion resistance can be further improved by a wide variety of surface treatments [8]. There are many studies regarding the possibility of enhancing titanium corrosion resistance by severe plastic deformation (SPD) [9–15]. Without doubt, grain refinement leads to an improvement in mechanical properties [16–20]. Assessing the effect of grain size on corrosion resistance is more complicated; however, due to the large number of factors that have to be taken into consideration. It is worth noting that contrary findings have been published, some claiming that nanostructure improves titanium corrosion resistance, others that it leads to a decrease. The difficulty in reaching general conclusions based on the literature relate to different methods of processing the nanomaterials. Technological processes affect not only grain size, but also other microstructural features such as crystallographic texture, internal stresses, and grain size distribution. All of these factors can have an impact on the overall behavior of nanomaterials in a corrosive environment [21–23]. Moreover, the large variety of corrosion characterization methods and measurements parameters (temperature, time from immersion, composition of the solution, and degree of aeration) make it difficult to compare data in the literature.

In this chapter, the following issues are discussed:

- Corrosion evaluation methods used for NC titanium
- The effects of nanostructure on the properties of a titanium oxide layer
- Factors affecting titanium corrosion resistance

8.2 EXPERIMENTAL METHODS FOR INVESTIGATION CORROSION RESISTANCE

This section provides a brief description of methods commonly used for evaluating the effects of grain refinement on titanium corrosion resistance. The first part deals with principles of electrochemical techniques such as potential monitoring, electrochemical impedance spectroscopy, and potentiodynamic polarization. The second part contains information about surface characterization after corrosion tests by microscopic and spectroscopic methods. These techniques made it possible to perform a qualitative analysis (using scanning electron microscopy (SEM)) of the surface topography and to evaluate the chemical composition and thickness of the oxide layer (using Auger electron spectroscopy (AES)). The introduction to the basics of the experimental methods used should help the reader fully understand the following sections relating to the factors affecting titanium corrosion resistance.

8.2.1 ELECTROCHEMICAL TECHNIQUES

According to the literature, electrochemical tests of titanium and its alloys are conducted in various solutions (from acidic to alkaline, containing or without chloride), at ambient or elevated temperatures, and for different exposure times (Table 8.1). These are critical to corrosion behavior. Unfortunately, in many cases a description of the experimental procedures is incomplete or simply lacking. Another factor that has a significant impact on the results of electrochemical tests is surface topography, which is determined by how the samples are prepared. Prior to the corrosion tests,

Table 8.1 Characterization of Corrosive Solutions Used in Cited Papers

Material	Solution	pH	Temperature (°C)	Aeration	Immersion Time Before Measurement (h)	References
Ti Grade 2	Ringer solution		37			[24]
	PBS	10.7			1	[25]
	0.9% NaCl				20	[12]
	SBF		37 ± 1		0.5	[26]
	0.9% NaCl		Room	Natural	24	[14]
Ti–Nb–Zr	0.9% NaCl		37	Natural	0.5	[27]
	SBF	7.4	37	Natural	0.5	[27]
	0.9% NaCl				0.5	[28]
	0.2% NaF				0.5	[28]
	Ringer sol.	6.9	37 ± 1	Aerated		[29]
Ti–Ni	1 M H_2SO_4		Room			[30]
	0.9% NaCl					[31]

Table 8.2 Preparation of Ti and Ti Alloys Samples Before Electrochemical Corrosion Tests

Grinding/Polishing	Cleaning	Electrochemical Pretreatment	References
Ground to 2000 grit	In deionised water		[24]
Mechanically and vibratorily polished	Sonicated in ethanol, acetone, and isopropanol for 20 min		[32]
Electro-polished in HCL/ethanol			[28]
Mechanically machined			[29]
Polished to remove 1–3 μm		Cathodic polarization at −1 V for 5 min	[26]
Polished	Cleaned and dried in cold air	Cathodic polarization at −0.8 V for 5 min	[33]
Polished			[34]
Polished up to 2400 and finished with a standard colloidal silica suspension			[35]
Lack of information			[12, 14, 27, 31]

samples are usually mechanically ground and polished. Intensive cathodic or anodic prepolarization, which can significantly change the conditions of growth on passive films and their structure, are also performed (Table 8.2). In a notable number of the papers we reviewed, information on sample preparation is very general or even lacking (Table 8.2). Differences in the conditions of the corrosion solutions and in the method of sample preparation create difficulties when trying to compare data from the literature.

8.2.1.1 Potential monitoring

When performing impedance or polarization measurements, the main requirement is to have a stable open circuit potential (OCP) value. For this reason, it is necessary to monitor the corrosion potential value prior to further electrochemical tests [2]. Moreover, changes in OCP yield important information about the process of passivation and the properties of the oxide layer. An increase in the OCP value with immersion time indicates constant inhibition of the anodic reaction (titanium passivation). This suggests that the passive layer has good protective properties under the given conditions. A decline in OCP value after immersion implies the existence of local breakdowns within the film, or the dissolution of the film. Rapid fluctuations in OCP value

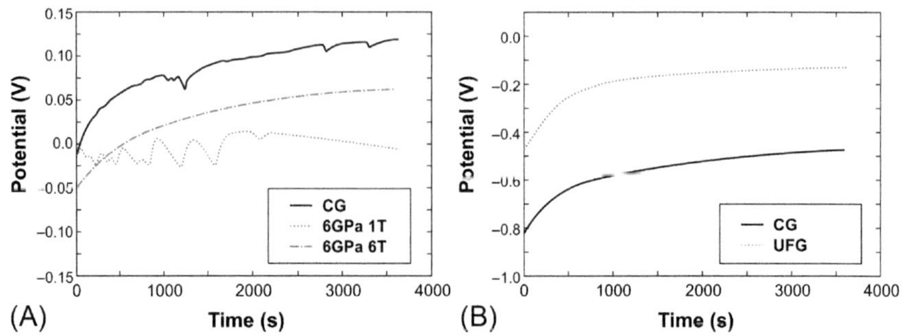

FIG. 8.2

Time evolutions OCP observed for both ultrafine-grained (UFG) and coarse-grained (CG) titanium in (A) 3.5% NaCl, (B) simulated body fluid (SBF) [9, 36].

are also worth mentioning (Fig. 8.2). These fluctuations suggest the formation of unstable local corrosion damages or a rebuilding of the structure of the passive layer.

8.2.1.2 Impedance measurements

Electrochemical impedance spectroscopy (EIS) involves a minimal perturbation in the measurement system. For that reason, it can be classified as a nondestructive electrochemical technique. The principle of EIS is based on stimulating a measurement system by an alternating potential signal of a known frequency and a small amplitude (usually 5–10 mV). The impedance is calculated by using the relationship between the value of the voltage signal and the current response. The EIS method makes it possible to calculate how the impedance value varies with the frequency. Measurements are performed for a number of frequencies (usually from 10^5 to 10^{-3} Hz). It is possible to separate processes that have different time constants during a single measurement. At the highest frequencies, quick processes such as charge transfer are observed, whereas slower, diffusive processes can be observed for the lower frequencies. The result of an EIS measurement is a sequence of amplitude and phase angle values for a number of frequencies [13]. There are two common ways of illustrating EIS data: Nyquist and Bode plots. A Nyquist plot includes the imaginary part vs the real part of the impedance (Fig. 8.3). The main disadvantage of a Nyquist plot is the lack of information about the frequency of each measurement point. A Bode plot shows two relationships, log (impedance modulus) and phase angle vs log frequency (Fig. 8.4). Analyzing the phase angle part of the plot is helpful in estimating the number of processes that have an impact on the overall corrosion response [2].

An interpretation of EIS data is based on a selection of an appropriate equivalent electric circuit. The individual elements of the circuit describe the electrical and electrochemical processes detected during the corrosion measurement. Fig. 8.5 presents an equivalent circuit with a single time constant, which is commonly used for analyzing EIS plots obtained for titanium immersed in physiological solutions.

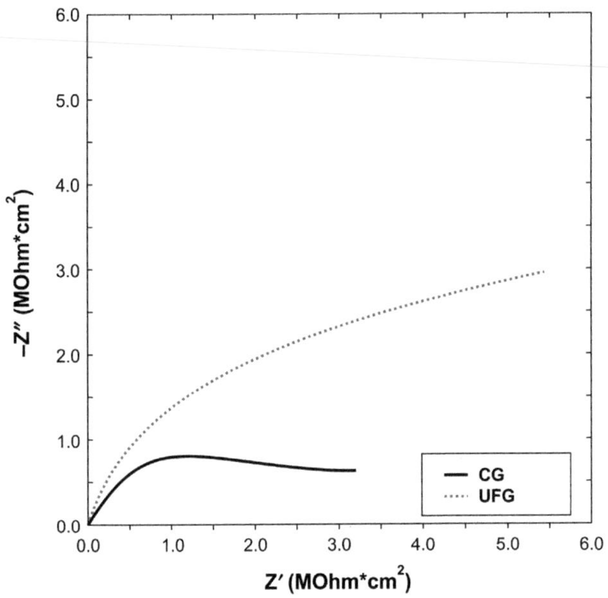

FIG. 8.3

Nyquist plot obtained for UFG and CG titanium in simulated body fluid (SBF) [9].

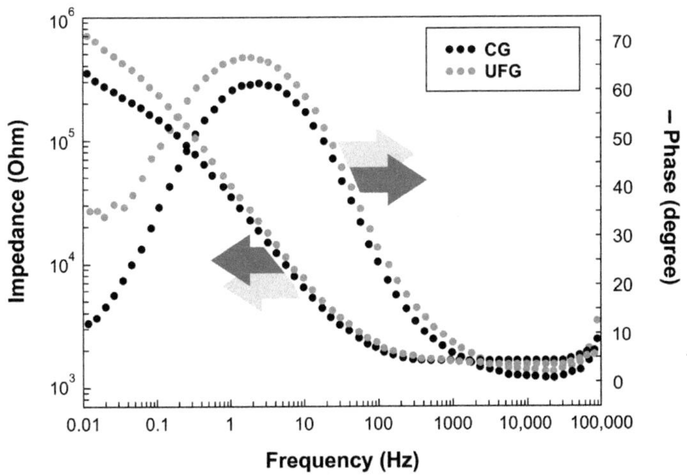

FIG. 8.4

Bode plot obtained for UFG and CG titanium in simulated body fluid (SBF) [9].

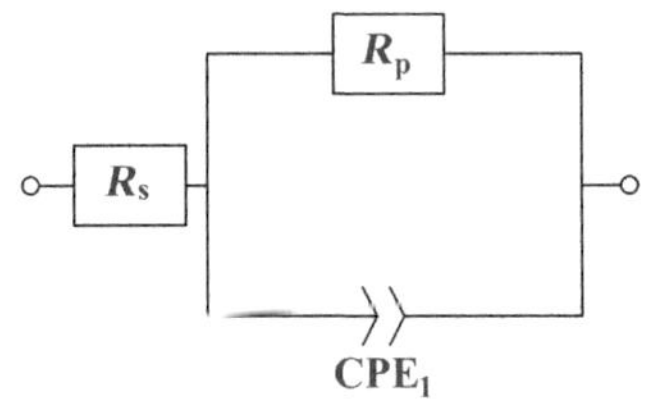

FIG. 8.5

One-time constant equivalent circuit commonly used for analyzing EIS measurements of titanium.

It contains three components: R_s—the solution resistance, R_p—polarization resistance, and CPE—the constant phase element for the interfacial capacitance. At the highest frequencies, the resistance of the solution is measured. It depends upon the distance between the sample and the reference electrode, or on the shape of the corrosion cell, and so is not an important parameter when comparing corrosion properties. Nevertheless, knowledge of the R_s value makes it possible to perform an Ohmic drop correction. At lower frequencies, the loading current of the material/electrolyte interface is measured. However, a simple capacitor is not appropriate in order to describe double layer ability to accumulate an electrical charge. Features of the metal/solution interface, such as structural defects or the nonuniform roughness of corrosion products, have to be taken into consideration. For this reason, the ability of an electrical charge to accumulate on the material/electrolyte interface must be described by the constant phase element rather than by a simple capacitor. The capacitance of the CPE element can be calculated from the following equation:

$$C_{CPE} = Y_0(j\omega)^{n-1} \tag{8.1}$$

where: Y_0—is CPE parameter (admittance), n—is a parameter which informs about surface properties, and ω—is frequency.

According to Eq. (8.1), when the n parameter is close to unity, capacitance can be approximated by the value of the CPE parameter (admittance). At the lowest frequencies, when the time of measurement is the longest, the polarization resistance (R_p) value is measured. The higher the value of polarization resistance, the better the properties of the passive layer, and consequently, the slower the corrosion process [13].

EIS data obtained for NC titanium can also be described by using an equivalent circuit with two time constants (Fig. 8.6). The reason for introducing an additional

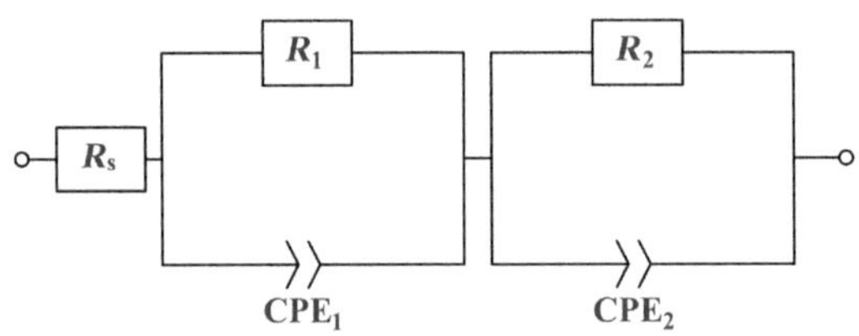

FIG. 8.6

Two-time constants equivalent circuit commonly used for analyzing EIS measurements of titanium.

time constant can be twofold. In some literature data, it is claimed that one time constant is attributed to the resistance and capacitance of the inner part of the oxide layer, whereas the second is related to the resistance and capacitance of the porous outer part or surface defects [36]. In other sources, the first time constant is attributed to the resistance and capacitance of the electrical double layer on the solution/oxide film interface, and the second describes the resistance and capacitance of the oxide film.

8.2.1.3 Polarization curves

Potentiodynamic polarization is a technique that provides an opportunity to estimate the potential range in which material is in the active or passive state [2]. Potentiodynamic measurements are performed, controlling the potential (whose value changes at a determined rate) and measuring the current response. Due to the large perturbation within the measurement system, potentiodynamic polarization can in general be classified as a destructive method. The polarization curves show the relationship between the potential applied and the current measured (Fig. 8.7). An analysis of the polarization curve makes it possible to estimate the corrosion potential value, the current density in the passive state, or the breakdown (pitting) potential. The higher the value of the breakdown potential, the less the susceptibility of the material to pitting corrosion.

An important element in interpreting potentiodynamic data is the estimation of the Tafel slopes (slopes of the linear region). This makes it possible to determine the density of the corrosion current, and consequently, to calculate the corrosion rate. The higher the corrosion current value, the faster the overall corrosion rate [2].

8.2.1.4 Comparison of electrochemical techniques

A summary of the comparison between the two most common electrochemical methods for evaluating corrosion is presented in Table 8.3. Table 8.4 gives some insight into the electrochemical methods most widely used in corrosion research.

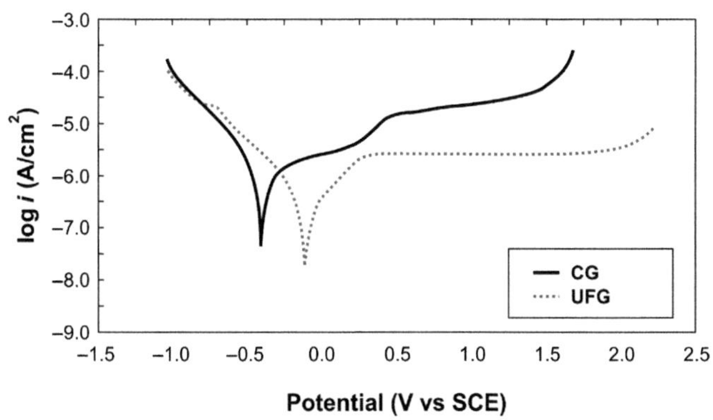

FIG. 8.7

Potential curve illustrating the result of potentiodynamic measurements obtained for UFG and CG titanium in simulated body fluid (SBF) [9].

Table 8.3 Comparison of the Most Common Electrochemical Measurements

Impedance Measurements	Potentiodynamic Measurements
Minimal perturbation in the measurement system: nondestructive method	Significant perturbation in the measurement system (materials polarization): destructive method
Opportunity to separate processes of different rates (electrode reactions, mass transport, adsorption)	Opportunity to observe only the slowest process (summing up processes and resultant corrosion response)
Opportunity to perform measurements in low-conductive systems (e.g., materials with organic coatings)	Performing measurements in low-conductive systems (such as nonconductive coatings) is difficult due to Ohmic potential drop
Greater accuracy of measurement (weak alternating currents)	Lower accuracy of measurements (weak direct currents)

For clarity, the parameters extracted relate to corrosion, and an interpretation of them is also provided.

8.2.2 SURFACE CHARACTERIZATION TECHNIQUES

An evaluation of corrosion resistance often requires a characterization of the morphology, structure, and composition of the corroded surface. Such an analysis is undoubtedly helpful in identifying structural features that could increase a material's susceptibility to corrosion. Electron- and photon-based techniques are commonly used for characterizing corroded surfaces. The selection of appropriate characterization methods depends mainly on the kind of information required (quantitative/qualitative), the technique resolution, the detection limit, and the possibility of destroying the analyzed element [2]. According to the literature, a typical method for evaluating surface morphology is scanning electron microscopy (SEM), and for analysing chemical composition, Auger Electron Spectroscopy (AES).

8.2.2.1 Scanning electron microscopy: Characterization of surface morphology

SEM makes it possible to carry out a quantitative analysis of different morphological aspects of a corroded surface, including estimating surface roughness, characterizing the corrosion pits (their dimension and shape), locating the corrosion and determining the morphology of corrosion products (such as oxides) (Fig. 8.8). The principle of SEM observation is based on scanning a specimen surface with a focused electron beam. The accelerating voltage applied in SEM is typically in a range of from a few to 30keV. SEM images are created by collecting secondary low-energy electrons (SEs) or backscattered electrons (BSEs). SEM observations made with a secondary electron detector provide information on topographic contrast, making this an appropriate method for estimating surface roughness and characterizing local corrosion damage. Backscattered electrons carry relatively high energies and emerge from a

Table 8.4 Corrosion Parameters Calculated From Selected Electrochemical Tests

Technique and Representative Plot	Parameter	Interpretation
Evolution of corrosion potential with exposure time 	E_{corr}—corrosion potential—evolution vs time	Also called open circuit potential (OCP); enoblement/increase of E_{corr} with exposure time suggests the formation of protective/passive film; decrease can be related to activation/breakdown of surface film and initiation of local corrosion
Evolution of anodic current with time at constant potential (potentiostatic run) 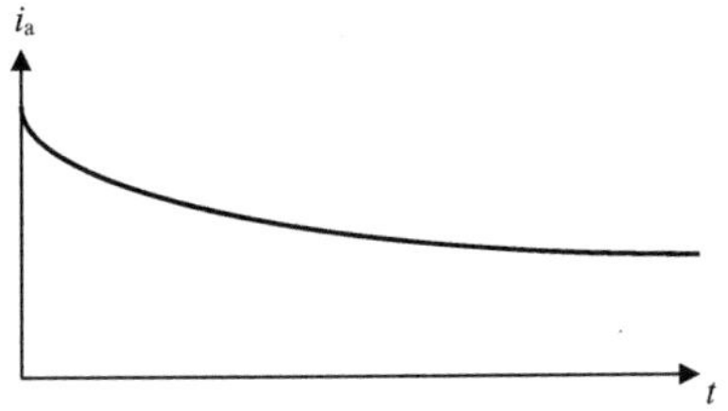	i_a—anodic (dissolution) current density—evolution vs time	Decrease of j_a with polarization time indicates the formation of protective/passive film; when measurement is taken in passive range, steady state values of passive current density i_{pass} can be determined
Potentiodynamic polarization curve 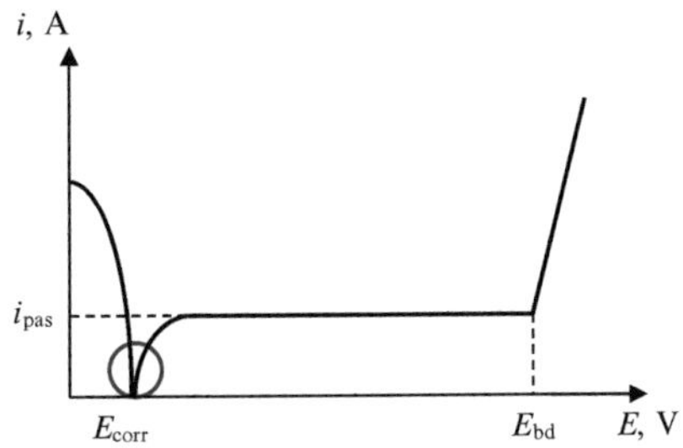	E_{corr}—corrosion potential	Potential corresponding to passage from cathodic to anodic current
	i_{corr}—corrosion current density	Electrochemical measure of corrosion rate, extracted from region close to E_{corr} marked with red circle
	i_{pass}—passive current density	measure of protective properties of passive film: good passive films exhibit very small values of i_{pass}
	E_{bd}—breakdown potential	Related to initiation of local corrosion (pitting), large value E_{bd}–E_{corr} indicates low susceptibility to local corrosion

Continued

Table 8.4 Corrosion Parameters Calculated From Selected Electrochemical Tests — cont'd

Technique and Representative Plot	Parameter	Interpretation
Electrochemical impedance spectroscopy 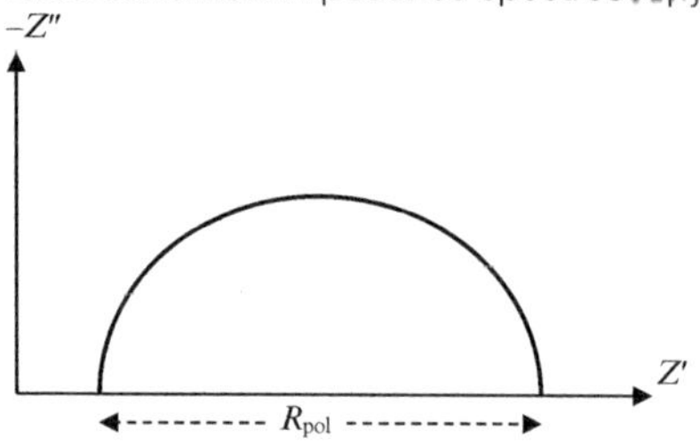	R_{pol}— polarization resistance	When measurement taken at E_{corr}, polarization resistance R_{pol} is inversely proportional to corrosion current density i_{corr}: $i_{corr} = B/R_{pol}$ (the proportionality factor B depends on the corrosion mechanism)

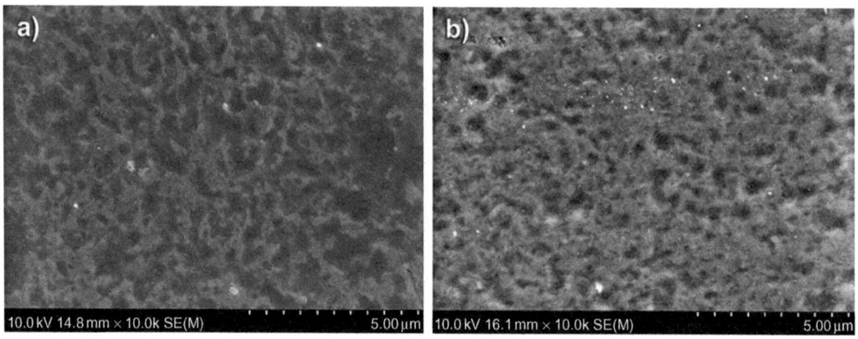

FIG. 8.8

Surface topography (after potentiodynamic polarization in artificial saliva with fluoride ions): (A) coarse-grained Ti Grade 2, (B) UFG Ti Grade 2 processed by multiple-pass rolling.

deeper part of the specimen [2]. For that reason, observations using a BSE signal detect the contrast between sites having different chemical compositions, and are thus a useful method for investigating how corrosion products are distributed.

8.2.2.2 X-ray photoelectron spectroscopy and Auger electron spectroscopy: Characterization of surface chemical composition

Analyzing the chemical and physical nature of surfaces is one of the main subjects of corrosion research. A passive layer formed spontaneously in air on a titanium surface is only a few nanometers thick. For this reason, characterizing their chemical composition and estimating their thickness requires accurate measurement techniques. Most of our present knowledge on the chemical composition of passive films comes from surface analytical methods. X-ray photoelectron spectroscopy (XPS) and AES provide fundamental information about the chemical composition and the state of components of surface layers, and particularly about surface atoms [12,37]. The composition of a surface layer often differs from that of the bulk material, due to contamination, oxidation, and processing. The above methods give insight into important problems regarding the nature of passive films.

Interesting information about the structure and chemical composition of passive films can be obtained by using a combination of ion sputtering and the AES method. Depth profiling is usually performed by inert gas (Ar^+) ion bombardment to remove successive layers of material from the surface. AES composition profiles are particularly useful, because they can be acquired via a high depth resolution. Sputter profiling, however, damages the sputtered oxide, and can lead to composition changes by preferential sputtering and/or ion mixing. Auger studies yield information on the first monolayers of materials [12]. Knowledge of the sputtering rate etching time makes it possible to estimate the thickness of the layer removed [38]. In the case of titanium, it is commonly accepted that the point at which titanium content reaches 75% can be classified as the passive layer/metal border.

Subtle changes in the chemistry of passive layers can be examined by XPS. This method makes it possible to detect many elements, even if they are present at the surface only in trace amounts (Fig. 8.9). Moreover, the high resolution of XPS measurements makes it possible to carry out a detailed analysis of the stoichiometry of the oxides on the surface. Analyzing the contribution of different oxidation states in the passive layer structure provides useful information about the kinetics of passive layer formation. Moreover, evaluating the strength of the signal from the metallic substrate makes it possible to compare the thickness of the passive layers formed on different samples.

8.3 PASSIVE FILM PROPERTIES

Most studies concerned with the corrosion resistance of NC titanium emphasize that the protective properties of a passive layer depend on the layer's stoichiometry, defect density, and diffusivity. The PDM makes it possible to describe the growth and breakdown of a passive layer at the atomic scale. This model was proposed for analyzing passive layer growth under steady-state and transient conditions, and is based on the point defects movement in the electrostatic field. The migration of oxygen anions or oxide ions is fundamental for passive layer formation, while the diffusion of cations vacancies or metal cations leads to passive layer dissolution. Oxide film growth involves solid-state diffusion through imperfections in the lattice [3]. The impact of grain refinement on the density and diffusivity of point defects in a passive layer formed on titanium has been studied recently. Up to now, the relationship between point defects density and grain size has been discussed mainly for titanium deformed by accumulative roll bonding (ARB) and friction stir processing (FSP) [24, 25, 39, 40]. It has been found that oxygen vacancies and/or cation interstitials act as electron donors and lead to an n-type passive film conduction. Studies indicate that grain refinement does not change the n-type conduction of

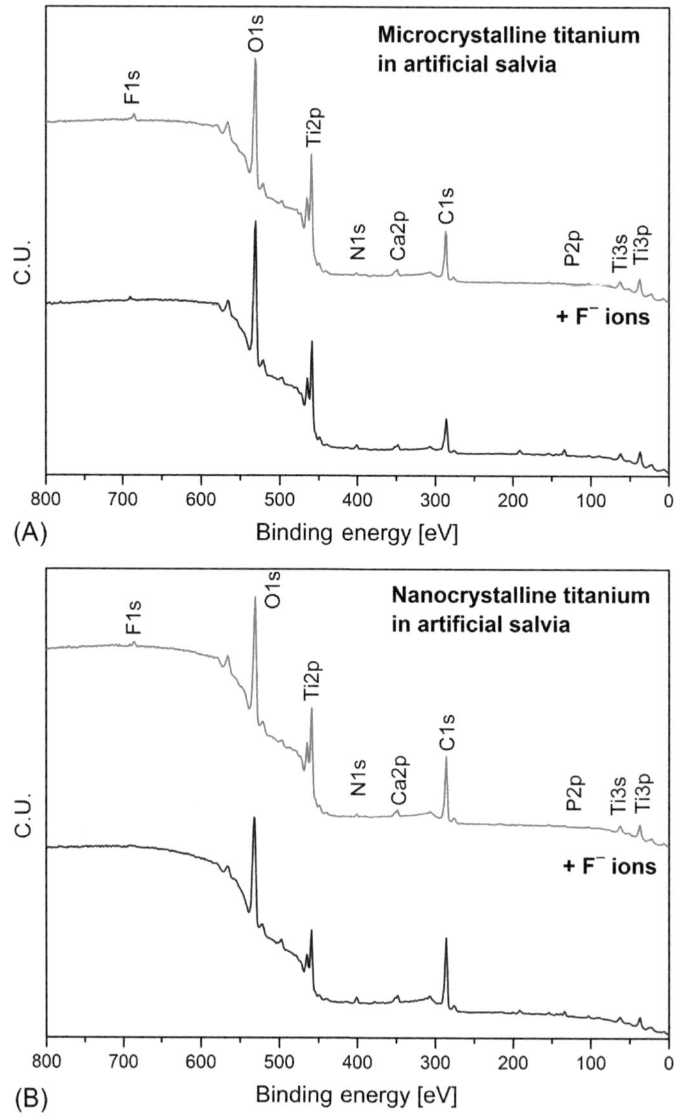

FIG. 8.9

XPS spectra showing the composition of passive layers formed in artificial saliva: (A) on MC titanium, (B) on NC titanium processed by multiple-pass rolling.

titanium. Calculations of donor density values (from the interfacial capacitance) have shown that grain refinement leads to the formation of a less defective, thicker passive layer on the titanium surface that results in better corrosion properties (Fig. 8.10). Nevertheless, the effect of nanostructure on point defect density and diffusivity in a passive film require further investigation.

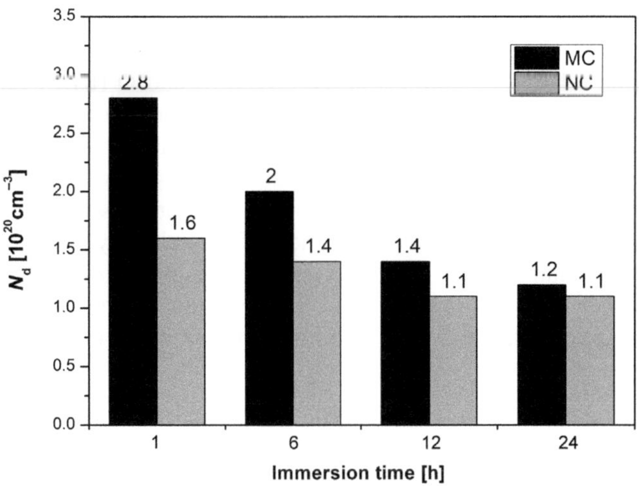

FIG. 8.10

Donor density in a passive film calculated in Ringer's physiological solution for annealed MC titanium and for NC titanium processed by ARB [39].

Studies of the effect of grain refinement on surface properties are also related to an analysis of the chemical composition, chemical state, and thickness of passive layers using surface analytical measurements. Typical composition profiles obtained by AES are shown in Fig. 8.11. Studies were performed on MC titanium and NC titanium processed by hydrostatic extrusion (HE). Three separate domains can be easily noticed within the profiles: an I-oxide film enriched in carbon (which is a contamination), a II-oxide layer containing Ti^{OX}, and a III-titanium substrate. The composition profiles revealed that the passive layers formed on the analyzed samples are

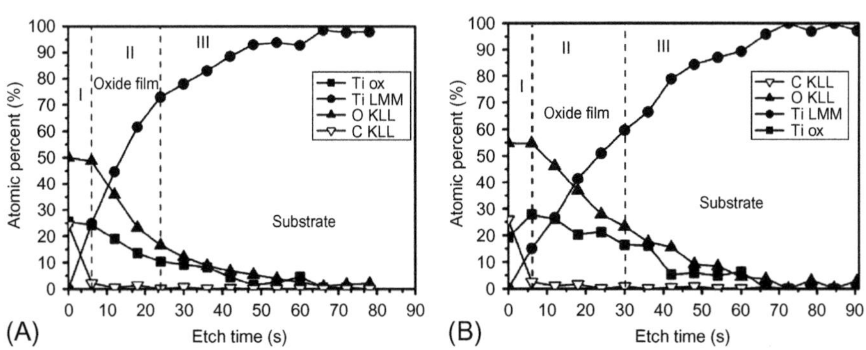

FIG. 8.11

AES results showing the composition of air-formed passive layers: (A) on MC titanium, (B) on NC titanium processed by HE [12].

very thin and have the same chemical composition. The signal from Ti^{OX} and O is stable only at the beginning of etching (Fig. 8.11). During the measurements, changes in the relative proportions of Ti^{OX} (LMM) and O (KLL) can be easily noticed. The time required to remove the oxide layer from micro- and NC titanium were virtually identical. Taking into consideration the constant value of ion etching, it can be concluded that the thickness of the air-formed passive layer is similar for both samples (5 nm for MC titanium and 6 nm for HE-processed titanium). Hence, the nanostructure does not affect the thickness of the oxide layer formed spontaneously on the titanium surface. The changes in titanium corrosion resistance after grain refinement would seem to be related more to the differences in point defect density in the passive layer, as well as to the diverse kinetics of their formation.

8.4 EFFECT OF GRAIN REFINEMENT ON TITANIUM CORROSION RESISTANCE

Physical processing such as SPD results not only in a significant increase in grain boundaries, but also in other phenomena such as changes in crystallographic texture, residual dislocation density, or grain size distribution. All of these changes can affect the resultant corrosion response, and it is difficult to state categorically what factor dominates in the grain size–corrosion resistance dependency.

8.4.1 THE EFFECT OF GRAIN BOUNDARIES

Structural defects such as grain boundaries possess higher energy, and an increase in their number enhances a material's reactivity. Hence, in accordance with thermodynamic principles, NC and UFG materials possess higher reactivity and higher susceptibility to corrosion. However, it has been observed that grain refinement can beneficially affect corrosion resistance, especially in the case of titanium and its alloys. The impact of increased reactivity caused by a greater number of grain boundaries is not straightforward [21–23, 41]. It can differ depending on the type of material corrosion (active or passive) in a particular solution. According to the literature, if the corrosion current density is lower than 10 μA/cm^2, it can be assumed that the material is in the passive state [42].

Corrosion in the active state relies on the metal atoms oxidizing into their ionic forms. This process includes the following stages: the removal of atoms from the crystal lattice, the ionization of those atoms, and their transfer to the solution in the form of hydrated ions. The overall rate of active corrosion can be estimated using the equation [43]:

$$i_a = i_0 \exp\left(\Delta E/2{,}303 b_a\right) \tag{8.2}$$

where i_a is anodic current density, i_0 is exchange current density, ΔE is overpotential, b_a is the anodic Tafel constant.

The energy needed to remove atoms differs depending on the location of those atoms in a crystalline material. Atoms situated within structural defects such as grain boundaries are less strongly bound, and it is easier for them to transfer into the solution than it is for atoms in nondefective sites of the crystal lattice. Hence, the increased number of grain boundaries in UFG and NC materials accelerates their active dissolution.

The enhanced reactivity of UFG and NC materials affects their corrosion resistance in the passive state in different ways. Due to their enhanced energy, structural defects are preferential sites for the nucleation of a protective passive layer. The large number of grain boundaries in materials processed by SPD methods can involve acceleration passivation and repassivation processes. The high density of grain boundaries can result in another beneficial effect in the case of UFG and NC alloys [44]. Grain boundaries are high-diffusivity paths, so a great number of them increases the diffusion of alloys elements to the surface of the material. This can be important for alloys containing elements with a strong affinity for oxygen. However, it is worth noting that a passive layer is only a few nanometers thick, and so its structure strongly depends on the structure of the metal substrate. The more defective structure of the passive layers formed on materials after SPD lead to reduce their resistance to local corrosion. Halogen ions can adsorb more easily on defect agglomerations due to their enhanced reactivity.

The main areas of application of titanium and its alloys, where their high corrosion resistance is crucial, are chemistry and medicine. For this reason, investigations of the impact of grain refinement on titanium corrosion behavior are usually performed in sulfuric or hydrochloric acid, or in biological environments such as physiological saline, or Hank's or Ringer's solution. Generally speaking, a nanostructure amplifies titanium corrosion resistance. It can be concluded that grain boundaries act as preferential sites for the nucleation of a passive layer, even in the presence of aggressive halogenic ions. Nevertheless, in the literature there are a number of contrary opinions. Tables 8.5–8.7 presents a summary of literature data concerning how a high proportion of grain boundaries affects titanium corrosion parameters such as corrosion current density, passive current density, and polarization resistance.

After a broad analysis of the influence of grain size on corrosion resistance, a classification which clarified the general trend was proposed. In 2010, Ralston and Birbilis created an equation connecting grain size with corrosion current density [42]. The following relationship was proposed based on experimental data in the literature:

$$i_{corr} = A + B^{*}d^{-0.5} \tag{8.3}$$

where i_{corr} is corrosion current density, d is average grain size, A is a constant that is a function of the environment, and B is a material constant that depends on composition or impurity level.

The values of B and A for given data can be calculated by using linear regression. This relationship separates the literature data into two classes: materials corroded in the passive state (when i_{corr} is less than $10\,\mu A/cm^2$) and materials that display a

Table 8.5 Impact of a High Number of Grain Boundaries on Ti Grade 2 Corrosion Current Density in Biomedical Solutions (Summary of Literature Data)

Processing Method	Grain Size Span (μm)	Solution	Change in Corrosion Current Density		References
			$\mu A/cm^2$	Difference (%)	
ARB	45 → 0.08	Ringer solution	0.21 → 0.14	33	[24]
FSP	25 → 0.042	Phosphate buffer	0.09 → 0.03	67	[25]
Sliding friction treatment (SFT)	42 → 0.09	Simulated body fluid	1.16 → 0.55	53	[26]
Multiple-pass rolling	35 → 0.420	Artificial saliva	0.34 → 0.21	38	[35]
Multiple-pass rolling	35 → 0.420	Artificial saliva + 1500 ppm F^-	0.45 → 0.25	44	[35]
HE	12 → 0.06	0.9% NaCl	0.015 → 0.01	34	[14]
Equal-channel angular pressing (ECAP)	20 → 0.183	Simulated body fluid	2.8 → 0.9	68	[9]
ECAP	12 → 0.173	0.9% NaCl	33 → 2	94	[45]
ECAP	15 → 0.2	Simulated body fluid	0.76 → 0.07	99	[46]
ECAP	20 → 0.26	0.16 NaCl	0.02 → 0.01	50	[34]

Table 8.6 Impact of a High Number of Grain Boundaries on Ti Grade 2 Passive Current Density in Biomedical Solutions (Summary of Literature Data)

Processing Method	Grain Size Span (μm)	Solution	Change in Passive Current Density		References
			$\mu A/cm^2$	Difference	
ARB	45 → 0.08	Ringer solution	47 → 41	13%	[24]
			0.27 → 0.13	52%	
FSP	25 → 0.042	Phosphate buffer	7.5 → 4.2	54%	[25]
HE	12 → 0.06	0.9% NaCl	Both about 10	0%	[11]

Table 8.7 Impact of a High Number of Grain Boundaries on Ti Grade 2 Polarization Resistance in Biomedical Solutions (Summary of Literature Data)

Grain Size Span (μm)	Solution	Equivalent Circuit Code[a]	Change in R_p, MΩcm^2	Difference in $R_{p^{-1}}$	References
25 → 0.042	Phosphate buffer	*R*(QR)	0.33 → 1	67%	[25]
21 → 0.09	0.9% NaCl	*R*(QR)(QR)	0.082 → 0.062	33%	[12]
42 → 0.09	Simulated body fluid	*R*(QR)	0.14 → 0.39	65%	[26]
12 → 0.06	0.9% NaCl	*R*(QR)	1.1 → 1.7	33%	[14]
35 → 0.420	Artificial saliva	*R*(QR)Q	0.11 → 0.13	15%	[35]
35 → 0.420	Artificial saliva +1500 ppm F^-	*R*(QR)Q	0.03 → 0.09	66%	[35]
20 → 0.183	Simulated body fluid	*R*(RQ)(RQ)	1.37 → 1.19	15%	[9]
20 → 0.26	0.16 NaCl	*R*(RQ)(RQ)	1.9 → 2.2	14%	[34]

[a]*Explanation of equivalent circuit code: R, resistance; C, capacitance; Q, constant phase element; ( ), parallel combination.*

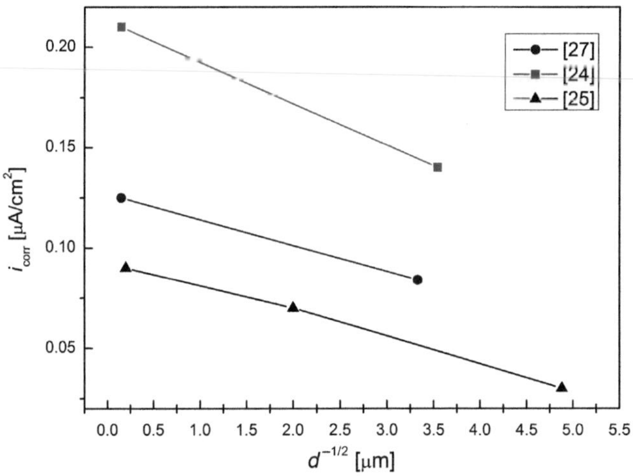

FIG. 8.12

Relationship between corrosion rate and grain size for materials in the passive state in Ringer solution [26], SBF [24], and phosphate buffer solution [25].

higher corrosion current density. In case of the passivated materials, the B constant is less than zero, and inversely for materials actively corroded. Typical behavior for titanium in the passive state (a decrease of i_{corr} with grain refinement) is illustrated in Fig. 8.12. The differences in the value of the A coefficient related to environmental conditions are clearly visible. Surprisingly, a similar slope (values of the B coefficient linked with material conditions) is observed despite the use of various processing methods.

According to Eq. (8.3), when a material has the ability to passivate in given conditions, the corrosion rate decreases with a reduction in grain size (Fig. 8.13B), but

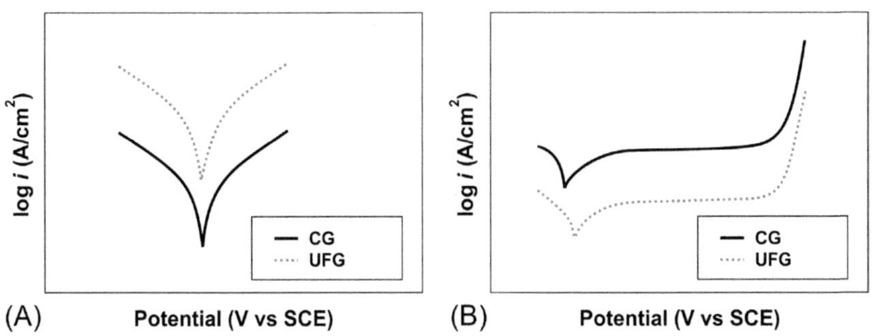

FIG. 8.13

Models illustrating the impact of grain size on a material's behavior during potentiodynamic measurements: (A) active conditions, (B) passive conditions [22].

when the dissolution rate is higher than 10 μA/cm^2, the fine-grained structure is expected to accelerate the corrosion process (Fig. 8.13A) [42].

The corrosion current density is inversely proportional to the polarization resistance (R_{pol}) taken at the corrosion potential, according to the formula:

$$i_{corr} = B/R_{pol}$$

where B is the proportionality factor.

Therefore, changes in corrosion current density are often compared with changes in the reciprocal of R_{pol} taken from EIS measurements. However, one must consider that the proportionality factor B depends on the mechanism of the corrosion process. When the transition from a coarse-grained metal to its NC counterpart is associated with a change in the kinetics of corrosion reactions, the changes observed in the $R_{pol^{-1}}$ can be only treated as a rough approximation of the corresponding variation in the i_{corr}. On the other hand, the polarization resistance seems to be easier to detect from impedance measurements than corrosion current density from the Tafel method.

The Ralston and Birbilis correlation is important when comparing the effect of grain size on corrosion resistance across a wide range of corrosion solutions. It helps to estimate how the impact of grain refinement changes with an increase in the aggressiveness of the environment. Moreover, it makes it possible to evaluate the impact of the impurity level or alloy additions on the slope of corrosion rate vs grain size. The phenomena reported are analogous to the classical Hall–Petch relationship. When a stable passive layer exists on a material's surface, then a reduction in grain size not only enhances the material's mechanical properties, but also its corrosion resistance.

8.4.2 EFFECT OF CRYSTALLOGRAPHIC TEXTURE

In order to obtain a nano- or UFG structure by physical deformation, a material had to be processed by multistage SPD processes. Initially, corrosion studies were limited to materials after the final stages of deformation. However, electrochemical measurements performed on materials in intermediate stages of processing revealed that the relationship between corrosion resistance and grain size is not linear, as had been predicted by Ralston and Birbilis [14, 34, 47]. Variations in crystallographic texture during deformation is one probable reason for the phenomena observed. This factor significantly affected the corrosion behavior in NaCl of Ti Grade 2 processed by ECAP. To evaluate the impact of crystallographic texture on corrosion resistance, it was necessary to check the orientation of the planes parallel to the surface that are exposed to the corrosive environment. It was noticed that the texture changed significantly after two passes of ECAP, and then (after four stages) returned to the crystallographic orientation prior to ECAP. In the case of titanium, after two stages of ECAP, the most privileged orientation is close to (0002), which is the basal (closed-packed) plane in the hcp system [34]. The modification of texture correlates with an enhancement in corrosion resistance. In the case of grains in which the atoms are oriented in closely-packed directions, the removal of those atoms to the corrosion solution is hampered in comparison with grains oriented in other, more loosely-packed directions [21]. In another work, in order to provide a quantitative description of texture, a texture index was calculated from orientation distribution functions. It was noticed that, for the strongest basal texture, corrosion resistance is also the greatest [47]. Nevertheless, in the literature related to the corrosion behavior of materials after SPD, the effect of texture is commonly neglected. The lack of available information makes it difficult to formulate a general correlation between crystallographic texture and corrosion resistance for titanium processed by SPD methods.

8.4.3 EFFECT OF GRAIN SIZE DISTRIBUTION

In recent years, systematic studies in the area of corrosion properties of materials processed by various SPD methods have been performed. It is commonly known that not only average grain size, but also grain size distribution, depends on the SPD method selected for grain refinement. According to the simulation performed, a broad grain size distribution leads to a decrease in the value of maximum yield strength and also promotes a reduction in critical grain size in which the Hall–Petch relationship is reversed. Thus, the influence of grain size distribution on corrosion resistance is also worth analyzing. It has been observed that, in the case of NC materials, the passivation kinetics are faster but the uniformity of the passive layer depends on the grain size distribution. In 2012, Gollapudi presented a modified Ralston–Birbilis correlation which considered a standard deviation in the number of weighted grain size distribution [48] according to the following equation:

$$i_{corr} = A + B^{*}gs^{-0.5}\exp\left(-\frac{9}{8}Sn^{2}\right) \tag{8.4}$$

where A is a constant whose value depends on the corrosive environment, B is a constant that represents a material (its value is related to the material's composition materials and impurity level), and Sn—is the standard deviation in the number weighted grain size distribution. The Sn can obtain values from 0 to 1. When Sn = 0, every grain in the analyzed part of the material has the same size. The higher the Sn value, the broader the grain size distribution. In typical materials, the grain size distribution is between 0.3 and 0.6.

The determination of this relationship was based on a methodology proposed by Phaniraj et al. [49], which was used to evaluate the impact of grain size distribution on the Hall–Petch equation. It is worth adding that Eq. (8.4) can be used only for pure metals. Moreover, the corrosion current density considers only uniform corrosion. In general, uniform corrosion can be considered as corrosion that occurs from the bulk of a material and is localized at grain boundaries [48]. Fig. 8.14 illustrates the corrosion current density vs grain size relationship for different values of the Sn parameter. It can be easily noticed that grain size distribution directly affects the corrosion behavior in both passivating and nonpassivating environments. Moreover, the impact of grain size distribution on corrosion behavior is substantially greater for NC materials. For materials in the passive state, a narrow grain size distribution favors the

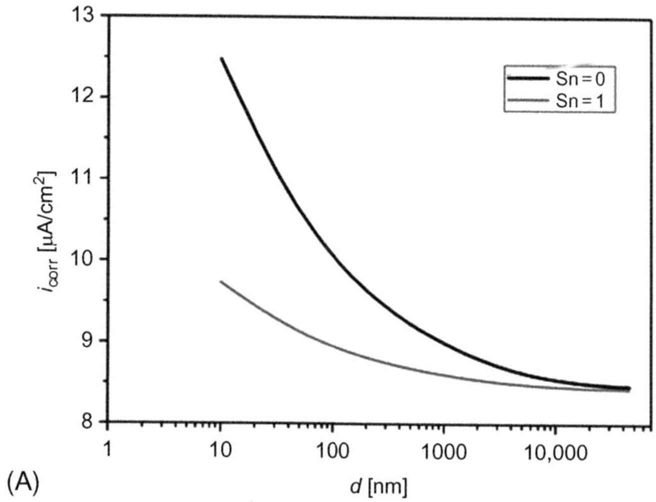

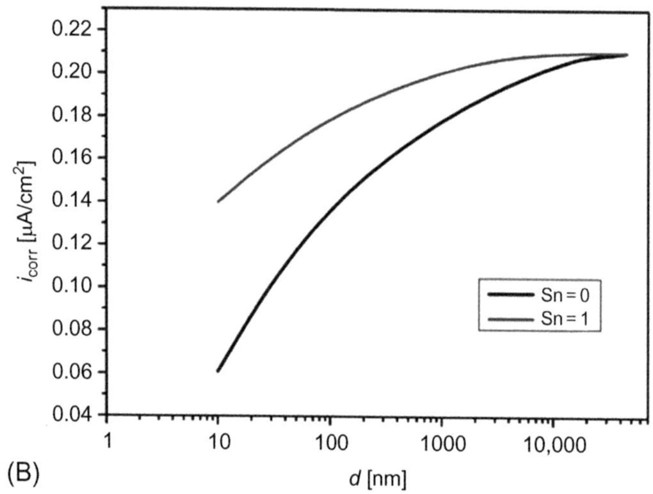

FIG. 8.14

Corrosion current density for different grain size distribution in a: (A) nonpassivating and (B) passivating environment [48].

formation of a homogenous and tight passive layer. The homogeneity of layer is the result of the close vicinity of grain boundaries and triple junctions. If the distance between sites with enhanced reactivity varies (in case of a broad grain size distribution), the oxide layer is nonuniform (Fig. 8.15). The reverse is to be expected for materials corroded in a nonpassivating environment. For the same average grain size, a material with a broader grain size distribution possesses better corrosion resistance in the active state. The probable reason for this is the smaller total volume of grain

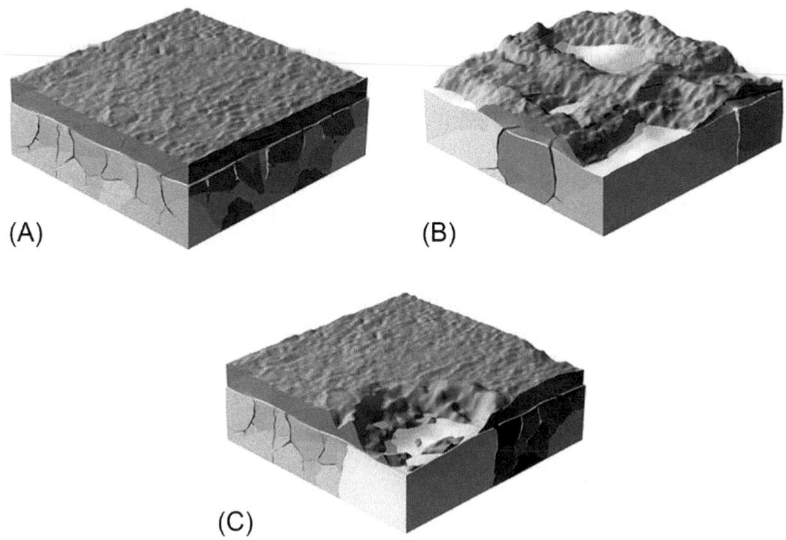

FIG. 8.15

Models of a passive layer structure in the case of: (A) fine-grained material with narrow grain size distribution, (B) coarse-grained material, (C) deformed material with large grain size distribution [48].

boundaries in the case of the material with a broader grain size distribution. However, this opinion remains to be confirmed by appropriate calculations. The calculations presented here indicate that grain size distribution is a substantial parameter that has an impact on corrosion resistance, especially in the case of nanomaterials. It should therefore be taken into consideration when designing NC materials whose corrosion properties are crucial for a specific application.

8.5 SUMMARY

In this chapter, the influence of grain refinement on the corrosion resistance of pure titanium was investigated. Both electrochemical and surface techniques of NC titanium were briefly described. The main conclusions can be summarized as follows:

- Processing by SPD techniques results not only in a significant increase in grain boundaries but also in an increase in residual stresses, changes in crystallographic texture, or grain size distribution. All of these factors can affect the corrosion resistance of titanium.
- Titanium corrosion resistance is usually evaluated in biological solutions such as physiological saline, simulated body fluid, or phosphate buffer solution. Generally, grain refinement enhances titanium's corrosion properties. It can be

concluded that grain boundaries act as preferential sites for the nucleation of passive layers, even in the presence of aggressive chloride or fluoride ions.

- A high number of grain boundaries resulted in a faster passivation and repassivation of NC titanium. However, it was found that the thickness of the passive layers formed on the titanium with micro- and nanometric grains are very similar. Thus, changes in titanium corrosion resistance after grain refinement are related more to differences in the point defect density in the passive layer, as well as to the diverse kinetics of their formation.
- The relationship between corrosion resistance and grain size is analogous to the Hall–Petch equation. For materials in the passive state, a reduction in grain size improves both the material's mechanical properties and its corrosion resistance. However, electrochemical tests performed for titanium in intermediate stages of deformation revealed that the relationship between corrosion resistance and grain size does not have to be linear. Changes that take place in the crystallographic texture during processing is a probable cause of this phenomenon.

REFERENCES

[1] G. Lütjering, J.C. Williams, Titanium, Springer, 2007.

[2] R.A. Cottis, Shreir's Corrosion, Elsevier, 2010.

[3] B. Munirathinam, R. Narayanan, L. Neelakantan, Electrochemical and semiconducting properties of thin passive film formed on titanium in chloride medium at various pH conditions. Thin Solid Films 598 (2016) 260–270, https://doi.org/10.1016/j.tsf.2015.12.025.

[4] Z. Jiang, X. Dai, H. Middleton, Investigation on passivity of titanium under steady-state conditions in acidic solutions. Mater. Chem. Phys. 126 (2011) 859–865, https://doi.org/10.1016/j.matchemphys.2010.12.028.

[5] Z. Jiang, X. Dai, T. Norby, H. Middleton, Investigation of pitting resistance of titanium based on a modified point defect model. Corros. Sci. 53 (2011) 815–821, https://doi.org/10.1016/j.corsci.2010.11.015.

[6] M.J. Muñoz-Portero, J. García-Antón, J.L. Guiñón, R. Leiva-García, Pourbaix diagrams for titanium in concentrated aqueous lithium bromide solutions at 25°C, Corros. Sci. 53 (2011) 1440–1450.

[7] S. Prasad, M. Ehrensberger, M.P. Gibson, H. Kim, E.A. Monaco, Biomaterial properties of titanium in dentistry. J. Oral Biosci. 57 (2015) 192–199, https://doi.org/10.1016/j.job.2015.08.001.

[8] A. Revathi, A.D. Borrás, A.I. Muñoz, C. Richard, G. Manivasagam, Degradation mechanisms and future challenges of titanium and its alloys for dental implant applications in oral environment. Mater. Sci. Eng. C 76 (2017) 1354–1368, https://doi.org/10.1016/j.msec.2017.02.159.

[9] H. Maleki-Ghaleh, K. Hajizadeh, A. Hadjizadeh, M.S. Shakeri, S. Ghobadi Alamdari, S. Masoudfar, E. Aghaie, M. Javidi, J. Zdunek, K.J. Kurzydlowski, Electrochemical and cellular behavior of ultrafine-grained titanium in vitro. Mater. Sci. Eng. C 39 (2014) 299–304, https://doi.org/10.1016/j.msec.2014.03.001.

[10] H.S. Kim, W.J. Kim, Annealing effects on the corrosion resistance of ultrafine-grained pure titanium. Corros. Sci. 89 (2014) 331–337, https://doi.org/10.1016/j.corsci.2014.08.017

[11] A. Balyanov, J. Kutnyakova, N.A. Amirkhanova, V.V. Stolyarov, R.Z. Valiev, X.Z. Liao, Y.H. Zhao, Y.B. Jiang, H.F. Xu, T.C. Lowe, Y.T. Zhu, Corrosion resistance of ultra fine-grained Ti. Scr. Mater. 51 (2004) 225–229, https://doi.org/10.1016/j.scriptamat.2004.04.011.

[12] H. Garbacz, M. Pisarek, K.J. Kurzydłowski, Corrosion resistance of nanostructured titanium. Biomol. Eng. 24 (2007) 559–563, https://doi.org/10.1016/j.bioeng.2007.08.007.

[13] A. Sotniczuk, M. Zwolińska, M. Sozańska, A. Królikowski, H. Garbacz, Nanostructure effect on the electrochemical response of titanium. Inżynieria Mater. Mater. Eng. 6 (2016) 311–316, https://doi.org/10.15199/28.2016.6.6.

[14] M. Zwolińska, A. Sotniczuk, K. Topolski, A. Królikowski, H. Garbacz, The influence of deformation the corrosion resistance of titanium grade 2, Solid State Phenom. 227 (2015) 27–30.

[15] H.S. Kim, S.J. Yoo, J.W. Ahn, D.H. Kim, W.J. Kim, Ultrafine grained titanium sheets with high strength and high corrosion resistance. Mater. Sci. Eng. A 528 (2011) 8479–8485, https://doi.org/10.1016/j.msea.2011.07.074.

[16] S. Tamimi, M. Ketabchi, N. Parvin, Microstructural evolution and mechanical properties of accumulative roll bonded interstitial free steel. Mater. Des. 30 (2009) 2556–2562, https://doi.org/10.1016/j.matdes.2008.09.039.

[17] W.J. Kim, S.J. Yoo, J.B. Lee, Microstructure and mechanical properties of pure Ti processed by high-ratio differential speed rolling at room temperature. Scr. Mater. 62 (2010) 451–454, https://doi.org/10.1016/j.scriptamat.2009.12.008.

[18] L. Mishnaevsky, E. Levashov, R.Z. Valiev, J. Segurado, I. Sabirov, N. Enikeev, S. Prokoshkin, A.V. Solov'yov, A. Korotitskiy, E. Gutmanas, I. Gotman, E. Rabkin, S. Psakh'e, L. Dluhoš, M. Seefeldt, A. Smolin, Nanostructured titanium-based materials for medical implants: modeling and development. Mater. Sci. Eng. R Rep. 81 (2014) 1–19, https://doi.org/10.1016/j.mser.2014.04.002.

[19] Y. Estrin, A. Vinogradov, Extreme grain refinement by severe plastic deformation: a wealth of challenging science. Acta Mater. 61 (2013) 782–817, https://doi.org/10.1016/j.actamat.2012.10.038.

[20] M. Geetha, A.K. Singh, R. Asokamani, A.K. Gogia, Ti based biomaterials, the ultimate choice for orthopaedic implants—a review. Prog. Mater. Sci. 54 (2009) 397–425, https://doi.org/10.1016/j.pmatsci.2008.06.004.

[21] K.D. Ralston, N. Birbilis, Effect of grain size on corrosion. Corrosion 66 (2010) 1–4, https://doi.org/10.5006/1.3462912.

[22] H. Miyamoto, Corrosion of ultrafine grained materials by severe plastic deformation, an overview. Mater. Trans. 57 (2016) 559–572, https://doi.org/10.2320/matertrans.M2015452.

[23] P. Herrasti, C. Ponce de León, F.C. Walsh, The corrosion behaviour of nanograined metals and alloys, Rev. Metal. 48 (2012) 377–394.

[24] G. Ansari, A. Fattah-alhosseini, On the passive and semiconducting behavior of severely deformed pure titanium in Ringer's physiological solution at 37 °C: a trial of the point defect model, Mater. Sci. Eng. C 75 (2017) 64–71.

[25] A. Fattah-alhosseini, M. Vakili-Azghandi, M. Sheikhi, M.K. Keshavarz, Passive and electrochemical response of friction stir processed pure titanium, J. Alloys Compd. 704 (2017) 499–508.

[26] J. Lu, Y. Zhang, W. Huo, W. Zhang, Y. Zhao, Y. Zhang, Electrochemical corrosion characteristics and biocompatibility of nanostructured titanium for implants. Appl. Surf. Sci. 434 (2018) 63–72, https://doi.org/10.1016/J.APSUSC.2017.10.168.

[27] R. Huang, Y. Han, The effect of SMAT-induced grain refinement and dislocations on the corrosion behavior of Ti–25Nb–3Mo–3Zr–2Sn alloy. Mater. Sci. Eng. C 33 (2013) 2353–2359, https://doi.org/10.1016/j.msec.2013.01.068.

[28] L. Jin, W. Cui, X. Song, G. Liu, L. Zhou, Effects of surface nanocrystallization on corrosion resistance of β-type titanium alloy. Trans. Nonferrous Met. Soc. Chin. 24 (2014) 2529–2535, https://doi.org/10.1016/S1003-6326(14)63379-3.

[29] D. Raducanu, E. Vasilescu, V.D. Cojocaru, I. Cinca, P. Drob, C. Vasilescu, S.I. Drob, Mechanical and corrosion resistance of a new nanostructured Ti–Zr–Ta–Nb alloy, J. Mech. Behav. Biomed. Mater. 4 (2011) 1421–1430.

[30] S. Mathur, R. Vyas, R. Jain, P. Kumar, K. Sachdev, S.K. Sharma, Electrochemical behavior of different structural states of the alloy Ti 60 Ni 40, J. Non Cryst. Solids 357 (2011) 3084–3087.

[31] T. Hu, Y.C. Xin, S.L. Wu, C.L. Chu, J. Lu, L. Guan, H.M. Chen, T.F. Hung, K.W. K. Yeung, P.K. Chu, Corrosion behavior on orthopedic NiTi alloy with nanocrystalline/amorphous surface. Mater. Chem. Phys. 126 (2011) 102–107, https://doi.org/10.1016/J.MATCHEMPHYS.2010.11.061.

[32] S. Faghihi, D. Li, J.A. Szpunar, Tribocorrosion behaviour of nanostructured titanium substrates processed by high-pressure torsion, Nanotechnology 21 (2010) 485703.

[33] J. Tang, H.Y. Luo, Y.B. Zhang, Enhancing the surface integrity and corrosion resistance of Ti-6Al-4V titanium alloy through cryogenic burnishing. Int. J. Adv. Manuf. Technol. 88 (2017) 2785–2793, https://doi.org/10.1007/s00170-016-9000-y.

[34] M. Hoseini, A. Shahryari, S. Omanovic, J.A. Szpunar, Comparative effect of grain size and texture on the corrosion behaviour of commercially pure titanium processed by equal channel angular pressing. Corros. Sci. 51 (2009) 3064–3067, https://doi.org/10.1016/j.corsci.2009.08.017.

[35] A. Sotniczuk, D. Kuczyńska, D. Kubacka, A. Królikowski, H. Garbacz, Influence of nanostructure on titanium corrosion resistance in fluoridated medium. Mater. Sci. Technol. (2017). https://doi.org/10.1080/02670836.2017.1407557.

[36] M. Nie, C.T. Wang, M. Qu, N. Gao, J.A. Wharton, T.G. Langdon, The corrosion behaviour of commercial purity titanium processed by high-pressure torsion. J. Mater. Sci. 49 (2014) 2824–2831, https://doi.org/10.1007/s10853-013-7988-z.

[37] M. Janik-Czachor, A. Jaskiewicz, M. Dolata, Z. Werner, Passivity and its breakdown in Al-based amorphous alloys, Mater. Chem. Phys. 92 (2005) 348–353.

[38] H. Garbacz, Microstructure and Properties of Nanocrystalline Titanium, Publishing House of the Warsaw University of Technology, Warsaw, 2012.

[39] A. Fattah-alhosseini, A.R. Ansari, Y. Mazaheri, M.K. Keshavarz, Effect of immersion time on the passive and electrochemical response of annealed and nano-grained commercial pure titanium in Ringer's physiological solution at 37°C. Mater. Sci. Eng. C 71 (2017) 771–779, https://doi.org/10.1016/j.msec.2016.10.057.

[40] A. Fattah-alhosseini, O. Imantalab, G. Ansari, The role of grain refinement and film formation potential on the electrochemical behavior of commercial pure titanium in Hank's physiological solution. Mater. Sci. Eng. C 71 (2017) 827–834, https://doi.org/10.1016/j.msec.2016.10.072.

[41] D. Orlov, K.D. Ralston, N. Birbilis, Y. Estrin, Enhanced corrosion resistance of Mg alloy ZK60 after processing by integrated extrusion and equal channel angular pressing. Acta Mater. 59 (2011) 6176–6186, https://doi.org/10.1016/j.actamat.2011.06.033.

[42] K.D. Ralston, N. Birbilis, C.H.J. Davies, Revealing the relationship between grain size and corrosion rate of metals, Scr. Mater. 63 (2010) 1201–1204.

[43] P. Marcus, Corrosion Mechanisms in Theory and Practice, CRC Press, 2011.

[44] A. Królikowski, Corrosion behaviour of amorphous and nanocrystalline alloys, in: Solid State Phenomena, Trans Tech Publications, 2015, pp. 11–14.

[45] C. Gode, S. Attarilar, B. Eghbali, M. Ebrahimi, Electrochemical behavior of equal channel angular pressed titanium for biomedical application, AIP Conference Proceedings, AIP Publishing, 2015, p. 20041.

[46] A. Balakrishnan, B.C. Lee, T.N. Kim, B.B. Panigrahi, Corrosion behaviour of ultra fine grained titanium in simulated body fluid for implant application, Trends Biomater. Artif. Organs 22 (2008) 58–64.

[47] N.P. Gurao, G. Manivasagam, P. Govindaraj, R. Asokamani, S. Suwas, Effect of texture and grain size on bio-corrosion response of ultrafine-grained titanium. Metall. Mater. Trans. A Phys. Metall. Mater. Sci. 44 (2013) 5602–5610, https://doi.org/10.1007/s11661-013-1910-9.

[48] S. Gollapudi, Grain size distribution effects on the corrosion behaviour of materials. Corros. Sci. 62 (2012) 90–94, https://doi.org/10.1016/j.corsci.2012.04.040.

[49] M.P. Phaniraj, M. Prasad, A.H. Chokshi, Grain-size distribution effects in plastic flow and failure, Mater. Sci. Eng. A 463 (2007) 231–237.

CHAPTER

Biological properties

9

Halina Garbacz
Faculty of Materials Science and Engineering, Warsaw University of Technology, Warsaw, Poland

9.1 INTRODUCTION

Biocompatibility is a feature of a substance or a material that determines its normal behavior with respect to living tissues [1,2]. The term "biocompatibility" is used to describe materials implanted into the body that produce a normal/healthy response and do not adversely affect cells (by being carcinogenic or toxic). The basic research that allows us to describe the behavior of materials in contact with foreign tissue are cell culture studies. An assessment of biocompatibility requires both in vitro (cell tests) and in vivo studies (animal trials), as well as clinical trials.

The nanomaterials used in medicine have a number of unique properties that distinguish them from conventional materials. They exhibit a high surface area to volume ratio (e.g., nanoparticles), which makes it possible to use them as drug delivery systems. Nanomaterials can be successfully designed to have specific sizes and shapes, and their chemical and surface properties can be controlled. This characteristic makes it possible to endow them with certain desirable optical, electrical, magnetic, and biological properties [3]. Nanoparticles or nanopowders would seem to be very efficient materials, yet they entail certain risks in use by causing toxicological and environmental effects [4–6]. It is true that individual nanosized particles can certainly enter the human body via the lungs and the intestines [4], but this problem does not arise in the case of bulk nanostructured metallic biomaterials.

9.2 INFLUENCE OF NANOSTRUCTURE ON BIOLOGICAL RESPONSE

Studies of the effects of nanostructured materials on biocompatibility are interdisciplinary, and require a combination of knowledge in medicine, biology and materials science and engineering. The adhesion of osteoblasts to nanocrystalline materials was first described in 1999 [7]. The results of in vitro studies of nanoceramics such as Al_2O_3, TiO_2, and hydroxyapatite revealed that the proliferation and other

Nanocrystalline Titanium. https://doi.org/10.1016/B978-0-12-814599-9.00009-2

long-term functions of osteoblasts, as well as the deposition of minerals containing calcium and phosphorus (calcium–phosphorous salts are the most important components of bone tissue) are significantly improved when materials with grains of fibers <100 nm are used. In addition, the capacity of osteoblasts with bone formation was found to increase on implants containing nanofibers in their structures as opposed to coarse grains (e.g., Al_2O_3). It is assumed that nanofibers are similar in shape to the hydroxyapatite and collagen crystals present in bone [8]. This feature facilitates the process of bone formation on an implant surfaces. Ceramics were produced by means of powder metallurgy. An additional factor other than grain size that affects the results of biological tests is their porosity. Porous biomaterials attract interest mainly because they enhance implant anchoring in the human body [2]. Yet, the mechanical strength of sinters is poor, and so the proper pore size and density is crucial.

In in vitro studies, especially those concerning protein adsorption and cell adhesion, the correct degree of surface roughness is essential. A suitable surface roughness of the artificial material can affect cell adhesion, growth, maturation, proliferation, differentiation [9], and even bone growth [10]. The acceptance or rejection of a biomaterial depends on the interface created between the implanted material and the surrounding tissue. The bulk structure must provide only mechanical stability for the implanted device, and must not interact with tissues, cells, and body fluids. The surface of the material plays the main role in contact with the surrounding tissue in respect of improving biological response [1,2,11]. In the case of ceramics and sinters, the degree of surface roughness is related to open porosity. The consolidation of nanopowders allows for a high degree of roughness on the sinter surface [12]. Surface topography often reproduces the shape and size of sintered nanopowders. Decreasing the size of TiO_2 particles from 2 μm to 32 nm has a significant impact on the roughness and architecture of the surface, and on the wettability of the sinters. The specific surface area of nanocrystalline TiO_2 is about 25% larger than that of microcrystalline titanium [7]. A similar effect can be obtained with solid implants obtained by conventional methods (casting, plastic processing) by modifying their surfaces by, for example, applying a layer of porous hydroxyapatite or a nanotube structure on the surface. It has been found that decreasing the surface roughness from the micro- to nanoscale results in an increase in the adhesion of osteoblasts by also increasing their biological activity [13–18]. Osteoblasts that adhere to the surface of nanocrystalline titanium oxides, such as Ti–6Al–4V and CoCrMo alloys, also undergo faster mineralization than microcrystalline substrates [19]. It is now believed that the fabrication of a hierarchical topography can be used to take advantage of beneficial micro- and nano-roughness. The primary advantage of a multimodal topography is that it mimics the bone hierarchical surface structure after the remodeling process. It has been observed that osteoclasts have the ability to dissolve bone, and create microcracks and gaps after the resorption of the bone surface (a diameter 30–100 μm). According to Gittens [20–22], osteoclasts acidify the calcium phosphate but do not produce collagenase, the enzyme needed to degrade collagen. As a result of this process, hierarchical structures with gaps after resorption (at the microscale) and bundles of collagen fibers (at the nanoscale) are formed on the surface of the bone. Hierarchical surface structures also promote protein adsorption [23–25].

A novelty is the use of nanotechnology to produce scaffolds, which are used in tissue engineering as a cell substrate [13]. Metals and their alloys having high porosity and varying pore sizes ensure the active growth of bone tissue. Titanium samples produced by multilayered laser sintering, containing very large pore sizes of between several dozen and several hundred nanometers in microstructure, exhibit excellent integration with bone tissue [17].

Lai and coworkers [26] investigated the possibility of using nanocrystalline titanium foils produced by electrodeposition in urology. The maximum grain diameter was 300 nm. The reference materials were titanium with an average particle size of about 80 μm, and polymers. The use of different biomaterials in an artificial urinary tract is limited by the deposition of calcium compounds on their surfaces, commonly called stones. Surface inclusions result in the closure of the artificial urinary tract and the necessity to replace it. Surface roughness also affects the deposition process: it provides a greater number of active nucleation areas, resulting in greater reactivity and improved particle adhesion to the surface. This effect has also been observed for nanocrystalline titanium. An analysis of nanocrystalline films of different surface topography revealed that smooth surfaces prevent incrustation. The main achievement of Lai's research is the demonstration that nanocrystalline titanium has greater resistance to incrustation than coarse titanium has, as well as to existing polymers such as polyvinyl chloride and polyethylene. This fact is explained by the greater chemical homogeneity of nanomaterials that have a natural tendency, for example, to reduce the pitting corrosion resistance of nanocrystalline aluminum [27]. Coarse materials are typically characterized by heterogeneous chemical compositions that, as shown by studies, can facilitate the heterogeneous nucleation of particles on the surface. The very favorable properties of nanocrystalline titanium in the context of applications in urology are currently being confirmed by in vivo studies.

Titanium and its alloys are used as prosthetics, dental implants, orthodontic devices, bridges, fixings for hearing aids, and in cardiology for the manufacture of mechanical heart valves [12,28–31]. Titanium exhibits high biocompatibility, and often, direct bone apposition [32]. Its low density and low Young's modulus result in less stress shielding and fewer artifacts on imaging techniques [33,34], and determines the biomechanical anchoring of the implant in the body. Studies have shown that titanium implants achieve more effective bone regeneration than implants made from titanium alloys. In addition, the results of surface examinations of implants used in bone surgery show that osteointegration in these metals is faster on developed surfaces than on smooth surfaces [35]. Significant improvements in the biological properties of titanium alloy implants have been achieved with great success in surface engineering [2,12,36,37]. Due to the relatively low mechanical strength and fatigue resistance of pure titanium, its biomedical use has limited to weakly loaded implants. A significant improvement in the mechanical properties of titanium can be achieved with the addition of alloying elements such as aluminum or vanadium [38]. However, despite the excellent corrosion resistance of a Ti–6Al–4V alloy, there have been several reports of adverse effects due to the release of ions (Al and V) into the surrounding tissue [39]. The risk of toxicity is particularly high under conditions of friction [32]. The main solution to this problem is based on eliminating the toxic elements from titanium alloys and replacing them with completely biocompatible elements such as niobium, zirconium, and tantalum [34,40,41]. An alternative approach to solving the problem of ion release is to stop using alloying elements while improving the mechanical properties of titanium through the grain reinforcement to the nanometer scale [32,42,43]. According to the literature, nanocrystalline titanium after severe plastic deformation (SPD) processes exhibits a higher affinity with oxygen [44] and higher corrosion resistance in physiological saline [44,45]. This strong oxygen affinity of titanium improves the metal's mechanical properties through solid solution effects [46,47]. The main advantage of bulk nanocrystalline materials [38], including titanium, is that they display surface properties that are completely different from those of conventional, coarse-grained titanium. Moreover, nanocrystalline titanium exhibits a high strength-to-weight ratio, superior wear resistance, and a high grain boundary area that allows for a reduction in implant thickness, which may reduce the flexural modulus of the implant [42].

Nowadays, a combination of these two approaches—eliminating toxic elements and refining grain—is being met with special interest. New biomedical beta titanium alloys are being designed, which form a new group of metallic biomaterials with superior mechanical properties, including ultra-low stiffness comparable to that of human bone [39,48,49]. The unique features of these beta titanium alloys are the result of the low thermodynamic stability of the beta phase formed after solution treatment. In the case of Ti–29Nb–13Ta–4.6Zr, the main disadvantage is its relatively low strength, which in fact is lower than that of the Ti–6Al–4V alloy. An improvement in mechanical properties and the conservation of a low elastic modulus may be achieved in metals by grain refinement and an enhancement of dislocation density [34,41,50,51].

9.3 SURFACE INTERACTIONS WITH SURROUNDING BODY ENVIRONMENT

In vitro tests are conducted in order to predict the impact of an implant surface on living tissues. The analysis takes place outside the living organism, using proteins, cell cultures, or bacteria. One of the most commonly anticipated processes with titanium is osseointegration, which is defined as direct contact between the bone and the implant surface, without a fibrous tissue at the interface [52]. As defined, the biological response of implants strongly depends on the first contact, when the implant surface reacts with the surrounding environment, proteins, cells, ions, etc. [53]. The process of osseointegration is very complex (Fig. 9.1). Cell adhesion (osteoblasts) is preceded by a protein adsorption process, which is dynamic, and enhanced by the physicochemical interactions between the implant surface and biological

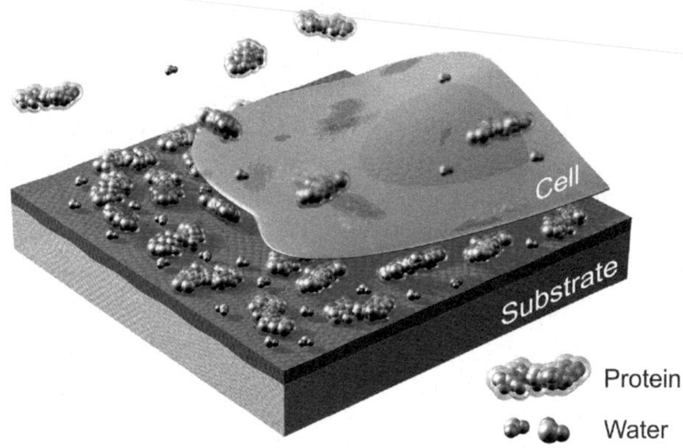

FIG. 9.1

Schematic illustration of interaction between biological environment and substrate.

macromolecules [54,55]. The adhesion of osteoblasts requires proteins, for example fibronectin, on the surface attached to them.

Cell–surface interactions depend on the interactions between proteins and the surface, and subsequently, between proteins and cells. Understanding these protein–cell and surface interactions makes it possible to predict what happens when the biological environment responds to surface topography or chemistry. Surface properties control the amount and quality of proteins and cells that adhere on the implant, and consequently, tissue growth. The surface properties that most strongly determine protein adsorption and cell adhesion are surface topography and surface chemical reactivity. Current strategies for designing biomaterials include improving biological properties such as rapid osseointegration, and antithrombogenic and antibacterial effects, through surface functionalization techniques that affect both incipient and long-term implant–body interactions [53]. Therefore, controlling wettability, surface roughness, texture, morphology, and chemical composition are of particular concern [35].

The topography or texture of a surface affects processes at the cellular level such as the orientation and migration of cells and the cellular production of organized cytoskeletal arrangements, or transitioning of cells. Surface texture has an impact on what signals are transmitted from focal adhesions to the cell nucleus, where the function of the cell can be changed. For example, a grooved topography causes an elongation of the cell by the contact guidance phenomenon [56] and promotes the osteogenic transition of the cell through changes in gene expression [57–59]. This elongation of the cell shape creates tension, which is transferred through actin filaments to the nucleus, causing a deformation thereof [60]. Changes in the morphology of the cytoskeleton and nucleus due to surface topography can have an impact on the differentiation of stem cells [57]. Modifying the surface topography also has a beneficial effect on protein adsorption; this is especially true of a nanotopography, because it reproduces the extracellular matrix (ECM), which consists of adhesion proteins [61].

Osteoblast differentiation, proliferation, and matrix production, as well as the production of local growth factors and cytokines, are all affected by surface roughness.

Surface chemistry can be used to control cell behavior in the following two ways: by hydrophobicity or surface charges, or by directly coupling receptor ligands to a surface. Hydrophobic surfaces favor protein adsorption, but also a strong unfolding of the molecules, which leads to a loss in the biological activity of the protein. For hydrophilic surfaces, the adsorption is weaker due to a strong interaction of water with the surface. Therefore, in case of wettability, the authors conclude that on flat surfaces moderate hydrophilicity is the most favorable for protein adsorption [1]. The surface treatment can be used to directly improve or prevent cell attachment to the implant surface. Surfaces modified with functional groups affect the orientation and conformation of adsorbed proteins, and thereby affect cell differentiation. A direct improvement in cell adhesion can be achieved by the covalent attachment of RGD peptides on the surface [1].

According to the literature, through severe plastic deformation (SPD) processing of titanium, changes can be made in surface characteristics such as roughness, wettability, and chemical composition. Different surface properties of nanocrystalline titanium affect the biological responses of the material. Because the successful integration of a biomaterial depends on its surface properties and its interaction with the physiological fluid, in this chapter recent advances in nanostructured titanium biomaterials are highlighted, with special focus on the impact of grain size on protein and cell–material interactions.

9.4 THE IMPACT OF GRAIN SIZE ON PROTEINS

Plasma protein adsorption is the first event that occurs when a foreign material comes into contact with foreign tissue [1,42,62]. A modification of the material's surface properties, such as wettability, chemical composition, or roughness, which are important for cell adhesion, also affects protein response [1]. There are two main methods of controlling the process of protein adsorption. The first is to select an appropriate substrate, and the second is to modify the surface of the material so as to change its chemical composition or structure [63]. Increasing the amount of albumin and fibronectin adsorbed on a titanium alloy after oxidation has been attributed to the increasing polar component of the surface energy [64]. According to the literature, defects such as dislocation that are introduced into the subsurface of biomedical alloys during manufacturing may affect the adsorption of protein on the surface of those alloys [63].

Taking that into consideration, an enhancement of the surface properties after SPD processing may also affect the protein adsorption process and conformational changes on the surface. According to the literature, the conformation of absorbed protein [65–67] yields more useful information than the amount of protein adsorbed. Diversified protein adsorption on ultrafine-grained surfaces can play a significant role in accelerating cell adhesion on those surfaces. According to the literature, adsorption on SPD materials take place in higher energy regions, such as grain boundaries, providing probable adsorption sites with partial surface coverage [42]. Researchers have found that the concentration of albumin was more than 5 times higher on nanocrystalline surfaces [42,62]. A description of a biomaterial surface having an ultrafine or nanograined structure and of its interaction with protein, makes it possible to select the desired response from the surrounding cells and tissues necessary for implant efficiency.

The authors attribute accelerated protein adsorption in ultrafine or nanocrystalline titanium mainly to the wettability and surface free energy. Nanograined titanium obtained by SPD processing exhibits a higher surface energy due to its small grain size and larger number of grain boundaries [68]. Moreover, higher surface energy materials exhibit high wettability because the resultant solid–liquid interface formed tends to lower its energy, resulting in a lower liquid contact angle [69]. This higher wettability of ultrafine-grained titanium can potentially affect the amount of protein that becomes adsorbed on the surface. Furthermore, along with the change in the grain size, the values of the Surface Free Energy components change [70]. For nanocrystalline titanium, the acid–base component (γ_s^{AB}) mainly consists of a Lewis base component (γ_s^-). The surface of nanocrystalline titanium is alkaline, and according to the Lewis acid–base theory is an electron donor that willingly reacts with acidic components such as proteins (albumin, integrin) [71]. According to the literature, the alkaline character of a surface after SPD processing is a phenomenon often observed [72].

9.5 THE IMPACT OF GRAIN SIZE ON CELL CULTURES

Grain refinement methods offer great potential for developing the microstructure and properties of titanium. By virtue of their high grain boundary length to grain size ratio and high strength to weight ratio, ultrafine-grained and nanograined materials provide the opportunity to decrease the thickness of an implant while promoting mechanical stability under physiological stresses and strains [42]. Severe plastic deformation methods significantly improve the strength characteristics of titanium to the level of titanium alloys [73,74]. One of the first works on the subject of SPD process titanium for medical applications was published in 2008 by Valiev [75]. The surface of micro- and nanocrystalline titanium was etched in hydrofluoric acid prior to cell exposure. The cell investigation shows that fibroblast colonization of the titanium surface dramatically increases after nanostructuring, suggesting more rapid osseointegration. The main difference between two analyzed samples was surface topography after etching. On nanocrystalline titanium a homogeneous and nanometer-sized roughness being apparent and for microcrystalline much coarser structure was observed.

Changes in surface topography at the nanoscale have been considered the most important factor determining the improved biocompatibility of nanocrystalline titanium [76]. The material was analyzed after equal-channel angular pressing (ECAP) in contact with mesenchymal stem cells derived from human bone marrow hMSCs (human bone marrow-derived mesenchyme stem cells). Investigations using an atomic force microscope (AFM) revealed that, after ECAP, the surface of the titanium was much more developed at the nanoscale when compared to microcrystalline titanium. This accelerated the adhesion and flattening of cells at an early stage of their growth. Similar conclusions were reached by the authors in Ref. [77], where nanocrystalline titanium obtained by ECAP was compared with a microcrystalline material and a Ti–6Al–4V alloy. In contact with murine 3T3 mouse fibroblasts, nanocrystalline titanium has been found to have a much better effect on their proliferation. Two explanations were given for these results. The first was based on the presence of nanogrooves in places of triple points. The grain boundaries are characterized by a relatively low atomic density and by the energetic instability of their atoms compared to the grain interiors. After the process of large deformation, some small voids or grooves (in the order of several nanometers) sometimes remain in the locations of the triple points as a result of incomplete diffusion compensation. In nanocrystalline and UFG materials, the number of grains is much higher, and increases the number of nanogrooves, as illustrated in Fig. 9.2. It was found that

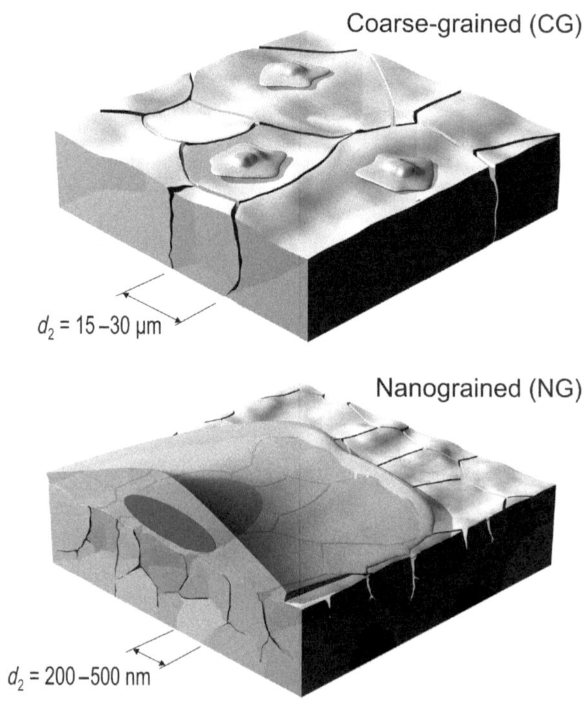

FIG. 9.2

Schematic illustration of cell adhesion on an implant surface as a function of grain size. As grain size decreases, the cells encounter more grooves and penetrate into them, contributing to stronger cell adhesion on an NG surface, whereas on a CG surface the cells seldom encounter a groove, and hence their morphology is round.

the high energy boundaries of the grains and nanogrooves filled active areas for cell adhesion. The second explanation suggested by the authors was based on an increase in surface energy after the SPD process. This is related to the increase in the share of grain boundaries, and consequently, to the surface area. High energy grain boundaries tend to release energy in every possible way. This increases the intensity of the material's interaction with the aqueous medium, leading to an increase in the wetting of nanomaterials compared to their coarse counterparts. This is due to an increase in the cell–matrix interactions of ECM proteins that mediate adsorption, which results in an increase in cell proliferation. Similar conclusions were drawn from an analysis of publication [78]. Yet another reason is that the SPD process causes a high density of surface discontinuities along the grain boundaries and junctions, possibly in the form of nanodefects. These defects could lead to the formation of focal adhesion points, which may help in the process of cell proliferation and adhesion to the substrate [69]. Recent studies have showed the nanocrystalline pure titanium improved combination of not only the in vitro bioactivity, bioaffinity, and long-term cellular functionalization in cytobiology but also the in vivo biostability and humanized bio-osseointegration in histomorphology [79].

The beneficial effect of the ECAP process on the biocompatibility of titanium can be also explained not by the effect of grain size distribution, but by changes that take place in the crystallographic texture of the material. Investigations demonstrate that the basal planes of those crystals exposed to the surface of the titanium substrate favor cell attachment. The study was conducted using different deformation paths. XPS analysis revealed differences in the chemical state of the surface of the materials tested. It was shown that, despite the presence of TiO_2 on all surfaces, they differed in the proportion of hydroxyl groups attached (−OH). According to the literature, the main factor promoting the formation of Ti–OH on the surface is the surface texture, which consists of more densely-packed basal planes that in turn enhance the cell–substrate interactions. The greater the –OH group's share of the titanium surface, the better the cellular response. It can therefore be said that a surface favoring the attachment of hydroxyl groups is responsible for the increase in cell adhesion [80].

According to the literature, there is a correlation between grain size and cell adhesion or viability, while the SPD processing of titanium improves its interaction with cells. The results obtained differ depending on how the nanostructures are obtained, the type of cells tested, the grain size obtained and the substrate material. Table 9.1 summarizes the impact of titanium grain size on cellular response.

9.6 SUMMARY

In this chapter the current knowledge of the influence of nanostructure on the biological activity of pure titanium was briefly summarized. Available literature indicated the tendency of a favorable impact of the grain refinement on the biological properties of the titanium.

Table 9.1 Summary of the Effect of SPD Processed Titanium on Cell Interaction

Substrate	SPD Process	Grain Size (nm)	Type of Cells	Impact of Grain Size	Surface Preparation	References
Ti Grade 2	Hydrostatic extrusion (HE)	87	Human bone derived cells	More advanced osteoblast colonization process	Polished	[70]
Ti Grade 2	High-pressure torsion (HPT)	~10–50	Mice preosteoblasts MC3T3-E1 Rat fibroblasts	Improved cell adhesion Higher spreading of osteoblasts	Polished	[81]
Ti Grade 2	Surface mechanical attrition (SMAT)	~60–80	Mesenchymal stem cells	Increased adhesion, cell proliferation and differentiation	Acid etched and sand-blasted	[82]
Ti Grade 2	SMAT	30	Osteoblast SaOS2	Increased adhesion	SMAT surface	[83]
Ti Grade 2	Ultrasonic shot peening (USP)	57–88	Osteoblast MG63	Increased adhesion, proliferation and differentiation	Peened	[84]
Ti Grade 2	ECAP	183	Mice fibroblast L929	Increased adhesion and accelerated proliferation	Grinded	[85]
Ti Grade 2	ECAP	240	Fibroblast 3T3	Improved fibronectin adsorption	Polished	[69]
Ti Grade 2	ECAP	200	Human mesenchymal stem cells	Improved adhesion and faster cell proliferation associated with nanotopography change	Polished	[86]

Ti Grade 2	ECAP	170–200	Human mesenchymal stem cells	Accelerated cell adhesion and cell growth	Polished	[76]
Ti Grade 2	ECAP	238	Mice fibroblast 3T3	Increased proliferation	Polished	[77,87]
Ti Grade 2	ECAP	400	Mice preosteoblasts MC3T3-E1	Improvement of cell adhesion associated with texture	—	[80]
Ti Grade 2	ECAP	280	Osteoblast MG63	Improved cell adhesion and proliferation	Blasted	[88]
Ti Grade 2	HE	~87–434	Osteoblast SaOS2	Nanostructure enhanced cellular growth on titanium Grade 2, especially after longer incubation periods	Polished	[89]
Ti Grade 4	ECAP + thermomechanical treatment	150	Fibroblasts L929	Increased colonization of cells	Etched	[75]
Ti Grade 4	ECAP	250	Fibroblasts L929, Osteoblast MG63, vascular smooth muscle cells VSMC and endothelial cells ECV304	Improved in vitro bioactivity, bioaffinity, and long-term cellular functionalization in cytobiology	Polished	[79]
Ti Grade 4	ECAP	200	Osteoblast-like U2OS	Higher cell concentration on the UFG titanium	Etched	[90]

The phenomenon of the improved biocompatibility of the pure titanium formed by SPD method is still under the discussion. The effect of the titanium grain refinement on biological properties can be explained by:

- The increasing value of the wettability and surface free energy of nanocrystalline titanium process by the SPD method.
- Changes in surface roughness, especially on the nanoscale. Increasing amount of the nanometric grooves on the nanocrystalline titanium due to the high number of the grain boundaries.
- Differences in chemical composition of the surface, especially adsorbed –OH groups on the surface.
- Changes of the crystallographic texture resulted from SPD processing of the titanium.

REFERENCES

[1] A.G. Harvey, E.W. Hill, A. Bayat, A.G. Harvey, E.W. Hill, A. Bayat, A.G. Harvey, E. W. Hill, A. Bayat, Designing implant surface topography for improved biocompatibility, Expert Rev, Med. Devices 10 (2) (2013) 257–267, https://doi.org/10.1586/erd.12.82.

[2] S. Bauer, P. Schmuki, K. Von Der Mark, J. Park, Engineering biocompatible implant surfaces. Part I: materials and surfaces, Prog. Mater. Sci. 58 (2013) 261–326, https://doi.org/10.1016/j.pmatsci.2012.09.001.

[3] M.A. Gatoo, S. Naseem, M.Y. Arfat, A.M. Dar, K. Qasim, S. Zubair, Physicochemical properties of nanomaterials: implication in associated toxic manifestations, Biomed. Res. Int. 2014 (2014) 1–9.

[4] P.H.M. Hoet, I. Brüske-hohlfeld, O.V. Salata, Nanoparticles—known and unknown health risks, J. Nanobiotechnol. 2 (2004) 1–15, https://doi.org/10.1186/1477-3155-2-12.

[5] M. Kääriäinen, C.K. Weiss, S. Ritz, S. Pütz, D.C. Cameron, V. Mailänder, K. Landfester, Zinc release from atomic layer deposited zinc oxide thin films and its antibacterial effect on *Escherichia coli*, Appl. Surf. Sci. 287 (2013) 375–380, https://doi.org/10.1016/j.apsusc.2013.09.162.

[6] C. Buzea, I.I. Pacheco, K. Robbie, Nanomaterials and nanoparticles: sources and toxicity, Biointerphases 2 (4) (2007) MR17–MR71, https://doi.org/10.1116/1.2815690.

[7] T.J. Webster, R.W. Siegel, R. Bizios, Osteoblast adhesion on nanophase ceramics, Biomaterials 20 (1999) 1221–1227.

[8] S. Oh, S. Jin, Titanium oxide nanotubes with controlled morphology for enhanced bone growth, Mater. Sci. Eng. C 26 (2006) 1301–1306, https://doi.org/10.1016/j.msec.2005.08.014.

[9] L. Prodanov, E. Lamers, M. Domanski, R. Luttge, J.A. Jansen, X.F. Walboomers, The effect of nanometric surface texture on bone contact to titanium implants in rabbit tibia, Biomaterials 34 (2013) 2920–2927, https://doi.org/10.1016/j.biomaterials.2013.01.027.

[10] K. Kubo, N. Tsukimura, F. Iwasa, T. Ueno, L. Saruwatari, H. Aita, W. Chiou, T. Ogawa, Cellular behavior on TiO_2 nanonodular structures in a micro-to-nanoscale hierarchy model, Biomaterials 30 (2009) 5319–5329, https://doi.org/10.1016/j.biomaterials.2009.06.021.

[11] L. Ponsonnet, K. Reybier, N. Jaffrezic, V. Comte, C. Lagneau, M. Lissac, C. Martelet, Relationship between surface properties (roughness, wettability) of titanium and titanium alloys and cell behaviour, Mater. Sci. Eng. C 23 (2003) 551–560, https://doi.org/10.1016/S0928-4931(03)00033-X.

[12] X. Liu, P.K. Chu, C. Ding, Surface modification of titanium, titanium alloys, and related materials for biomedical applications, Mater. Sci. Eng. R 47 (2005) 49–121, https://doi.org/10.1016/j.mser.2004.11.001.

[13] U. Müller, T. Imwinkelried, M. Horst, M. Sievers, Do human osteoblasts grow into open-porous titanium?, Eur. Cells Mater. 11 (2006) 8–15, https://doi.org/10.22203/eCM.v011a02.

[14] H.O.S. Fo, A.B. Novaes, L.M.S. De Castro, A.L. Rosa, P.T. De Oliveira, In vitro osteogenesis on a microstructured titanium surface with additional submicron-scale topography, Clin. Oral Implants Res. 18 (2007) 333–344, https://doi.org/10.1111/j.1600-0501.2006.01313.x.

[15] C. Galli, M.C. Coen, R. Hauert, V.L. Katanaev, M.P. Wymann, Protein adsorption on topographically nanostructured titanium, Surf. Sci. Lett. 474 (2001) L180–L184.

[16] C. Galli, M.C. Coen, R. Hauert, V.L. Katanaev, P. Gro, Creation of nanostructures to study the topographical dependency of protein adsorption, Colloids Surf. B Biointerfaces 26 (2002) 255–267.

[17] I. Shishkovsky, Y. Morozov, I. Smurov, Nanofractal surface structure under laser sintering of titanium and nitinol for bone tissue engineering, Appl. Surf. Sci. 254 (2007) 1145–1149, https://doi.org/10.1016/j.apsusc.2007.09.021.

[18] B.C. Ward, T.J. Webster, Increased functions of osteoblasts on nanophase metals, Mater. Sci. Eng. C 27 (2007) 575–578, https://doi.org/10.1016/j.msec.2006.05.032.

[19] T.J. Webster, J.U. Ejiofor, Increased osteoblast adhesion on nanophase metals: Ti, Ti6Al4V and CoCrMo, Biomaterials 25 (2004) 4731–4739, https://doi.org/10.1016/j.biomaterials.2003.12.002.

[20] R.A. Gittens, R. Olivares-Navarrete, Z. Schwartz, B.D. Boyan, Implant osseointegration and the role of microroughness and nanostructures: lessons for spine implants, Acta Biomater. 10 (2014) 3363–3371, https://doi.org/10.1016/j.actbio.2014.03.037.

[21] R.A. Gittens, T. McLachlan, R. Olivares-Navarrete, Y. Cai, S. Berner, R. Tannenbaum, Z. Schwartz, K.H. Sandhage, B.D. Boyan, The effects of combined micron-/submicron-scale surface roughness and nanoscale features on cell proliferation and differentiation, Biomaterials 32 (2011) 3395–3403, https://doi.org/10.1016/j.biomaterials.2011.01.029.

[22] R.A. Gittens, L. Scheideler, F. Rupp, S.L. Hyzy, J. Geis-Gerstorfer, Z. Schwartz, B. D. Boyan, A review on the wettability of dental implant surfaces II: biological and clinical aspects, Acta Biomater. 10 (2014) 2907–2918, https://doi.org/10.1016/j.actbio.2014.03.032.

[23] D. Kuczyńska, P. Kwaśniak, J. Marczak, J. Bonarski, J. Smolik, H. Garbacz, Laser surface treatment and the resultant hierarchical topography of Ti grade 2 for biomedical application, Appl. Surf. Sci. 390 (2016) 560–569, https://doi.org/10.1016/j.apsusc.2016.08.105.

[24] D. Kuczyńska, P. Kwaśniak, M. Pisarek, P. Borowicz, H. Garbacz, Influence of surface pattern on the biological properties of Ti grade 2, Mater. Charact. 135 (2018) 337–347 https://doi.org/10.1016/j.matchar.2017.09.024.

[25] P. Kwaśniak, J. Pura, M. Zwolińska, P. Wieciński, H. Skarżyński, L. Olszewski, J. Marczak, H. Garbacz, K.J. Kurzydłowski, Laser and chemical surface modifications of titanium grade 2 for medical application, Appl. Surf. Sci. 336 (2015) 267–273, https://doi.org/10.1016/j.apsusc.2014.11.178.

[26] L. Lai, D.R. Tyson, R.V. Clayman, J.C. Earthman, Encrustation of nanostructured Ti in a simulated urinary tract environment, Mater. Sci. Eng. C 28 (2008) 460–464, https://doi.org/10.1016/j.msec.2007.04.019.
[27] J.E. Sweitzer, G.J. Shiflet, J.R. Scully, Localized corrosion of Al90Fe5Gd5 and Al87Ni8.7Y4.3 alloys in the amorphous, nanocrystalline and crystalline states: resistance to micrometer-scale pit formation, Electrochim. Acta 48 (2003) 1223–1234, https://doi.org/10.1016/S0013-4686(02)00830-7.
[28] K. Wang, The use of titanium for medical applications in the USA, Mater. Sci. Eng. A 213 (1996) 8–11, https://doi.org/10.1016/0921-5093(96)10243-4.
[29] H.J. Rack, J.I. Qazi, Titanium alloys for biomedical applications, Mater. Sci. Eng. C 26 (2006) 1269–1277, https://doi.org/10.1016/j.msec.2005.08.032.
[30] Y. Okazaki, S. Rao, Y. Ito, T. Tateishi, Corrosion resistance, mechanical properties, corrosion fatigue strength and cytocompatibility of new Ti alloys without Al and V, Biomaterials 19 (1998) 1197–1215.
[31] D. Banerjee, J.C. Williams, Perspectives on titanium science and technology, Acta Mater. 61 (2013) 844–879, https://doi.org/10.1016/j.actamat.2012.10.043.
[32] H. Mora-Sanchez, I. Sabirov, M.A. Monclus, E. Matykina, J.M. Molina-Aldareguia, Ultra-fine grained pure titanium for biomedical applications ultra-fine grained pure titanium for biomedical applications, Mater. Technol. 31 (2016) 756–771, https://doi.org/10.1080/10667857.2016.1238131.
[33] M. Geetha, A.K. Singh, R. Asokamani, A.K. Gogia, Ti based biomaterials, the ultimate choice for orthopaedic implants—a review, Prog. Mater. Sci. 54 (2009) 397–425, https://doi.org/10.1016/j.pmatsci.2008.06.004.
[34] M. Niinomi, Mechanical biocompatibilities of titanium alloys for biomedical applications, J. Mech. Behav. Biomed. Mater. 1 (2008) 30–42, https://doi.org/10.1016/j.jmbbm.2007.07.001.
[35] D. Deligianni, N. Katsala, S. Ladas, D. Sotiropoulou, J. Amedee, Y. Missirlis, Effect of surface roughness of the titanium alloy Ti-6Al-4V on human bone marrow cell response and on protein adsorption, Biomaterials 22 (2001) 1241–1251.
[36] S.E. Kim, J.H. Lim, S.C. Lee, S.C. Nam, H.G. Kang, J. Choi, Anodically nanostructured titanium oxides for implant applications, Electrochim. Acta 53 (2008) 4846–4851, https://doi.org/10.1016/j.electacta.2008.02.005.
[37] T. Xu, N. Zhang, H.L. Nichols, D. Shi, X. Wen, Modification of nanostructured materials for biomedical applications, Mater. Sci. Eng. C 27 (2007) 579–594, https://doi.org/10.1016/j.msec.2006.05.029.
[38] W.T. Huo, L.Z. Zhao, S. Yu, Z.T. Yu, P.X. Zhang, Y.S. Zhang, Significantly enhanced osteoblast response to nano-grained pure tantalum, Sci. Rep. 7 (2017) 40868, https://doi.org/10.1038/srep40868.
[39] S. Guo, Q. Meng, X. Zhao, Q. Wei, H. Xu, Design and fabrication of a metastable β-type titanium alloy with ultralow elastic modulus and high strength, Sci. Rep. 5 (2015) 1–8, https://doi.org/10.1038/srep14688.
[40] Y. Li, C. Yang, H. Zhao, S. Qu, X. Li, Y. Li, New developments of Ti-based alloys for biomedical applications, Materials (Basel) 7 (2014) 1709–1800, https://doi.org/10.3390/ma7031709.
[41] T. Akahori, M. Niinomi, Y. Koyanagi, T. Kasuga, H. Toda, H. Fukui, M. Ogawa, Mechanical properties of biocompatible beta-type titanium alloy coated with calcium phosphate invert glass-ceramic layer, Mater. Trans. 46 (2005) 1564–1569.

[42] K.C. Nune, R.D.K. Misra, Biological activity of nanostructured metallic materials for biomedical applications, Mater. Technol. Adv. Perform. Mater. 31 (2016) 772–781, https://doi.org/10.1080/10667857.2016.1225148.

[43] B.R. Sunil, A. Thirugnanam, U. Chakkingal, Nano and ultra fine grained metallic biomaterials by severe plastic deformation techniques, Mater. Technol. 31 (2016) 743–755, https://doi.org/10.1080/10667857.2016.1249133.

[44] M. Zwolińska, A. Sotniczuk, K. Topolski, A. Królikowski, H. Garbacz, The influence of deformation the corrosion resistance of titanium grade2, Solid State Phenom. 227 (2015) 27–30, https://doi.org/10.4028/www.scientific.net/SSP.227.27.

[45] H. Garbacz, M. Pisarek, K.J. Kurzydłowski, Corrosion resistance of nanostructured titanium, Biomol. Eng. 24 (2007) 559–563, https://doi.org/10.1016/j.bioeng.2007.08.007.

[46] P. Kwasniak, M. Muzyk, H. Garbacz, K.J. Kurzydlowski, Influence of oxygen content on the mechanical properties of hexagonal Ti—first principles calculations, Mater. Sci. Eng. A 590 (2014) 74–79, https://doi.org/10.1016/j.msea.2013.10.004.

[47] P. Kwasniak, M. Muzyk, H. Garbacz, K.J. Kurzydlowski, Influence of C, H, N, and O interstitial atoms on deformation mechanism in titanium—first principles calculations of generalized stacking fault energy, Mater. Lett. 94 (2013) 92–94, https://doi.org/10.1016/j.matlet.2012.12.002.

[48] S. Ozan, J. Lin, Y. Li, C. Wen, New Ti-Ta-Zr-Nb alloys with ultrahigh strength for potential orthopedic implant applications, J. Mech. Behav. Biomed. Mater. 75 (2017) 119–127, https://doi.org/10.1016/j.jmbbm.2017.07.011.

[49] J. Stráský, P. Harcuba, K. Václavová, K. Horváth, M. Landa, Increasing strength of a biomedical Ti-Nb-Ta-Zr alloy by alloying with Fe, Si and O, J. Mech. Behav. Biomed. Mater. 71 (2017) 329–336, https://doi.org/10.1016/j.jmbbm.2017.03.026.

[50] T. Akahori, M. Niinomi, M. Nakai, T. Kasuga, M. Ogawa, Characteristics of biomedical beta-type titanium alloy subjected to coating, Mater. Trans. 49 (2008) 365–371, https://doi.org/10.2320/matertrans.MRA2007207.

[51] K.Y. Xie, Y. Wang, Y. Zhao, L. Chang, G. Wang, Z. Chen, Y. Cao, X. Liao, E.J. Lavernia, R.Z. Valiev, B. Sarrafpour, H. Zoellner, S.P. Ringer, Nanocrystalline ß-Ti alloy with high hardness, low Young's modulus and excellent in vitro biocompatibility for biomedical applications, Mater. Sci. Eng. C 33 (2013) 3530–3536, https://doi.org/10.1016/j.msec.2013.04.044.

[52] T. Albrektsson, P. Brånemark, H. Hansson, J. Lindström, Osseointegrated titanium implants: requirements for ensuring a long-lasting, direct bone-to-implant anchorage in man, Acta Orthop. Scand. 52 (1981) 155–170, https://doi.org/10.3109/17453678108991776.

[53] K. Cai, J. Bossert, K.D. Jandt, Does the nanometre scale topography of titanium influence protein adsorption and cell proliferation?, Colloids Surf B Biointerfaces 49 (2006) 136–144, https://doi.org/10.1016/j.colsurfb.2006.02.016.

[54] D.A. Puleo, A. Nanci, Understanding and controlling the bone-implant interface, Biomaterials 20 (1999) 2311–2321.

[55] D. Costa, P. Garrain, M. Baaden, Understanding small biomolecule-biomaterial interactions: a review of fundamental theoretical and experimental approaches for biomolecule interactions with inorganic surfaces, J. Biomed. Mater. Res. A 101A (2012) 1210–1222, https://doi.org/10.1002/jbm.a.34416.

[56] X.F. Walboomers, J.A. Jansen, Cell and tissue behavior on micro-grooved surfaces, Odontology 89 (2001) 2–11, https://doi.org/10.1007/s10266-001-8178-z.

[57] V. Dumas, A. Rattner, L. Vico, E. Audouard, J.C. Dumas, P. Naisson, P. Bertrand, Multi-scale grooved titanium processed with femtosecond laser influences mesenchymal stem cell morphology, adhesion, and matrix organization, J. Biomed. Mater. Res. A 100A (2012) 3108–3116, https://doi.org/10.1002/jbm.a.34239.
[58] K.A. Kilian, B. Bugarija, B.T. Lahn, M. Mrksich, Geometric cues for directing the differentiation of mesenchymal stem cells, Proc. Natl. Acad. Sci. U.S.A. 107 (2010) 4872–4877, https://doi.org/10.1073/pnas.0903269107.
[59] E. Martinez, E. Engel, J.A. Planell, J. Samitier, Effects of artificial micro- and nano-structured surfaces on cell behaviour, Ann. Anat. 191 (2009) 126–135, https://doi.org/10.1016/j.aanat.2008.05.006.
[60] E.K.F. Yim, S.W. Pang, K.W. Leong, Synthetic nanostructures inducing differentiation of human mesenchymal stem cells into neuronal lineage, Exp. Cell Res. 313 (2007) 1820–1829, https://doi.org/10.1016/j.yexcr.2007.02.031.
[61] P. Solar, O. Kylian, A. Marek, M. Vandrovcova, L. Bacakova, J. Hanus, J. Vyskocil, D. Slavinska, H. Biederman, Particles induced surface nanoroughness of titanium surface and its influence on adhesion of osteoblast-like MG-63 cells, Appl. Surf. Sci. 324 (2015) 99–105, https://doi.org/10.1016/j.apsusc.2014.10.082.
[62] C. Nune, R.D.K. Misra, M.C. Somani, L.P. Karjalainen, Dependence of cellular activity at protein adsorbed biointerfaces with nano- to microscale dimensionality, J. Biomed. Mater. Res. A 102 (2013) 1663–1676, https://doi.org/10.1002/jbm.a.34831.
[63] Z. Wang, Y. Yan, L. Qiao, Protein adsorption on implant metals with various deformed surfaces, Colloids Surf. B Biointerfaces 156 (2017) 62–70, https://doi.org/10.1016/j.colsurfb.2017.05.015.
[64] A. Michiardi, C. Aparicio, B.D. Ratner, J.A. Planell, J. Gil, The influence of surface energy on competitive protein adsorption on oxidized NiTi surfaces, Biomaterials 28 (2007) 586–594, https://doi.org/10.1016/j.biomaterials.2006.09.040.
[65] B. Sivaraman, K.P. Fears, R.A. Latour, Investigation of the effects of surface chemistry and solution concentration on the conformation of adsorbed proteins using an improved circular dichroism method, Langmuir 25 (2009) 3050–3056, https://doi.org/10.1021/la8036814.
[66] D.M. Hylton, S.W. Shalaby, R.A. Latour, Direct correlation between adsorption-induced changes in protein structure and platelet adhesion, J. Biomed. Mater. Res. A 73 (2005) 349–358, https://doi.org/10.1002/jbm.a.30295.
[67] S.N. Rodrigues, I.C. Gonçalves, M.C.L. Martins, M.A. Barbosa, B.D. Ratner, Fibrinogen adsorption, platelet adhesion and activation on mixed hydroxyl-/methyl-terminated self-assembled monolayers, Biomaterials 27 (2006) 5357–5367, https://doi.org/10.1016/j.biomaterials.2006.06.010.
[68] P. Jojibabu, B. Ratna Sunil, T.S. Sampath Kumar, U. Chakkingal, V. Nandakumar, M. Doble, Wettability and in vitro bioactivity studies on titanium rods processed by equal channel angular pressing, Trans. Indian Inst. Metals 66 (2013) 299–304, https://doi.org/10.1007/s12666-013-0281-7.
[69] S. Bindu, K.P. Sanosh, K. Smetana, A. Balakrishnan, T.N. Kim, An in vivo evaluation of ultra-fine grained titanium implants, J. Mater. Sci. Technol. 25 (2009) 556–560.
[70] H. Garbacz, Mikrostruktura i właściwości nanokrystalicznego tytanu, Warsaw University of Technology, 2010.
[71] E. Gongadze, D. Kabaso, S. Bauer, J. Park, P. Schmuki, A. Iglič, Adhesion of osteoblasts to a vertically aligned TiO2 nanotube surface, Mini-Rev. Med. Chem. 13 (2013) 194–200.

[72] V.K. Truong, S. Rundell, R. Lapovok, Y. Estrin, J.Y. Wang, C.C. Berndt, D.G. Barnes, C.J. Fluke, R.J. Crawford, E.P. Ivanova, Effect of ultrafine-grained titanium surfaces on adhesion of bacteria, Appl. Microb. Cell. Physiol. 83 (2009) 925–937, https://doi.org/10.1007/s00253-009-1944-5.
[73] H. Garbacz, M. Lewandowska, W. Pachla, K.J. Kurzydłowski, Structural and mechanical properties of nanocrystalline titanium and 316LVM steel processed by hydrostatic extrusion, J. Microsc. 223 (2006) 272–274, https://doi.org/10.1111/j.1365-2818.2006.01646.x.
[74] K. Topolski, H. Garbacz, W. Pachla, K.J. Kurzydlowski, Bulk nanostructured titanium fabricated by hydrostatic extrusion, Phys. Status Solidi C 7 (2010) 1391–1394, https://doi.org/10.1002/pssc.200983396.
[75] R.Z. Valiev, I.P. Semenova, E. Jakushina, V.V. Latysh, H.J. Rack, T.C. Lowe, et al., Nanostructured SPD processed titanium for medical implants, Mater. Sci. Forum. 584–586 (2008) 49–54 https://doi.org/10.4028/www.scientific.net/MSF.584-586.49.
[76] Y. Estrin, E.P. Ivanova, A. Michalska, V.K. Truong, R. Lapovok, R. Boyd, Accelerated stem cell attachment to ultrafine grained titanium, Acta Biomater. 7 (2011) 900–906, https://doi.org/10.1016/j.actbio.2010.09.033.
[77] T.N. Kim, A. Balakrishnan, B.C. Lee, W.S. Kim, K. Smetana, J.K. Park, B.B. Panigrahi, In vitro biocompatibility of equal channel angular processed (ECAP) titanium, Biomed. Mater. 2 (2007) S117–S120, https://doi.org/10.1088/1748-6041/2/3/S06.
[78] J.W. Park, Y.J. Kim, C.H. Park, D.H. Lee, Y.G. Ko, J.H. Jang, C.S. Lee, Enhanced osteoblast response to an equal channel angular pressing-processed pure titanium substrate with microrough surface topography, Acta Biomater. 5 (2009) 3272–3280, https://doi.org/10.1016/j.actbio.2009.04.038.
[79] F.L. Nie, Y.F. Zheng, S.C. Wei, D.S. Wang, Z.T. Yu, G.K. Salimgareeva, A.V. Polyakov, R.Z. Valiev, In vitro and in vivo studies on nanocrystalline Ti fabricated by equal channel angular pressing with microcrystalline CP Ti as control, J. Biomed. Mater. Res. A 101A (2013) 1694–1707, https://doi.org/10.1002/jbm.a.34472.
[80] M. Hoseini, P. Bocher, A. Shahryari, F. Azari, J.A. Szpunar, H. Vali, On the importance of crystallographic texture in the biocompatibility of titanium based substrate, J. Biomed. Mater. Res. A 102 (2014) 3631–3638, https://doi.org/10.1002/jbm.a.35028.
[81] S. Faghihi, F. Azari, A.P. Zhilyaev, J.A. Szpunar, H. Vali, M. Tabrizian, Cellular and molecular interactions between MC3T3-E1 pre-osteoblasts and nanostructured titanium produced by high-pressure torsion, Biomaterials 28 (2007) 3887–3895, https://doi.org/10.1016/j.biomaterials.2007.05.010.
[82] S. Faghihi, H. Vali, M. Tabrizian, Effects of crystal size and orientation of substrates on cell adhesion: implication for medical implants, Int. J. Mod. Phys. B 22 (2008) 3069–3081.
[83] M. Lai, K. Cai, Y. Hu, X. Yang, Q. Liu, Regulation of the behaviors of mesenchymal stem cells by surface nanostructured titanium, Colloids Surf. B Biointerfaces 97 (2012) 211–220, https://doi.org/10.1016/j.colsurfb.2012.04.029.
[84] M. Wen, C. Wen, P. Hodgson, Y. Li, Improvement of the biomedical properties of titanium using SMAT and thermal oxidation, Colloids Surf. B Biointerfaces 116 (2014) 658–665, https://doi.org/10.1016/j.colsurfb.2013.10.039.
[85] H. Maleki-Ghaleh, K. Hajizadeh, A. Hadjizadeh, M.S. Shakeri, S. Ghobadi Alamdari, S. Masoudfar, E. Aghaie, M. Javidi, J. Zdunek, K.J. Kurzydlowski, Electrochemical and cellular behavior of ultrafine-grained titanium in vitro, Mater. Sci. Eng. C 39 (2014) 299–304, https://doi.org/10.1016/j.msec.2014.03.001.

[86] Y. Estrin, C. Kasper, S. Diederichs, R. Lapovok, Accelerated growth of preosteoblastic cells on ultrafine grained titanium, J. Biomed. Mater. Res. A 90 (2009) 1239–1242, https://doi.org/10.1002/jbm.a.32174.

[87] T.N. Kim, A. Balakrishnan, B.C. Lee, W.S. Kim, B. Dvorankova, K. Smetana, J.K. Park, B.B. Panigrahi, In vitro fibroblast response to ultra fine grained titanium produced by a severe plastic deformation process, J. Mater. Sci. Mater. Med. 19 (2008) 553–557, https://doi.org/10.1007/s10856-007-3204-5.

[88] C.Y. Zheng, F.L. Nie, Y.F. Zheng, Y. Cheng, S.C. Wei, R.Z. Valiev, Enhanced in vitro biocompatibility of ultrafine-grained titanium with hierarchical porous surface, Appl. Surf. Sci. 257 (2011) 5634–5640, https://doi.org/10.1016/j.apsusc.2011.01.062.

[89] D. Kubacka, A. Yamamoto, P. Wiecinski, H. Garbacz, Biological behavior of titanium processed by severe plastic deformation, Appl. Surf. Sci. (2018), https://doi.org/10.1016/j.apsusc.2018.04.120.

[90] A.V. Polyakov, I.P. Semenova, R.Z. Valiev, High fatigue strength and enhanced biocompatibility of UFG CP Ti for medical innovative applications, IOP Conf. Ser. Mater. Sci. Eng. 63 (2014) 0–6, https://doi.org/10.1088/1757-899X/63/1/012113.

CHAPTER

Tribology 10

Halina Garbacz*, Maciej Motyka[†]

Faculty of Materials Science and Engineering, Warsaw University of Technology, Warsaw, Poland Faculty of Mechanical Engineering and Aeronautics, Rzeszow University of Technology, Rzeszow, Poland*[†]

10.1 INTRODUCTION

Bulk materials cannot be considered totally independently of their surface properties. Tribology is an important interdisciplinary field which involves the design of components with static and dynamic contacts for a required performance and reliability. It is believed that nanostructuring of metallic materials (e.g. by severe plastic deformation (SPD) methods), causing significant increase of their hardness, should also lead to improvements in the wear resistance. However, the problem is more complex, in that a material's wear resistance depends on many tribological factors such as the material of counter specimen, loading mode, and environment (lubricants).

There are three wear mechanisms that are considered (Fig. 10.1) [1,2]:

- Elastic deformation: after deformation, protrusions on the surface return to their original shape. This phenomenon may lead to the nucleation and growth of fatigue cracks;
- Protrusion shearing: this occurs when the shear stress concentration is high enough to nucleate and propagate cracking;
- Plastic deformation: protrusions change their shape, due to local stress concentration. Repeated deformation leads to strain accumulation, local strengthening, and may cause further brittle cracking. Plastic deformation also leads to the smoothing/flattening of the surface, which increases risk of adhesive wear.

It is generally accepted to consider the following wear types (Fig. 10.2) [1,2]:

- Adhesive wear: this process occurs when atomic bonding forces occurring between the materials are stronger than the strength of one of the materials. Most of removed material stays attached to surface of the other one;
- Abrasive wear: often occurs when a hard rough surface slides across a softer one. Process of material displacement or removal through a cutting—occurs when a

Nanocrystalline Titanium. https://doi.org/10.1016/B978-0-12-814599-9.00010-9

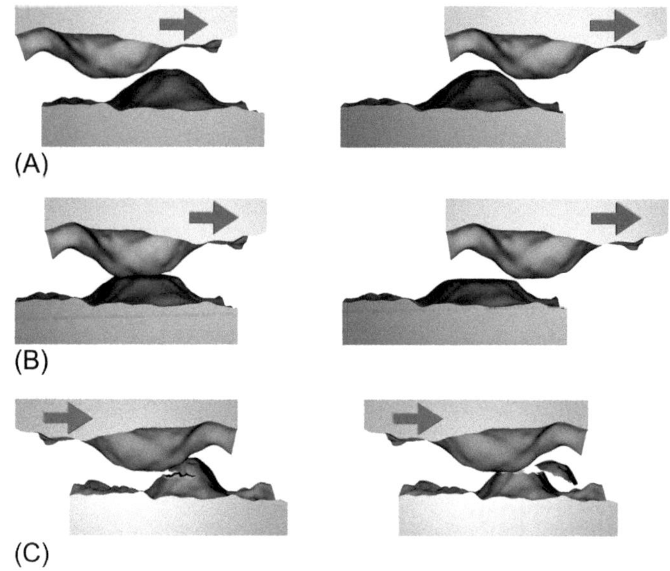

FIG. 10.1

Diagrams of wear mechanisms: (A) elastic deformation, (B) plastic deformation, (C) protrusion shearing.

material separates from a surface in a form of chips, or debris—of plowing—occurs when there is material displacement to the sideways;

- Fatigue wear: either through pitting—fatigue failure of the material due to nucleation and growth of the subsurface cracks (common wear process of ball bearings), or spalling—a fatigue failure of the material due to nucleation and growth of the surface cracks (common wear process of railways).

Generally, the investigations of wear behavior include measurement of friction coefficient and wear loss (volume, mass/weight, linear wear), wear rate, examination of wear surface and its morphology by various analytical techniques, and studies of the microstructure and mechanical properties. Many authors in their research also highlight the significant role of elastic properties and determine the ratio of hardness and modulus of elasticity (H/E or H^3/E^2) as a parameter by which tribological wear can be better predicted.

10.2 WEAR RESISTANCE

Tribological properties of nanomaterials have not been described comprehensively in the literature. Usually the nanolayers that are produced by various surface treatment are characterized in tribology [3–11]. The analysis of grain refinement effect on wear and friction coefficient of nanocrystalline (NC) layers (Table 10.1) indicates

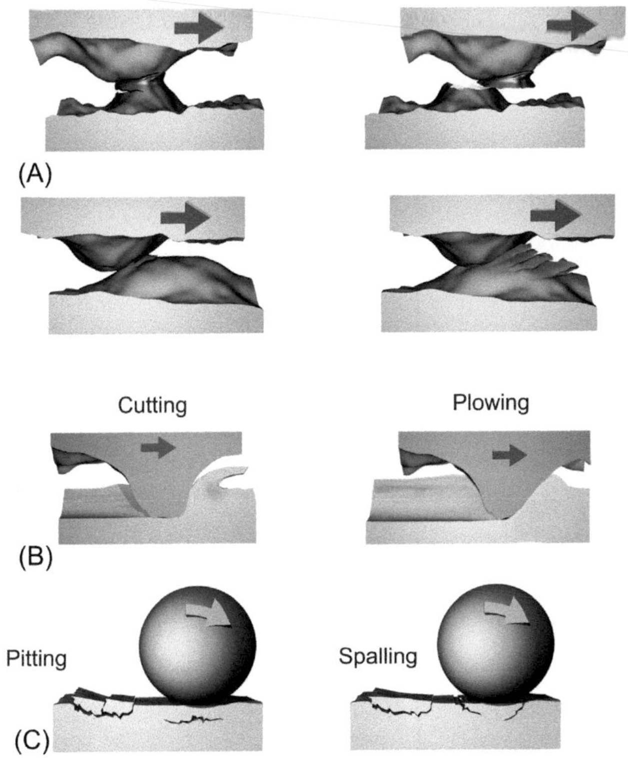

FIG. 10.2

Diagrams of wear types: (A) adhesive wear, (B) abrasive wear, (C) fatigue wear.

Table 10.1 Tribological Properties of NC Metals and Layers Compared to Coarse-Grained Ones

Material	Production Method	Tribological Test Parameters: Counter Specimen/ Environment	Wear Resistance	Friction Coefficient
Layers				
Low-carbon steel [3]	Surface mechanical attrition (SMAT)	Diamond needle/ dry	↑	↓
Al [4]	Magnetron sputtering	Steel arbor/dry	↑	↓
Co [5]	Electrochemical deposition	Steel ball/dry	↑	↓
Cu [6]	SMAT	Steel ball/dry	↑	↓

Continued

Table 10.1 Tribological Properties of NC Metals and Layers Compared to Coarse-Grained Ones —cont'd

Material	Production Method	Tribological Test Paramcters: Counter Specimen/ Environment	Wear Resistance	Friction Coefficient
Ni	Electrochemical deposition	Steel ball/dry [5]	↑	↓
		Al_2O_3 particle flux/ dry and wet [7]	↑	Const.
		Steel ball/dry [8]	Const.	↓
Ni [8]	Shear extrusion	Steel ball/fretting	No data	↓
Carbon steel [9]	Mechanical-pulse treatment	Steel insert/oil with fine quartz sand	↑	↓
Ti and Ti–6Al–4V [10]	Ultrasonic technique	Si_3N_4 ball/dry (fretting test)	↑	↓
Ti [11]	SMAT	Steel ball	↑	↓
Bulk materials				
Ti [12]	Shear extrusion	High frequency reciprocating rig (HFRR)	Const.	Const.
Ti [13]	Equal-channel angular pressing (ECAP)	Steel ball/dry	↑	Const.
Ti [14]	Large-strain machining (LSM)	Ball on disc AISI 52100/dry	↑	↑
Ti [15]	ECAP	Ball on disc counterpart: WCo8	No data	↓
Ti [16]	ECAP	Steel disc/dry	↑	No data
Ti [17]	High-pressure torsion (HPT)	Alumina ball/PBS	↑	No data
Ti [18]	HE	Steel disc/dry, NaCl, paraffin	↓	Const.
Ti [19]	HPT	Micro-scratch testing	↑	Const.
Cu [20]	LSM	Steel ball/dry	↑	Const.
Cu [20]	Large-strain extrusion machining (LSEM)	Steel pin/dry	↑	No data

that they mainly exhibit higher wear resistance when compared to microcrystalline layers.

On the basis of Archard's wear model, hardness increase that is caused by grain boundary strengthening should favorably impact on wear resistance [21]. It is expressed by the equation:

$$Q = K\frac{LN}{H} \tag{10.1}$$

where: Q—quantity of wear, K—coefficient of wear, L—friction distance, N—pressure used, and H—hardness.

Significant influence on tribological wear, beside hardness, is surface roughness [5]. The topography of any NC layers surface depends on surface treatment condition. Wear mechanisms of materials are determined by their crystal structure and phase composition and friction conditions. Wang and coauthors [5] evaluated the effect of grain size on wear resistance of NC layers on Ni and Co substrates with similar hardness, produced by the electrodeposition method. It was indicated that the value of friction coefficient for Ni decreased with an increase in surface hardness, caused by grain refining. Nevertheless, NC layers on a Co substrate were characterized by lower friction coefficient than that produced on the Ni substrate. Better wear resistance of NC layers on Co substrate was justified by higher adhesive wear resistance of hexagonal close packed (hcp) metals comparing to face centered cubic (fcc) Ni.

The reference materials considered when evaluating the tribological properties of NC layers are the same ones, but in a coarse-grained state. Usually the effect of grain size on wear resistance is analyzed, whereas other meaningful microstructural features of layers developed by surface treatment methods, distinguished them from bulk materials, are omitted. These are: grain shape, microstructural uniformity (grain size distribution), and the state of stress caused by technological process. The complex evaluation of the effect of material nanostructure on its tribological properties is not possible without taking into consideration all factors mentioned earlier. Determined wear resistance depends also on various additional factors: the examination method used, properties of counterspecimen's material, and the test conditions (load, friction distance, temperature, and environment). Due to very little literature data on wear of bulk NC materials, this problem needs to be much more extensively analyzed.

The influence of grain size on the tribological properties of titanium has been investigated (Table 10.1). The majority of presented results refer to the wear resistance of ultrafine-grained (UFG) titanium [12–19]. The examinations of NC titanium obtained by hydrostatic extrusion (HE) were performed using friction wear testing apparatus T-11 in a pin-on-disc configuration [22]. NC and microcrystalline (MC) titanium in the form of cylinders that were 3 mm in diameter were studied. The X17CrNi16-2 steel was used as the counterspecimens (hardness of discs surface—5.8 GPa) [23]. Surfaces in contact were previously grinded—the average roughness R_a of titanium sample's surface was 0.1 μm. For comparison, the ultra-high-molecular weight polyethylene (UHMWPE) discs were also used. The wear tests were carried out in the following conditions: normal loads in the range of 2–15 N (pressures 0.28, 0.75 and 2.0 MPa), friction distance 540 m and linear velocity of the pin 0.15 m/s. Tests were performed in conditions of dry friction and with addition of 0.9% water solution of NaCl or paraffin oil.

The performed examinations determined the change of friction coefficient value (μ), total linear wear Δl of tribological pairs (the sum of height loss of titanium pin and depth of wear path on the disc surface), and mass loss Δm of pins after a whole friction cycle. Surfaces in contact, after friction wear tests, were observed by the scanning electron microscopy (SEM) method (Hitachi S-3500 N equipped with EDS detector).

The results obtained for tribological pairs titanium pin/steel disc in dry friction tests (Fig. 10.3) confirmed the low wear resistance of titanium. An expected

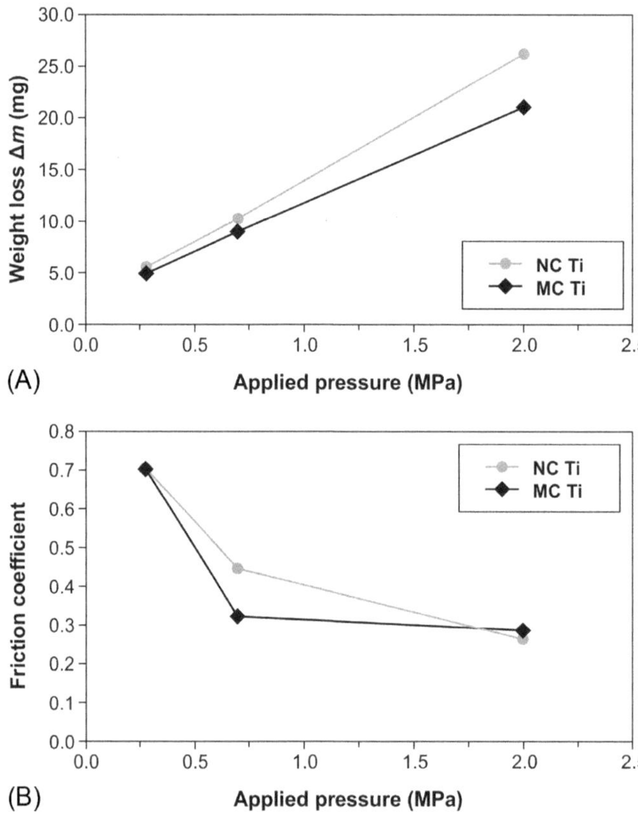

FIG. 10.3

Results of tribological examinations in dry friction condition using steel disc as a counter specimen: (A) weight loss Δm, (B) friction coefficient μ.

improvement of tribological properties was not found in NC titanium produced by the HE method. On the contrary, wear of NC titanium pins was higher than for MC ones—in the whole range of loads used. Similar results were obtained by Budinski [24] for pure titanium and Ti–6Al–4V alloy. This phenomenon can be explained on the basis of analysis of the wear mechanism occurring during the friction process.

Microscopic observations of wear paths confirmed that the friction process between examined surfaces is related to following phenomena: adhesion, transport of materials between titanium pin and steel disc, tribochemical reaction, and abrasive wear (Figs. 10.4–10.6). Observations of the surface of NC titanium pins showed that during the tests, the material "flows" from the contact surface and it breaks down chipping (Figs. 10.4 and 10.5). During tests performed the wear process of titanium pin dominated and its parts were transported onto the surface disc (Fig. 10.6). Tribochemical reactions between contact surfaces caused the formation of oxide layers, torn off during the friction wear tests. In the consequence of such processes, highly dispersed hard wear's products appeared in the surface contact zone and caused abrasive wear of titanium surface. This wear mechanism was confirmed by EDS analysis

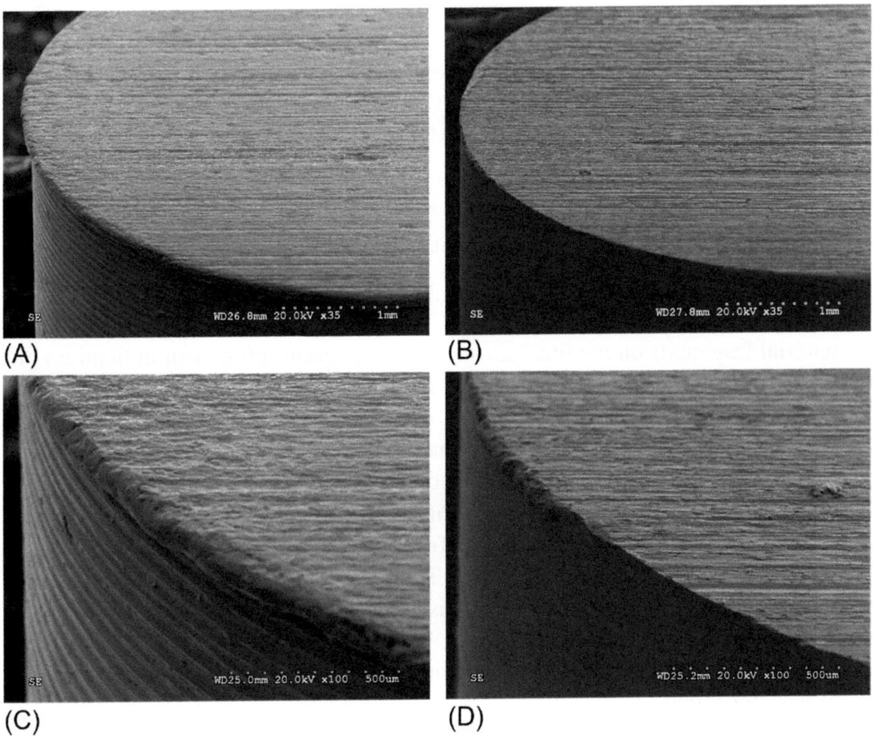

FIG. 10.4

Surface topography (SEM) of MC (A, C) and NC (B, D) titanium pins after dry friction process vs steel disc.

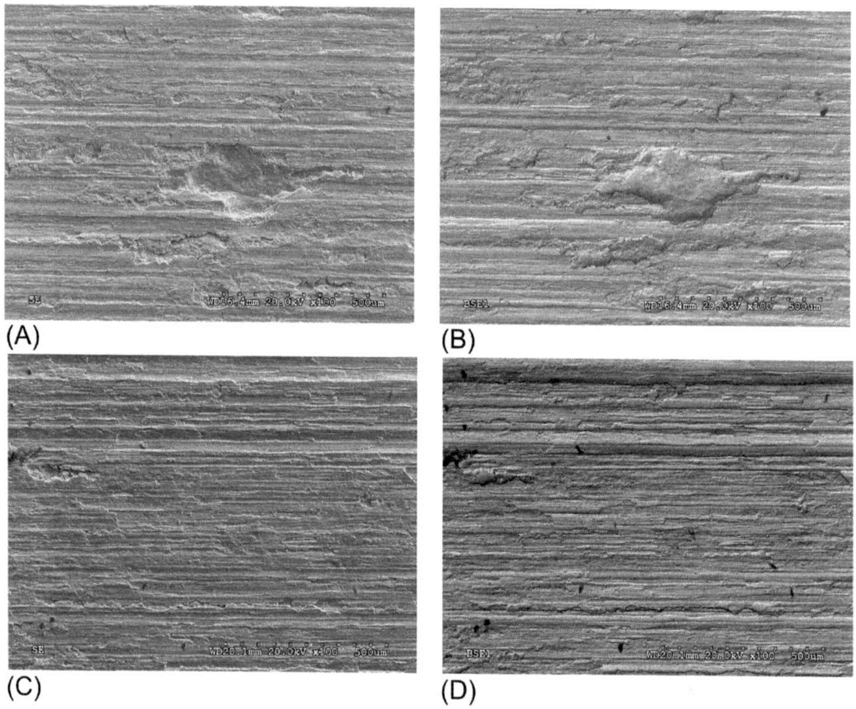

FIG. 10.5

Surface topography of NC (A, B) and MC (C, D) titanium pins after dry friction process vs steel disc—(A, C) SEM/SE, (B, D) SEM/BSE observation methods.

of material fragments on the titanium surface [22]. Grain refinement in titanium did not improve its wear resistance in dry friction condition. Whereas the presence of internal stresses and crystal defects in NC titanium could be a reason for intensified processes of oxide layers formation and surfaces tacking, therefore giving a decreased wear resistance in the examined material.

On the basis of chemical composition of wear's products from NC titanium pins it was found that the mechanism of such friction processes are complex. Among a wear's products, fine particles containing Ti and Fe formed as a result of tribochemical reactions. The presence of numerous large particles of pure Ti and isolated chips—products of the micro-machining process—was observed. Any qualitative difference between the wear's products from dry friction process of NC and MC titanium was not found.

Low tribological wear resistance of pure titanium is confirmed by the literature data. The rate of wear seems to be determined by oxidation reactions and deformation of surface layer during the friction wear test [4,5]. Titanium is characterized by

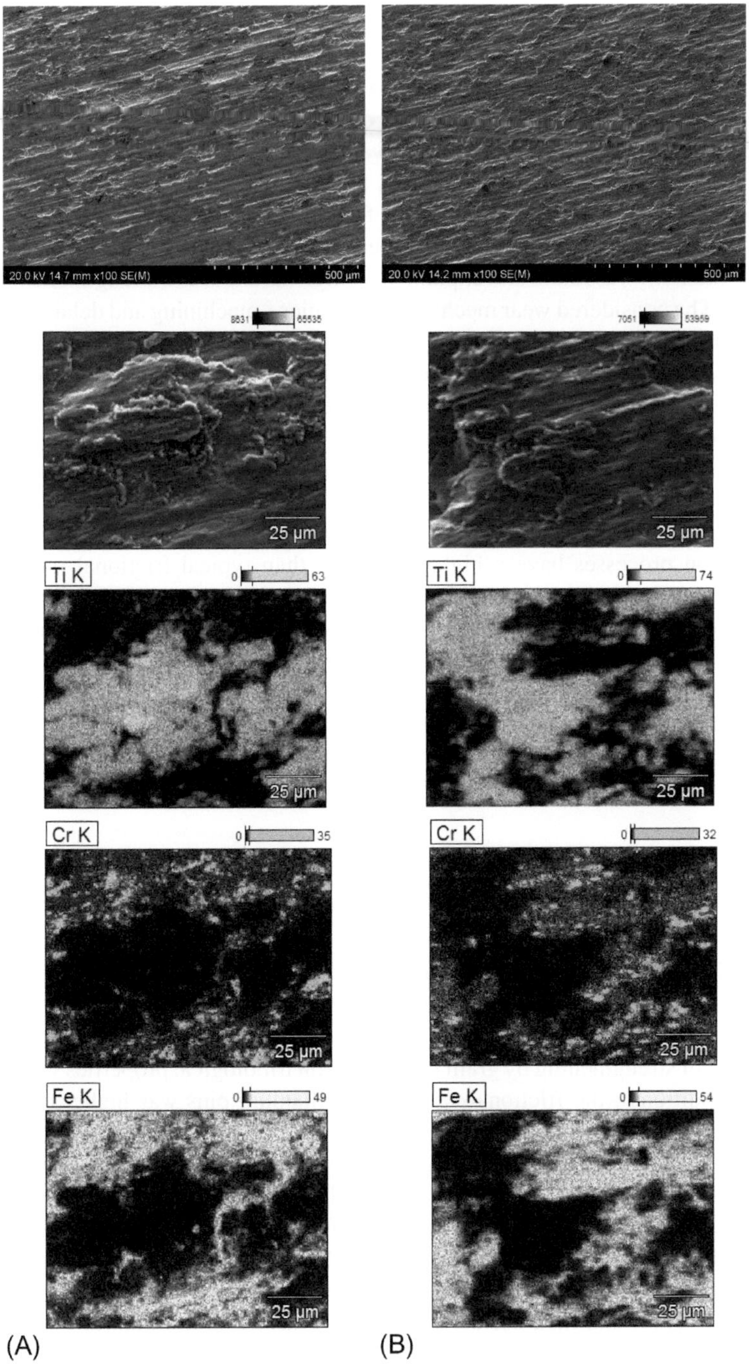

FIG. 10.6

Morphology and EDS mapping (elements distribution of Ti, Cr, and Fe) on the surface of wear products deposited on steal counter specimen during the dry friction process of NC (A) and MC (B) titanium.

relatively low values of yield and shear stresses, typical for hcp metals. Simultaneously values of friction coefficient are quiet large.

Many publications indicate the positive effect of strengthening by grain boundaries on tribological properties of metals [6,7,15]. Additionally, La and coauthors [13] have written about the reduction of frictional resistance in dry conditions for UFG titanium produced by the ECAP method. The smaller rate of frictional wear of UFG titanium was found and explained by high hardness and strength of examined material. The considered wear mechanism was micro-machining and delamination—tearing of the material's particles. The surface of coarse-grained titanium was also susceptible to delamination in the same test condition but the main wear mechanism was an abrasive one. It should be emphasized that the examinations mentioned earlier were conducted in different conditions than those for NC titanium. The ball-on-disc friction test was used (instead of pin-on-disc) using hard (ceramic or steel) balls and titanium discs (as a specimen).

The contact area between specimen and counterspecimen is quite small, therefore deformation processes have a higher meaning than typical frictional ones. Work hardening of disc surface causes justified increase of wear resistance in such conditions. Frictional process conditions using the pin-on-disc method are essentially different because the contact area between wearing surfaces was much larger, at around 7 mm^2. The differences of test results indicate the necessity of further research on wear mechanisms in NC titanium, especially in terms of evaluation of factors responsible for tribological properties.

10.3 THE EFFECT OF ENVIRONMENT ON THE WEAR PROCESS

Research on the wear resistance of NC titanium produced by HE method was also extended to friction tests with lubricants (Fig. 10.7). It was found that lubricants negatively affected the wear resistance of titanium during the pin-on-disc tests. Significant wear increase was determined for friction in the environment of paraffin oil. The effect of strengthening by grain refinement on tribological properties was similar to dry conditions—the frictional wear of NC titanium pins was higher comparing with MC ones, for all variants used.

Microscopic observations of wear paths revealed that during lubricated friction process the tribochemical reactions occurred, however their mechanism differed from ones found in dry conditions. Surfaces of steel discs were smooth after friction tests, without signs of abrasive wear (Fig. 10.8). Also in this case the fragments of titanium pins were transported to counterspecimen's surface, however titanium clusters were smaller and their surface distribution was much more uniform. Different tribological properties in environment of 0.9% water solution of NaCl could be caused by the action of thin oxide layers formed on the wearing surfaces. Additionally, the presence of liquid solution ensured better heat transfer from the contact zone and decreased deformation effects and surface tacking during the friction process.

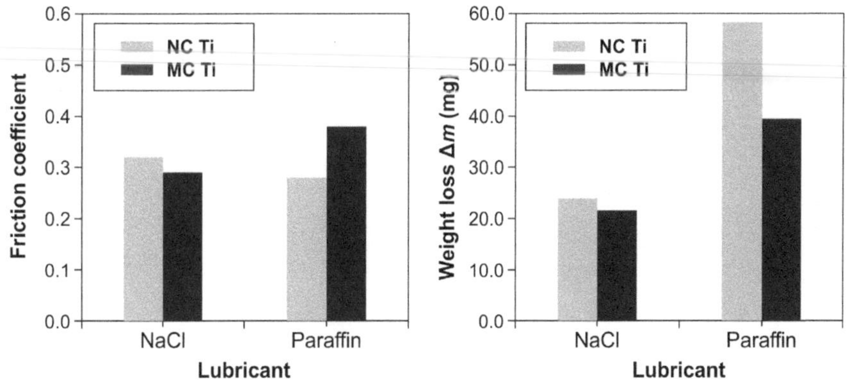

FIG. 10.7

Results of tribological examinations in lubricated friction condition using steel disc as a counter specimen (pressure 2 MPa).

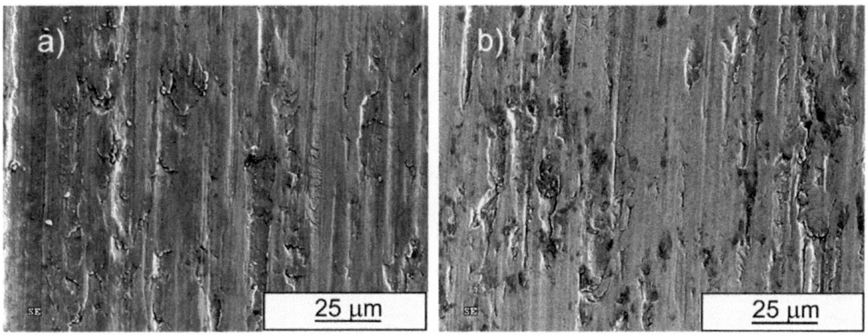

FIG. 10.8

Surface topography (using SEM) of steel disc (counter specimen) after lubricated friction process (in water solution of NaCl) against NC (A) and MC (B) titanium pins [18].

Much more intensive material transport between wearing surfaces was found after lubricated friction wear tests using paraffin oil. Among the wear's products some isolated large particles in form of scales were found (Fig. 10.9), which testified to the intensification of the adhesive process. They remained in a frictional zone and underwent partially spreading on wearing surfaces, leading to formation of secondary layers (Fig. 10.10). The presence of crystal defects and internal stresses in the surface layer of NC titanium, similar to tests in dry conditions, could intensify surface tacking and oxide layer formation processes, resulting in its lower wear resistance.

Note that that different tribological behavior was found in 316LVM steel that was HE processed [25]. Better wear resistance was found for NC steel examined in the lubricated friction test using paraffin oil—distinct lower values both friction coefficient and surface roughness were determined (compared to MC steel).

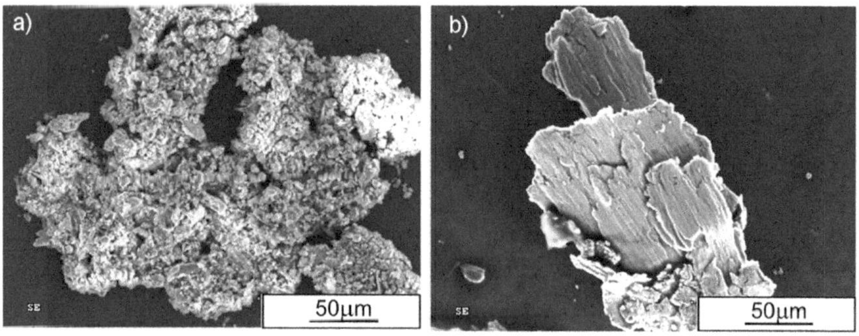

FIG. 10.9

SEM images of wear products formed during lubricated friction process of NC titanium pin/steel disc pair in water solution of NaCl (A) and paraffin oil (B) [18].

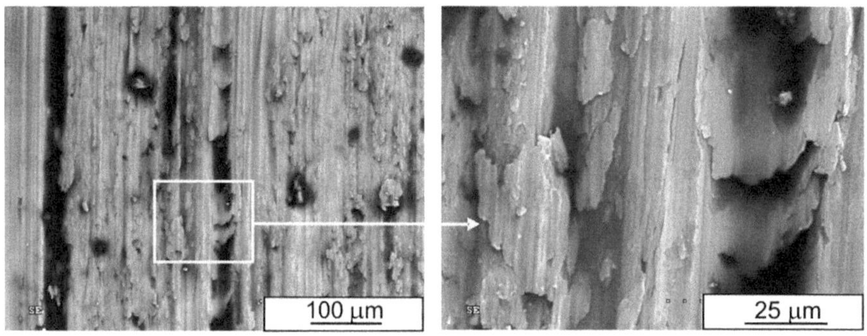

FIG. 10.10

Surface topography (using SEM) of steel disc after lubricated friction process in paraffin oil [18].

10.4 FRICTION IN NC TITANIUM/POLYETHYLENE PAIR

The analysis of results obtained from our own research was the basis of the statement that dominant wear mechanisms of MC and NC titanium in a tribological pairing with steel are tribochemical reactions and adhesive processes. For reduction of adhesive forces between titanium and steel surfaces, additional tests using UHMWPE discs were performed. Polyethylene inserted into the tribological set eliminated the interaction between metallic surfaces during the friction process. It was found, based on results obtained from dry friction wear tests, that values of linear wear Δl of NC titanium were significantly lower for higher pressures (2 MPa) compared to MC titanium (Fig. 10.11). It indicates the possibility of using of NC titanium in tribological sets with polyethylene, for example in biomedical applications.

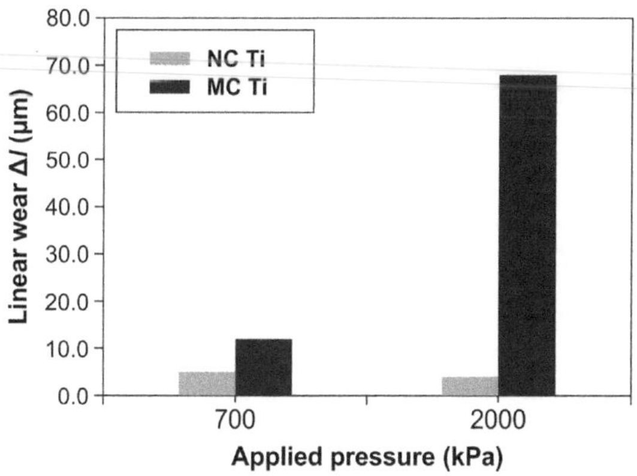

FIG. 10.11

Linear wear of titanium pins in dry friction process in set with polyethylene disc.

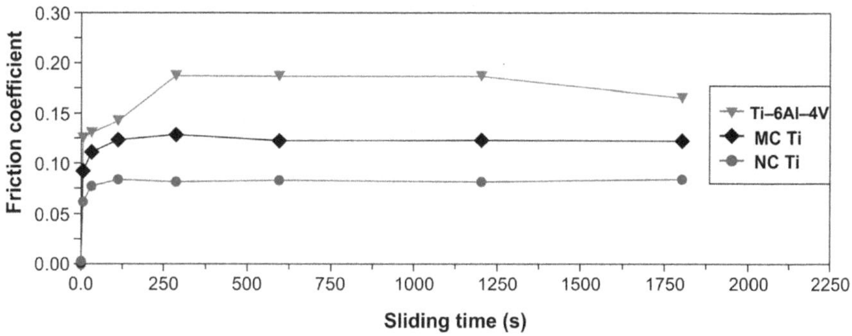

FIG. 10.12

Dependence of friction coefficient of NC and MC titanium and Ti–6Al–4V alloy on time of wear test in physiological saline [27].

Titanium alloys and polyethylene in tribological pair are often used in medicine, e.g. in hip replacement implants [26]. Motor systems "femoral head–acetabular shell' of such implants have relatively complex dimensions and shapes. Due to technological limitations it is not possible presently to make these components from NC titanium.

The biomedical application of NC titanium in a tribological set with polyethylene requires friction wear tests in environment of physiological saline. As reference materials, MC titanium and Ti–6Al–4V alloy were applied. It was revealed that NC titanium is characterized by the lowest value of friction coefficient ($\mu = 0.08$) in test conditions used (Fig. 10.12). Grain refinement in titanium seems to meaningfully reduce the friction coefficient because the value for MC titanium is much higher ($\mu = 0.125$).

The highest frictional wear resistance ($\mu = 0.18$) was determined for the Ti–6Al–4V alloy. It is worth adding that stabilization of friction coefficient value of pure titanium occurred after 2 min from the beginning of test, whereas this state for titanium alloy was achieved after 5 min (Fig. 10.12).

10.5 SUMMARY

The tribological properties of UFG and NC titanium are important in terms of its potential for biomedical applications. SPD-induced strong strengthening of titanium by grain boundaries and dislocations results in an increase of its hardness and strength. The results we have discussed demonstrate that the correlation between hardness/strength and wear resistance of titanium is limited. This fact can be explained by following properties, which reduce wear resistance of NC metals: higher activity–high oxidation rate, decreased ductility, low strain hardening capability, strain induced grain coalescence, and the presence of nonequilibrium grain boundaries [12,19].

Grain refinement can have a varied effect on the titanium wear resistance (Fig. 10.13). The variability in experimental results and various factors affecting tribological wear demonstrate that more systematic experiments are required to explain the effect of SPD on wear resistance of titanium. In order to improve the wear resistance of nanostructured titanium, well-known surface engineering methods that are employed for MC metals can be used. The criteria for selecting the surface treatment methods should take into account the thermal stability of nanostructure in the examined material [28].

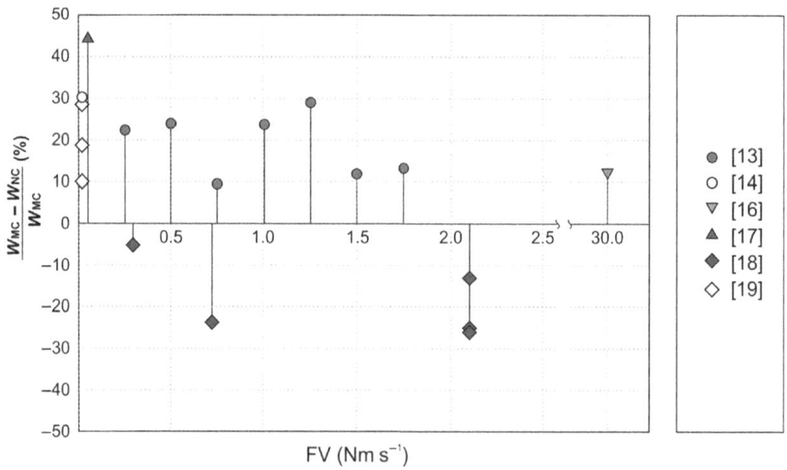

FIG. 10.13

Relation between relative (%) wear resistance changes of SPD-processed titanium and FV parameter (load times linear velocity).

REFERENCES

[1] K. Kato, Classification of wear mechanisms/models, Proc. Inst. Mech. Eng. J J. Eng. Tribol. 216 (6) (2002) 349–355.
[2] N. Gao, C.T. Wang, R.J.K. Wood, T.G. Langdon, Tribological properties of ultrafine-grained materials processed by severe plastic deformation, J. Mater. Sci. 47 (12) (2012) 4779–4797.
[3] Z.B. Wang, N.R. Tao, S. Li, W. Wang, G. Liu, J. Lu, K. Lu, Effect of surface nanocrystalization on friction and wear properties in low carbon steel, Math. Sci. Eng. A352 (2003) 144–149.
[4] Z.N. Farhat, Y. Ding, D.O. Northwood, A.T. Alpas, Effect of grain size on friction and wear of nanocrystalline aluminum, Math. Sci. Eng. A206 (1996) 302–313.
[5] L. Wang, Y. Gao, T. Xu, Q. Xue, A comparative study on the tribological behavior of nanocrystalline nickel and cobalt coatings correlated with grain size and phase structure, Mater. Chem. Phys. 99 (2006) 96–103.
[6] Y.S. Zhang, Z. Han, K. Wang, K. Lu, Friction and wear behaviors of nanocrystalline surface of pure copper, Wear 260 (2006) 942–948.
[7] D.H. Jeong, F. Gonzalez, G. Palumbo, K.T. Aust, U. Erb, The effect of grain size on the wear properties of electrodeposited nanocrystalline nickel coatings, Scr. Mater. 44 (2001) 493–499.
[8] R. Mishra, B. Basu, R. Balasubramaniam: Effect of grain size on the tribological behavior of nanocrystalline nickel. Math. Sci. Eng. A373 (2004), 370–373.
[9] H. Nykyforchyn, V. Kyryliv, O. Maksymiv, Wear resistance of steels with surface nanocrystalline structure generated by mechanical-pulse treatment, Nanoscale Res. Lett. 12 (2017) 150.
[10] A. Amanov, I.-S. Cho, D.-E. Kim, Y.-S. Pyun, Fretting wear and friction reduction of CP titanium and Ti-6Al-4V alloy by ultrasonic nanocrystalline surface modification, Surf. Coat. Technol. 207 (2012) 135–142.
[11] M. Wen, C. Wen, P. Hodgson, Y. Li, Wear behaviour of pure Ti with a nanocrystalline surface layer, Appl. Mech. Mater. 66-68 (2011) 1500–1504.
[12] J.E. Mogonye, T.W. Scharf, Tribological properties and mechanisms of self-mated ultrafine-grained titanium, Wear 376–377 (2017) 931–939.
[13] P. La, J. Ma, Y.T. Zhu, J. Yang, W. Liu, Q. Xue, R.Z. Valiev, Dry-sliding tribological properties of ultrafine-grained Ti prepared by severe plastic deformation, Acta Mater. 53 (2005) 5167–5173.
[14] P. Igliesias, M.D. Bermudez, S. Chandrasekar, B.C. Rao, A.E. Jimenez, T. Perdigon, A study of the wear behaviour of nanocrystalline titanium created by large strain machining, Proceedings of the STLE/ASME 2006 International Joint Tribology Conference, San Antonio, TX, 2006.
[15] V.V. Stolyarov, L.S. Shuster, M.S. Migranov, R.Z. Valiev, Y.T. Zhu, Reduction of friction coefficient of ultrafine-grained CP titanium, Math. Sci. Eng. A 371 (2004) 313–317.
[16] G. Purcek, O. Saray, O. Kul, I. Karaman, G.G. Yapici, M. Haouaoui, H.J. Maier, Mechanical and wear properties of ultrafine-grained pure Ti produced by multi-pass equal-channel angular extrusion, Math. Sci. Eng. A517 (2009) 97–104.
[17] S. Faghihi, D. Li, J.A. Szpunar, Tribocorrosion behaviour of nanostructured titanium substrates processed by high-pressure torsion, Nanotechnology 21 (2010) 485703.

[18] H. Garbacz, M. Gradzka-Dahlke, K.J. Kurzydlowski, The tribological properties of nano-titanium obtained by hydrostatic extrusion, Wear 263 (2007) 572–578.
[19] C.T. Wang, N. Gao, M.G. Gee, R.J.K. Wood, T.G. Langdon, Effect of grain size on the micro-tribological behavior of pure titanium processed by high-pressure torsion, Wear 280–281 (2012) 28–35.
[20] P. Iglesias, M.D. Bermudez, W. Moscoso, B.C. Rao, M.R. Shankar, S. Chandrasekar, Friction and wear of nanostructured metals created by large strain extrusion machining, Wear 263 (2007) 636–642.
[21] J.F. Archard, Contact and rubbing of flat surfaces, J. Appl. Phys. 24 (8) (1953) 981–988.
[22] H. Garbacz, M. Gradzka-Dahlke, K.J. Kurzydlowski, The tribological properties of nano-titanium obtained by hydrostatic extrusion, Wear 263 (2007) 572–578.
[23] Standard Test Method G-99-95a, ASTM Annual Book of Standards, vol.03.02, American Society for Testing and Materials, West Conshohocken, PA, 1999, pp. 401–405.
[24] K.G. Budinski, Tribological properties of titanium alloys, Wear 151 (1992) 203–217.
[25] J. Budniak, M. Lewandowska, W. Pachla, M. Kulczyk, K.J. Kurzydlowski, The influence of hydrostatic extrusion on the properties of an austenitic stainless steel, Solid State Phenom. 114 (2006) 57–62.
[26] R.T. Trousdale, D.J. Berry, J.L. Gilbert, Fracture of an acetabular component inserted without cement: a case report, J. Bone Joint Surg. 79A (1997) 901–905.
[27] H. Garbacz, Microstructure and Properties of Nanocrystalline Titanium, Warsaw University of Technology Publishing House, Warsaw, 2010.
[28] H. Garbacz, P. Wieciski, D. Kuczynska, D. Kubacka, K.J. Kurzydlowski, The effect of grain size on the surface properties of titanium grade 2 after different treatments, Surf. Coat. Technol. 335 (2018) 13–24.

CHAPTER

Machinability

11

Witold Habrat*, Angelos P. Markopoulos[†], Maciej Motyka*, Jan Sieniawski*

Faculty of Mechanical Engineering and Aeronautics, Rzeszow University of Technology, Rzeszow, Poland Manufacturing Technology Section, School of Mechanical Engineering, National Technical University of Athens, Zografos, Greece[†]*

11.1 INTRODUCTION

Nowadays, titanium and its alloys are mostly used in the aerospace and automotive industry, cosmonautics, bioengineering, shipbuilding, chemical plants, sport and recreation, and architecture. Moreover, nanocrystalline (NC) materials have unique mechanical and physical properties that are derived from their unique structures. These properties make them attractive for biomedical application [1] as orthopedic and dental implants and give them a significant potential for energy applications [2]. All of these applications need to be processed by manufacturing technology methods to produce relevant parts. Among all the manufacturing techniques, machining is most often used.

Favorable properties of titanium such as low density, low coefficient of thermal conductivity, high specific strength, low elastic modulus, excellent corrosion resistance, excellent biocompatibility, and durability at elevated temperatures make titanium one of the most advantageous and useful metals. The same properties, however, classify titanium and its alloys as difficult-to-cut materials.

11.2 MACHINABILITY OF MICRO- AND NANOCRYSTALLINE TITANIUM

The machinability depends on many factors related to the work material, cutting tool used, and the cutting parameters (Fig. 11.1). The workpiece material determines selection of cutting tool—geometry and material of cutter—and cutting parameters—cutting speed, feed rate, and depth of cut. There is correlation between the cutting tool and cutting parameters, because the material of the cutting tool is the fundamental factor determining the cutting speed and the geometry of the cutter affects the depth of the cut and feed rate.

Nanocrystalline Titanium. https://doi.org/10.1016/B978-0-12-814599-9.00011-0

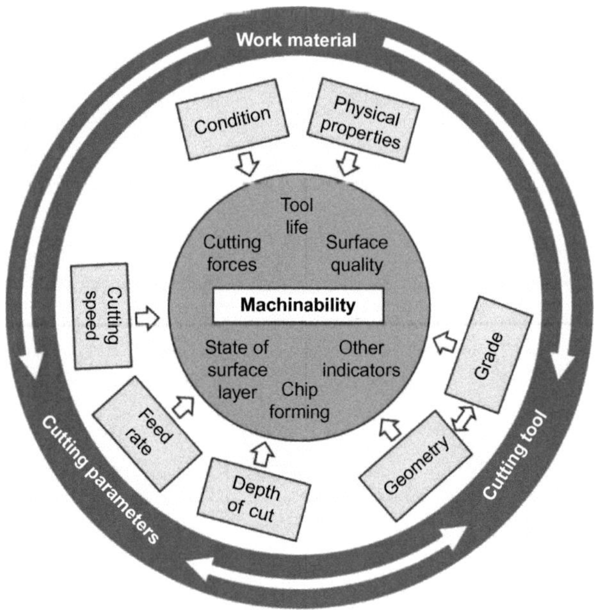

FIG. 11.1

Factors affecting the machinability of materials.

The condition and physical properties of the work material have a direct influence on its machinability. Among the factors which determine the condition of the work material microstructure, grain size, heat treatment, chemical composition, fabrication, and mechanical properties [3] can be distinguished.

The microstructure of a metal refers to its crystal or grain structure. NC materials with grain sizes of a few nanometers exhibit unusual properties and have high potential for use in structural and device applications, in which enhanced mechanical and physical properties are required. In Fig. 11.2 the microstructure of the high purity (99.99%) titanium before and after multipass hydrostatic extrusion (HE) process are compared. HE is one of the severe plastic deformation (SPD) methods, which permits obtaining the ultrafine and nanosized grains. Metals whose microstructures are similar have comparable machining properties. The variations in the microstructure of the same work material affect its machinability. NC pure titanium, in comparison to microcrystalline (MC) titanium, exhibits better mechanical properties, which influence its possibility of use and machinability.

The selection of fabrication method seems to be essential for obtaining the required properties of semi-finished products, especially for NC material. The manufacturing method, with which we obtain ultrafine-grained (UFG) or NC structures affects the properties of the workpiece material. Naseri et al. [5] analyzed the static mechanical properties and ductility of biomedical UFG commercially pure titanium produced by another one of SPD methods, that is, the equal-channel angular

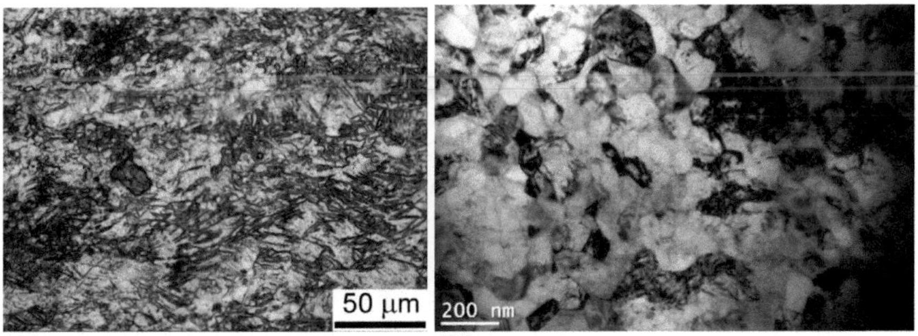

(A) (B)

FIG. 11.2

Microstructure of high purity (99.99%) titanium rod: (A) before (light microscopy) and (B) after HE process (transmission electron microscopy) [4].

pressing (ECAP) process [6]. The research has revealed that the ECAP process, when used for production of UFG and NC titanium, significantly increases the yield and ultimate tensile strengths, bending strength, hardness, and fracture toughness of the commercially pure titanium. It is observed that the processing method affects the strength and ductility. For example, NC materials produced by SPD techniques are contamination-free and porosity-free, so that they usually have high strength and good ductility. It can be also shown that SPD processing decreases the ductility to a smaller extent than conventional deformation processes such as rolling, drawing, and extrusion [7].

The properties of NC titanium are decisive, in relation to its machinability (Fig. 11.3). Smaller grains are not always desirable in order to obtain the required combination of high strength and high ductility for structural applications. Ductility usually decreases with decreasing grain size in NC materials.

NC metals and alloys with grain sizes less than 20 nm, or amorphous alloys, may have both lower strength and lower ductility than materials with larger grain sizes. There exists an optimum grain size range, in which a nanostructured material has both high strength and good ductility [7].

The stiffness of a part is affected by the modulus of elasticity of the workpiece material. It is a significant factor in machining of titanium as the thrust force, which deflects the part being machined, is considerably greater for this metal than for steel. Modulus of elasticity is important when designing machining conditions. Since the modulus of elasticity for titanium is only about half of that of steel, a titanium part may deflect several times as much as a similar steel part during machining [8].

One of the most important physical properties that have influence on machinability is work hardening, the process in which metal becomes stronger and harder by plastic deformation, due to changes in the crystal structure of the material. It is observed that a significant improvement in the yield strength of pure titanium is made by obtaining the UFG structure without alloying, with an absence of work hardening behavior compared to MC material [9]. In the case of cutting, the work

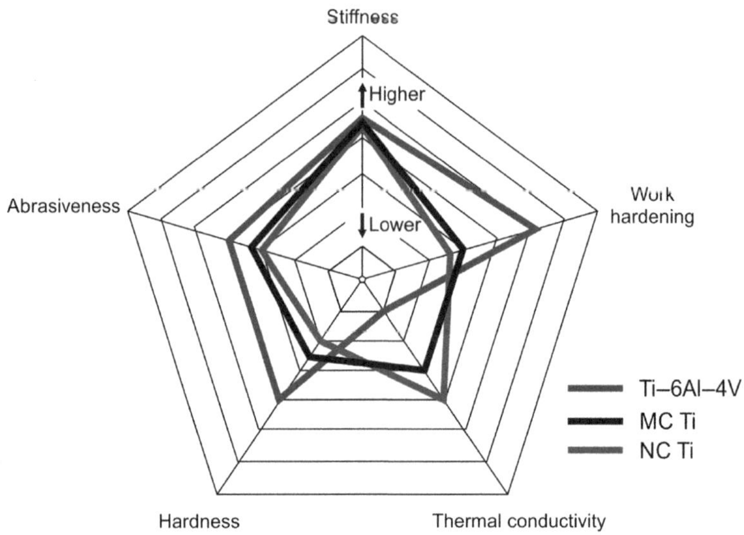

FIG. 11.3

Variation of machinability related properties of MC and NC titanium and conventional Ti–6Al–4V alloy.

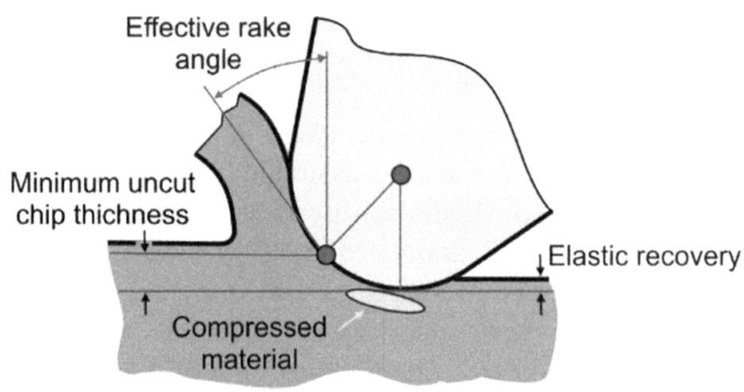

FIG. 11.4

Factors affecting the work hardening during cutting.

hardening results mainly from the roundness or sharpness of the cutting edge. It results in some of the material getting compressed instead of being cut off. Moreover, machining with a very small depth of cut results to work hardening and working material gets compressed and springs back instead of being cut. Such repeated cuts cause repeated deformation and work hardening (Fig. 11.4).

NC metals are often strong but have a low thermal stability [10]. The coefficient of thermal conductivity for pure titanium is reduced with the grain size, as the increased number of grain boundaries create the additional impediment of phonon and electron heat transport through the crystal lattice [11].

Thermal conductivity with high chemical reactivity of NC titanium has a significant influence on limiting its machinability. It can be assumed that wear of cutting tool at both flank and crater equally at high cutting speeds results from high heat accumulation at tooltip and adhesion with the serrated chips, respectively. For lower cutting speeds, the crater wear could be lower, due to absence of high cutting temperatures at the tool–chip interface.

In the case of NC titanium, it is necessary to limit the chemical reactivity. Titanium reacts with nearly all metals and refractory materials, and this, of course, includes cutting tools [8]. It is important to correctly select the material of cutting tool and keep the cutting temperature low to avoid a chemical reaction. The high heat generated during the cutting process also causes a work-hardening phenomenon that affects the surface integrity of titanium and could lead to geometric inaccuracies in the part and severe reduction in its fatigue strength [12].

Cutting tools for machining processes of NC titanium need certain special properties to ensure a good quality of cutting:

- geometry with a positive rake angle to minimize cutting forces, heat generation, and part deflection;
- sharp cutting edge—grinded cutting inserts without coating or with a small thickness of the coating;
- dedicated grade;
- limited tendency to vibrations;
- rigid construction of tooling system.

These requirements are met by tools that are dedicated not only to the machining of titanium and its alloys, but also to aluminum alloys. An example of such cutting inserts (Mitsubishi) are shown in Fig. 11.5.

Also, the thin coatings used on the cutting inserts ensure cutting edge sharpness while reducing the abrasive, thermal, and chemical impact on the cutting tool (Fig. 11.6).

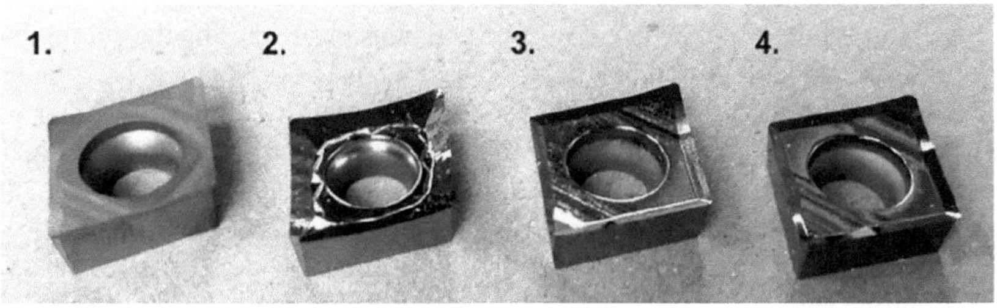

FIG. 11.5

Examples of uncoated cutting inserts with positive geometry: 1. CCGX120408-AL H10 (Sandvik), 2. CCGT120404-ASIC20 (Iscar), 3. CCGT120404-AZ HTi10, 4. CCGT120408-AZ HTi10 (Mitsubishi).

FIG. 11.6

Examples of thin coated cutting inserts (from *left*) VBGT160402-UM 1105, VBGT160402-UM 1115 and DNMG150608-SMC 1115 for titanium and titanium alloys (Sandvik).

FIG. 11.7

Example of solid end mill for standard milling operations.

An example of a solid carbide cutter is the end mill for standard milling operations from Kennametal company, shown in Fig. 11.7.

Cutting tools made of superhard materials can also be used in the machining of NC titanium. First of all, the polycrystalline diamond (PCD) cutting tool can be used. Typically, cubic boron nitride (CBN) tools are of limited use in application for the machining of titanium and its alloys. However, it is possible to use a modern superhard CBN tool with the "binderless" structure, made by Sumitomo, that provides high machining efficiency, while ensuring high durability of the tool and surface quality of the workpiece (Fig. 11.8) [13].

Saini et al. analyzed developments in cutting tool technology in improving machinability of the Ti–6Al–4V alloy. Their conclusions regarding the cutting tools used for the machining of titanium alloys can be adapted for UFG and NC titanium. They showed that the tungsten carbide (WC) tools with 6% cobalt as binder with grain size between 0.8 and 1.4 μm are found to be the most suitable. Although diamond inserts perform better than CBN tools in respect of chemical inertness, WC tools can be the better choice for machining titanium alloys due to their satisfactory performance at low and medium speeds as well as their cost-effectiveness [14].

The cutting parameters for application to NC titanium should be selected depending on the chosen material of cutting tool and the manufacturer's recommendations. However, the selection should not be the same as for pure titanium (Grade 2) but closer to the recommendations for two-phase titanium alloys (e.g., Ti–6Al–4V). This will ensure better tool durability and a good quality of the surface layer.

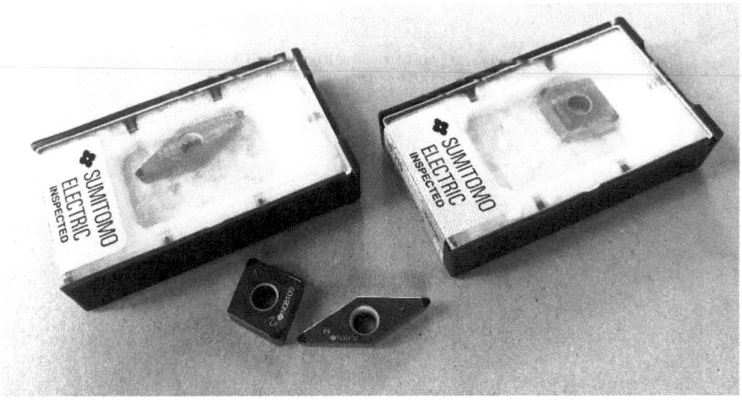

FIG. 11.8

CBN cutting inserts with the "binderless" structure (NCB100) for titanium alloys (Sumitomo).

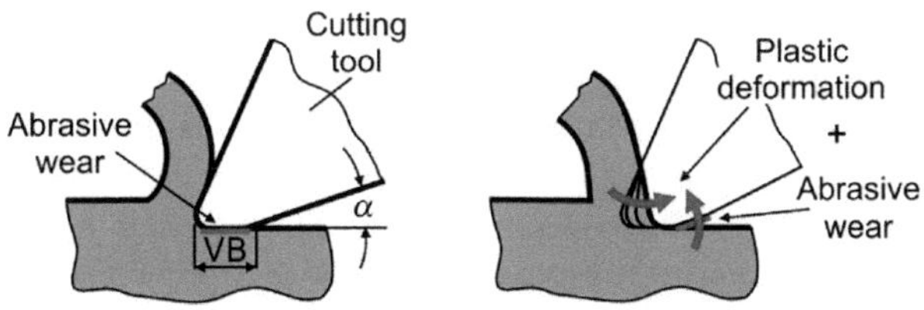

FIG. 11.9

Effect of tool geometry changes due to alone flank wear (*left*) and flank wear together with plastic deformation (VB—flank wear).

To assess the tool wear, appropriate indicators, such as wear on the flank surface or the dimensions of the crater on the rake face, are used. Many variables, such as workpiece material, cutting tool material, cutting tool geometry, and cutting parameters impact on the mechanism of tool wear in relation to the cutting time. The most desirable is the occurrence of minimal abrasive flank wear because of changes of tool geometry (Fig. 11.9). Flank wear together with plastic deformation adversely affects geometry due to work hardening.

One problem arising from the use of tool life as a machinability index is its sensitivity to the other machining variables. Of particular importance is the effect of tool material. Tool wear is very important machinability indicator in machining titanium, because of the significant cost of the cutting tool and its effect on the cutting forces and temperature in the cutting zone.

Morehead [15] studied tool wear in UFG and MC pure titanium with the use of PCD insert. The research was realized with small depth of cut (0.01; 0.03 mm) and cutting speeds of 60 and 180 m/min. It was observed that the tool wear rate increased significantly with the increase of cutting speed from 60 to 180 m/min. The UFG and MC titanium showed similar flank wear progressions when machining at 60 m/s.

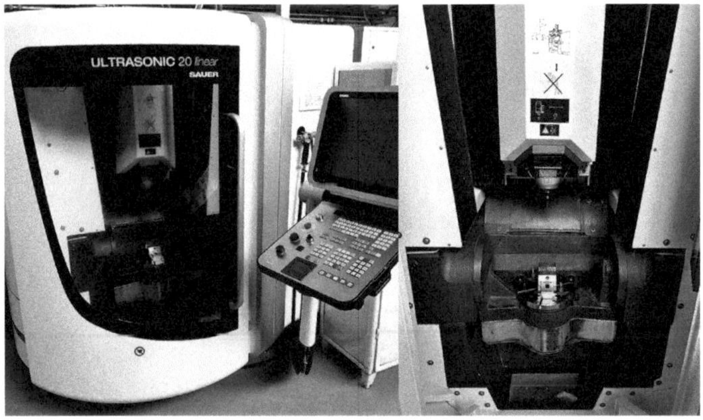

FIG. 11.10

5-axis precision milling machine Ultrasonic 20 linear with the possibility of ultrasonic assisted grinding.

There is a high probability of creating built-up edges during machining of NC titanium because this workpiece material promotes strain-hardening, the chip formation is stable and largely stationary, and a stagnant zone in the material flow in front of the cutting edge can occur. Built-up edges influence the cutting edge geometry. When they move off, they can drag along workpiece particles (adhesive wear). Sometimes strain-hardened parts of the built-up edges are integrated into the newly formed workpiece surface. Therefore, the formation of built-up edges is undesirable. However, it does not occur at higher cutting speeds and results in higher temperatures in the chip formation zone, because there is no strain-hardening if the recrystallization temperature is exceeded during the deformation process.

Khan et al. [16] applied cryogenic treatment on carbide inserts of cutting tool and observed significant improvement in wear resistance of carbide tools in the turning process of pure titanium. Moreover, this treatment was helped in preventing the formation of built-up edges and chipping of cutting tool surfaces.

Due to the method of producing of NC titanium, only machining of small parts is possible. Therefore machining of such parts can be considered as precision machining. An example of the precision machine tool is the 5-axis Ultrasonic 20 linear milling machine with the possibility of ultrasonic assisted grinding (Fig. 11.10). Kumar et al. [17] effectively realized an experimental study of ultrasonic machining of pure titanium.

11.3 CUTTING FORCES IN TURNING AND MILLING PROCESSES

The components of total cutting force in machining processes, especially the tangential cutting force, is one of the main parameters providing information on machinability of materials. Conventional MC–titanium alloys require high cutting forces, which result in deformation on the cutting tool during machining process. High dynamic shear strength induces abrasive saw-tooth edges, generating tool notching [18].

The cutting force components measure the resistance of machined material, which has to be overcome during the machining process. Graphical dependence of the measured values of components of dynamic cutting forces during turning of MC and NC titanium depends on the feed rate and are shown in Figs. 11.11 and 11.12.

The test results show that the structure of titanium has little effect on main force. A significant influence was observed in the case of thrust and feed forces. For the NC titanium, these forces are significantly lower than for MC Ti.

In the research [19] the components of cutting force for milling process were determined. The aim of this research was a comparison of cutting force components during machining of MC and NC titanium. The total cutting force F can be decomposed into tangential F_t, radial F_r, and axial component F_a. Based on the results, the components F_t and F_r were selected for the comparison. The tangential and axial component of cutting force were calculated by measurement results obtained from dynamometer (F_X, F_Y, F_Z) and geometric relations presented in Fig. 11.13.

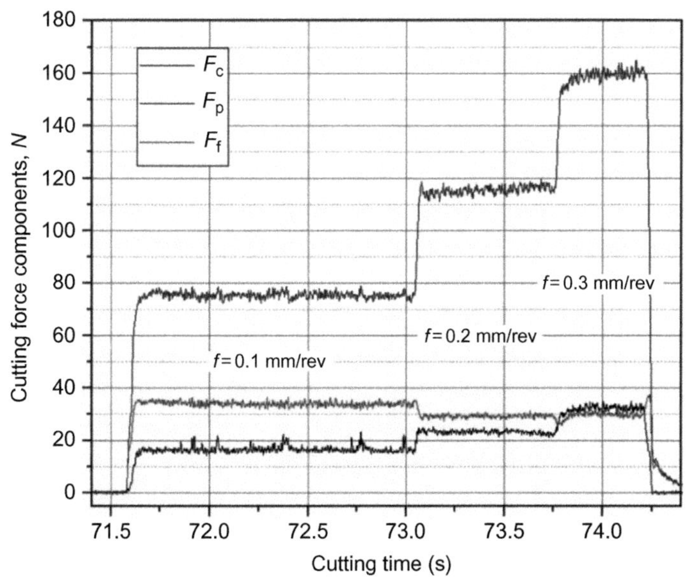

FIG. 11.11

Components of cutting forces during the turning of MC titanium Grade 2 for $v_c = 50$, $a_p = 0.4$ mm with different feed rates (cutting insert CCGT120402-AZ); where: F_c—main (tangential) force, F_p—thrust force, F_f—feed force.

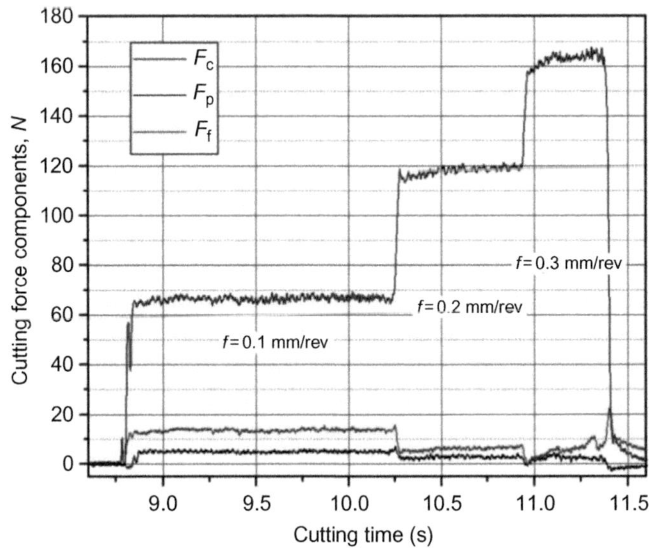

FIG. 11.12

Components of cutting forces during the turning of NC titanium Grade 2 with $v_c = 50$, $a_p = 0.4$ mm with different feed rates (cutting insert CCGT120402-AZ); where: F_c—main (tangential) force, F_p—thrust force, F_f—feed force.

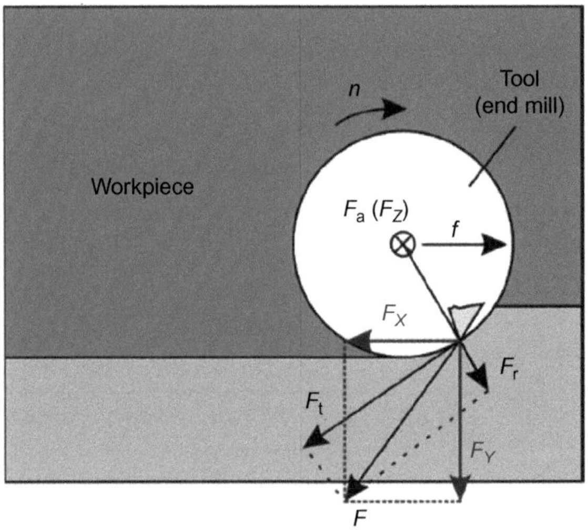

FIG. 11.13

The relationship between the components of the cutting forces.

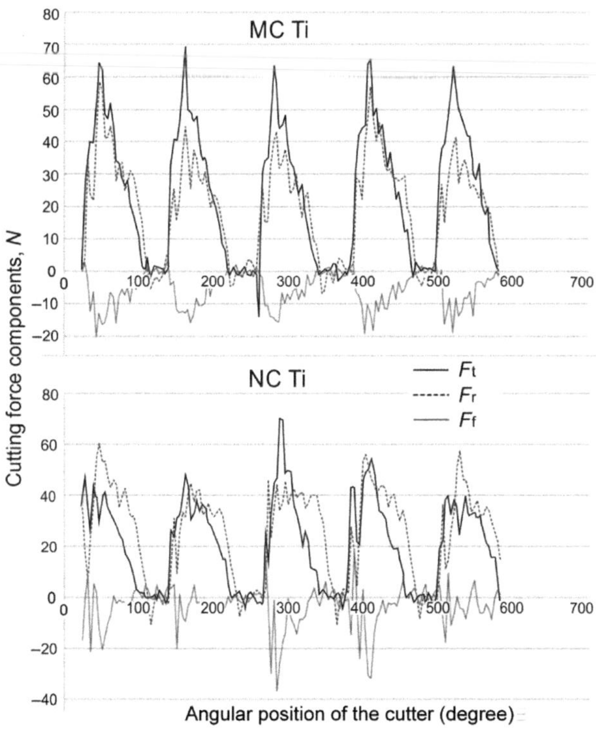

FIG. 11.14

Example of components of cutting forces as a function of angular position of cutter for milling of MC and NC titanium with $v_c = 40$ m/min, $f = 0.29$ mm/tooth and $a_p = 1$ (end mill UGDE0600A5ARA KC643M).

Fig. 11.14 shows the components of cutting force during milling of MC and NC titanium with cutting speed $v_c = 40$ m/min, feed rate $f = 0.29$ mm/tooth and depth of cut $a_p = 1$ mm, calculated according to relationship presented in Fig. 11.13.

The results of measurement of cutting force components during the milling process of MC and NC titanium are shown in Figs. 11.15–11.18.

The obtained results show that the values of tangential component of cutting force F_t during the milling process of NC titanium are lower (in the range from 5% to 35%) than in the case of its MC form. No significant differences in the range of radial component values F_r were found. This is a result of difference in so-called "cutting action," which is more "negative" for machining of MC titanium (as in case of tool with smaller rake angle).

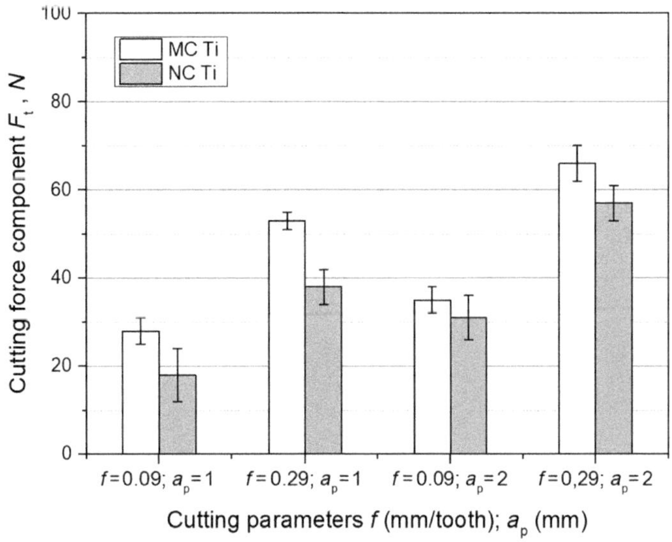

FIG. 11.15

Values of the tangential component of cutting forces F_t for the selected cutting parameters (a_p, f) obtained for cutting speed $v_c = 40$ m/min [19].

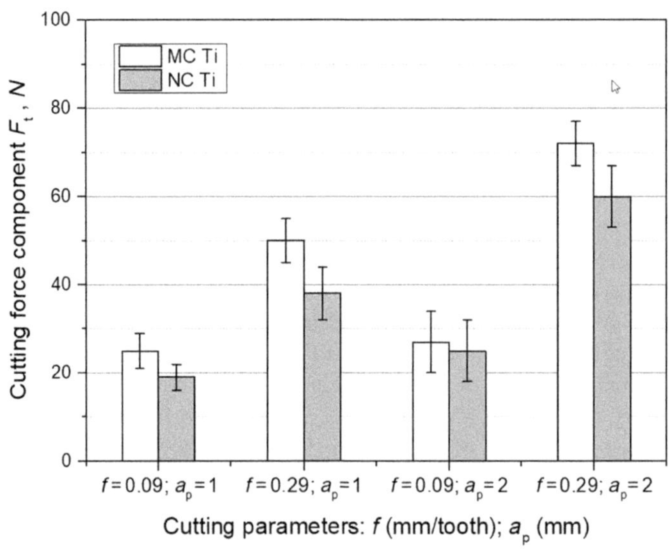

FIG. 11.16

Values of the tangential component of cutting forces F_t for the selected cutting parameters (a_p, f) obtained for cutting speed $v_c = 60$ m/min [19].

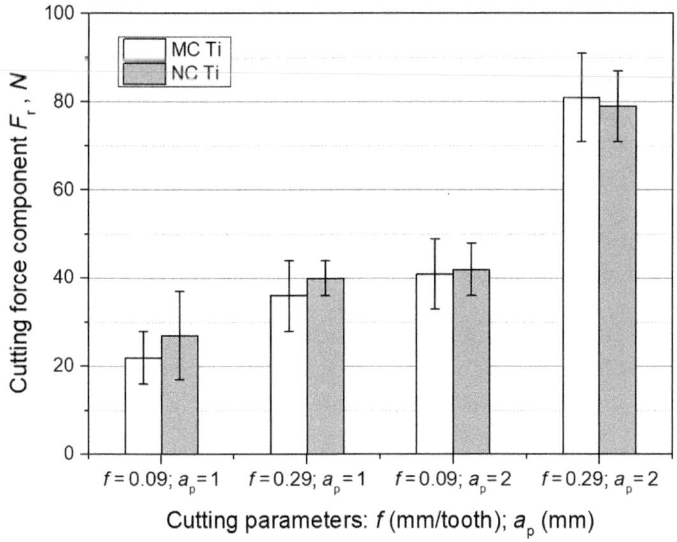

FIG. 11.17

Values of the tangential component of cutting forces F_r for the selected cutting parameters (a_p, f) obtained at cutting speed $v_c = 40$ m/min [19].

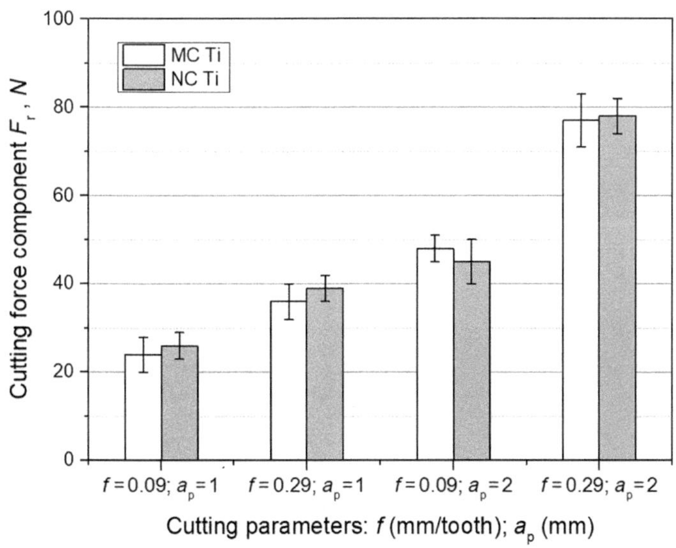

FIG. 11.18

Values of the tangential component of cutting forces F_r for the selected cutting parameters (a_p, f) obtained for cutting speed $v_c = 60$ m/min [19].

11.4 QUALITY OF MACHINED SURFACE

The geometric structure of surface and obtained values of roughness parameters indicate differences in machinability between analyzed materials as well. In Figs. 11.19 and 11.20, the profiles and geometric structure of the surface after turning for MC and NC titanium are compared. The visual comparison showed no significant differences.

The dependencies of the roughness parameters R_a and R_z on the feed rate for turning of MC and NC titanium are shown in Figs. 11.21 and 11.22. The obtained results of roughness parameters after turning of NC titanium are greater (about 10% for R_a) than for MC Ti. Based on the research, it can be noted that the roughness significantly depends on the feed.

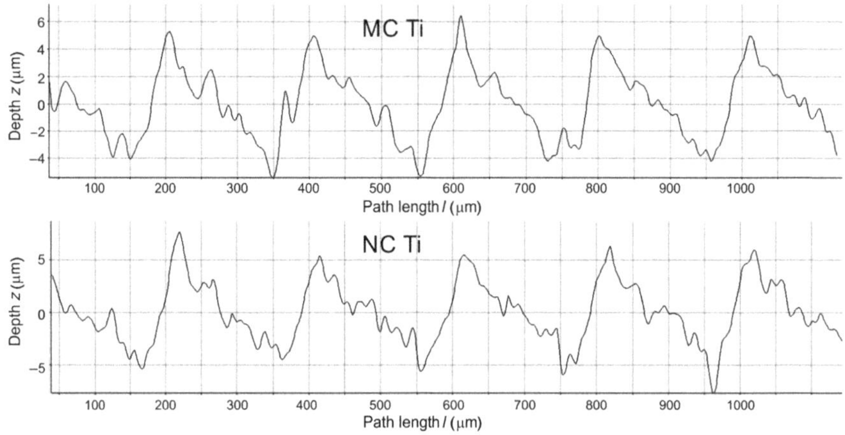

FIG. 11.19

Comparison of profiles of the surface roughness after turning for MC and NC titanium.

FIG. 11.20

Comparison of geometric structure of the surface after turning for MC and NC titanium.

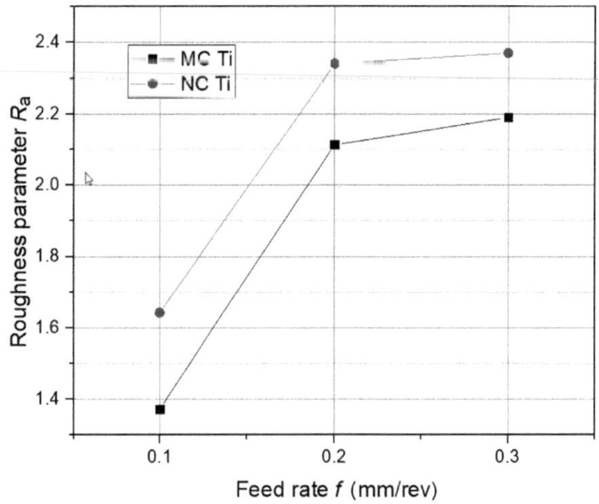

FIG. 11.21

The impact of feed rate on the roughness parameter R_a for turning on MC and NC titanium.

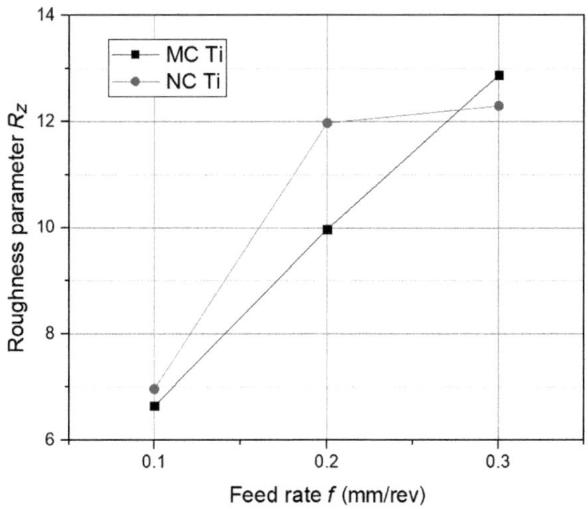

FIG. 11.22

The impact of feed rate on the roughness parameter R_z for turning on MC and NC titanium.

In research [4], the feed rate effect on selected parameters of surface roughness after milling of MC and NC pure titanium was determined. In Figs. 11.23 and 11.24, the images of surface topography and roughness parameters after side milling of MC and NC titanium for different feed rates are presented. The vibrating effects during

Feed Rate

	0.005	0.009	0.019	0.029

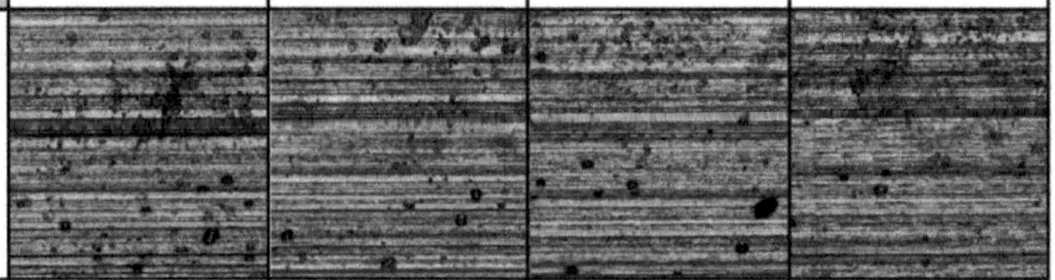

Parameters	0.005	0.009	0.019	0.029
S_a	1.879 μm	1.917 μm	1.9751 μm	1.8939 μm
S_q	2.4299 μm	2.4983 μm	2.6943 μm	2.446 μm
S_p	17.177 μm	12.389 μm	19.06 μm	15.112 μm
S_v	14.643 μm	11.939 μm	17.012 μm	10.483 μm
S_z	31.82 μm	24.328 μm	36.073 μm	25.596 μm

FIG. 11.23

The surface topography and roughness parameters after side milling of MC commercially pure titanium for different feed rates.

Feed Rate

	0.005	0.009	0.019	0.029

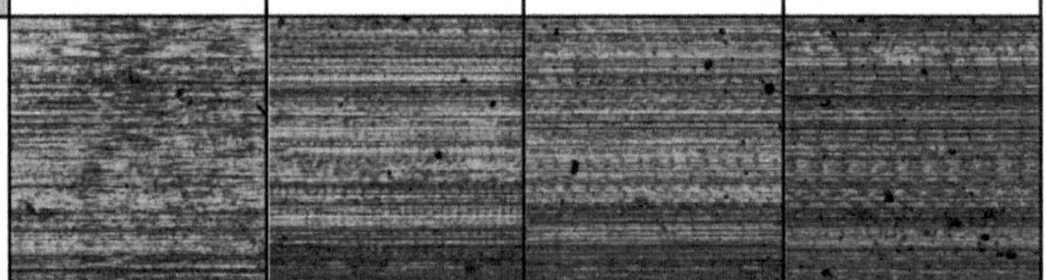

Parameters	0.005	0.009	0.019	0.029
S_a	2.1325 μm	2.2525 μm	2.4366 μm	2.4267 μm
S_q	2.7384 μm	2.9083 μm	3.0782 μm	3.1619 μm
S_p	16.304 μm	12.349 μm	16.96 μm	16.278 μm
S_v	21.576 μm	14.341 μm	11.87 μm	13.403 μm
S_z	37.88 μm	26.689 μm	28.83 μm	29.681 μm

FIG. 11.24

The surface topography and roughness parameters after side milling of NC titanium for different feed rates.

machining can be observed on the images of surface topography after side milling of NC titanium. This shows the differences in the mechanics of the cutting process of MC and NC titanium.

The dependences of the roughness parameters S_a and S_z on the feed rate for MC and NC titanium after side milling are shown in Figs. 11.25 and 11.26.

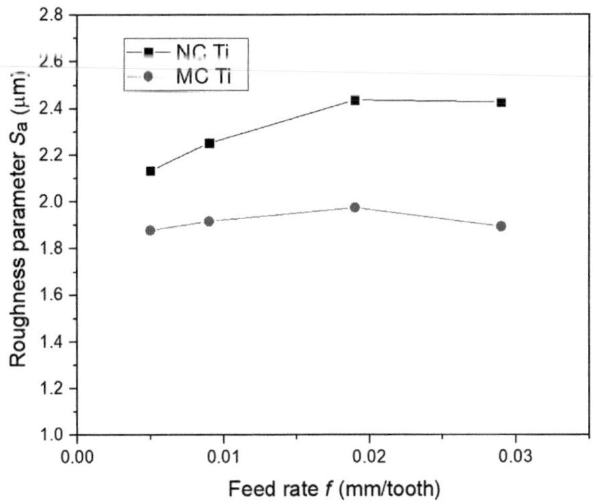

FIG. 11.25

The impact of feed rate on the roughness parameter S_a for side milling [4].

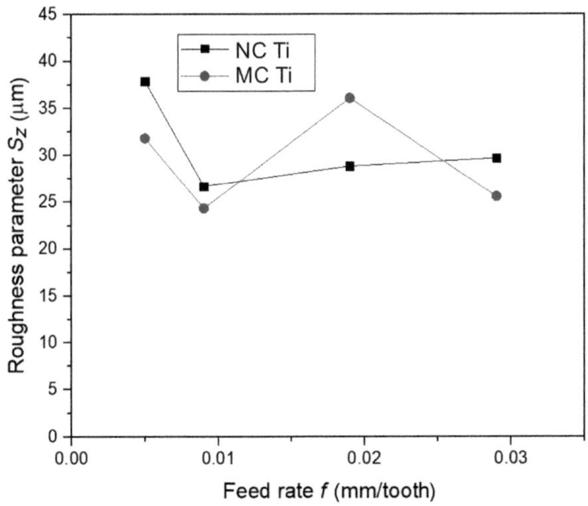

FIG. 11.26

The impact of feed rate on the roughness parameter S_z for side milling [4].

The obtained roughness parameters after milling of NC titanium are greater than for MC Ti. The significant impact of feed rate on the roughness parameters for side milling was not determined.

11.5 EVALUATION OF CHIP FORMATION

Davis et al. [20] were researching the chip formation mechanism during machining for UFG titanium. It was observed that transitions from cyclic shear localization within the low cutting speed regime to uniform shear localization within the moderately high cutting speed regime and to cyclic shear localization were occurring. The shear band spacing increases with the cutting speed and is always lower than that of the coarse-grained titanium. If the shear strain rate distribution contains a shift in the chip flow direction, the chip morphology appears saw-toothed, and cyclic shear localization is the chip formation mechanism. If no such shift occurs, the chip formation is considered continuous, and uniform shear localization is the chip formation mechanism.

Swaminathan et al. [21] determined in commercially pure titanium that, a shear strain of ~3, obtained during machining with +20 degrees rake angle tool, led to significant grain refinement. The microstructure of this chip is seen to be comprised of sub-100 nm grains that are considerably misoriented against one another.

Many factors influence chip formation. Therefore, it can be used to assess machinability because workpiece material is one of the major factors. Different combinations of shear strength and ductility have direct results on the chip formation process. The chip morphology after turning with various parameters is presented in Fig. 11.27.

Based on more detailed images of the chip morphology after turning process (Figs. 11.28 and 11.29), it can be concluded that there are significant differences in the chip forming process. For NC titanium, segmented chips and continuous (microsegmented) wear is observed. This indicates a change in the material's properties along with grain refinement.

Differences were also observed in the case of the milling process [19] (Figs. 11.30–11.32).

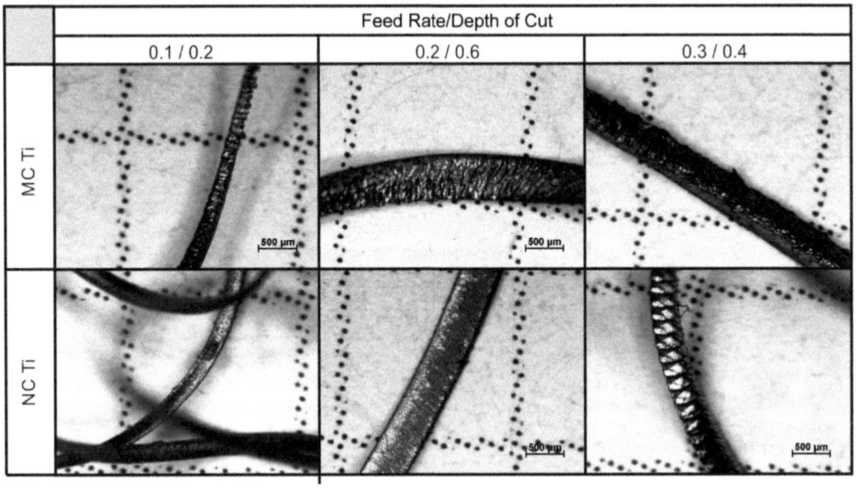

FIG. 11.27

Chip morphology for MC and NC titanium after turning with various parameters.

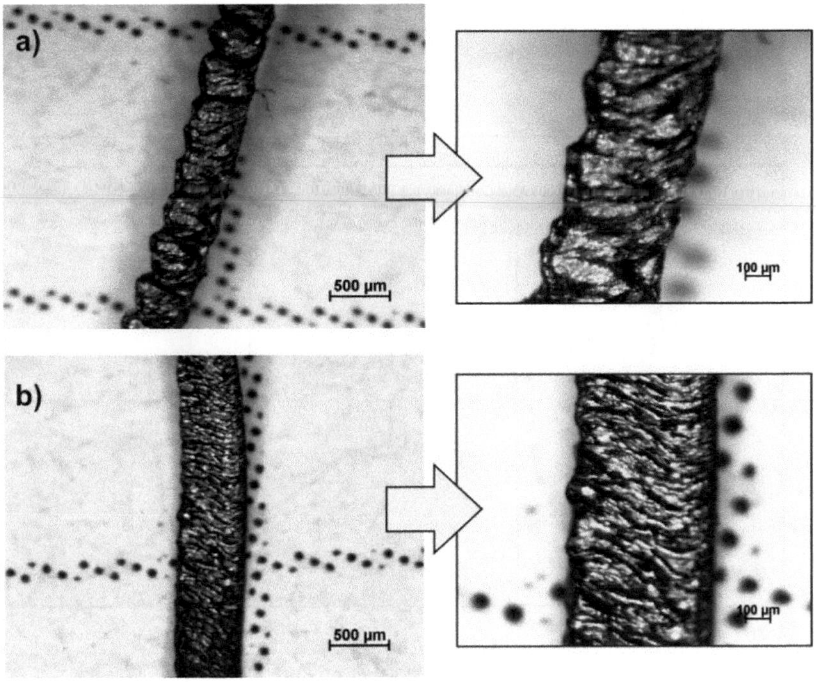

FIG. 11.28

Shape of the chips after turning of MC titanium with cutting parameters: (A) $v_c = 50$ m/min, $f = 0.3$ mm/rev, $a_p = 0.4$ mm; (B) $v_c = 50$ m/min, $f = 0.2$ mm/rev, $a_p = 0.6$ mm.

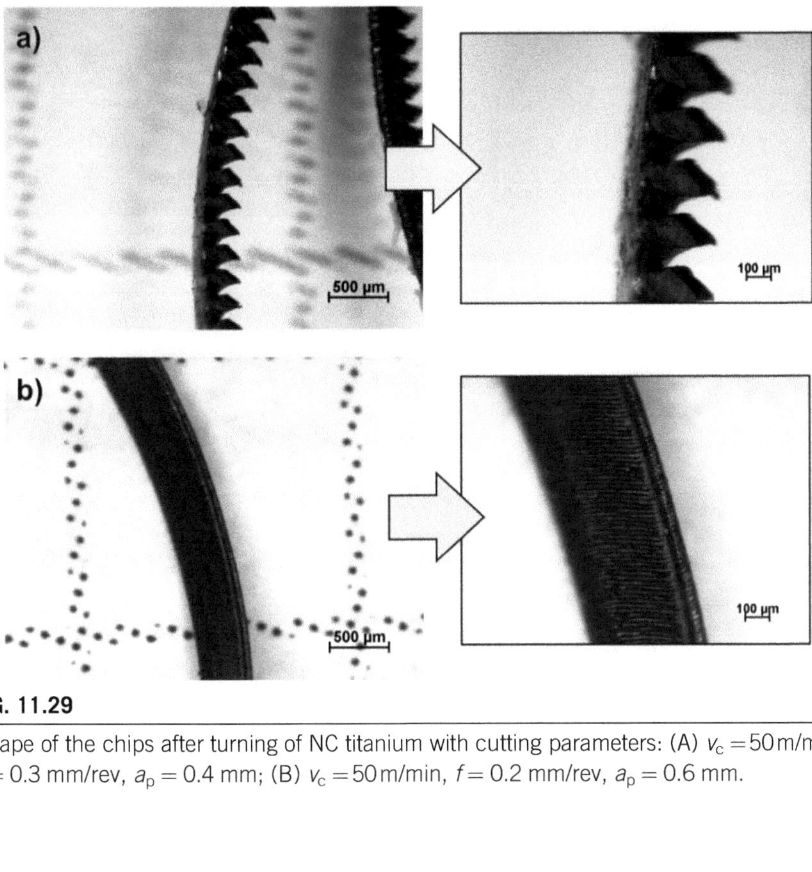

FIG. 11.29

Shape of the chips after turning of NC titanium with cutting parameters: (A) $v_c = 50$ m/min, $f = 0.3$ mm/rev, $a_p = 0.4$ mm; (B) $v_c = 50$ m/min, $f = 0.2$ mm/rev, $a_p = 0.6$ mm.

FIG. 11.30

Chip forms for MC and NC titanium after milling with different feed rate.

FIG. 11.31

Shape of the chip after milling of MC titanium with cutting parameters $v_c = 50$ m/min, $a_p = 1.5$ mm and $a_e = 1.2$ mm and feed rate: (A) $f = 0.029$, (B) $f = 0.009$ mm/tooth.

FIG. 11.32

Shape of the chip after milling of MC titanium with cutting parameters $v_c = 50\,m/min$, $a_p = 1.5$ mm and $a_e = 1.2$ mm and feed rate: (A) $f = 0.029$, (B) $f = 0.009\,mm/tooth$.

11.6 MODELING AND SIMULATION OF MACHINING OF NANOCRYSTALLINE TITANIUM

Modeling and simulation are commonly employed in manufacturing technology in order to predict the outcome of a process, aid the validation of a theory, and/or achieve process optimization. Especially in machining, the finite element method (FEM) has a history of more than five decades and is considered as the most popular method used for this purpose [22]. In general, the basic principle of this technique is the replacement of a continuum by finite elements forming a mesh in a procedure called discretization. As each finite element is simpler in geometry than the actual structure, it is easier to analyze. Every finite element possesses nodes, where the problem of initial and boundary conditions is applied and the degrees of freedom are calculated. The finite elements are connected to each another in nodes, where problem variables as well as properties are applied. The nodes of each element are assembled and global relations are formatted. The problem involves the selection of model formulation, mesh and element types, contact parameters, and the use of a remeshing strategy in most cases. However, the most challenging tasks are the application of the friction and the material model [23].

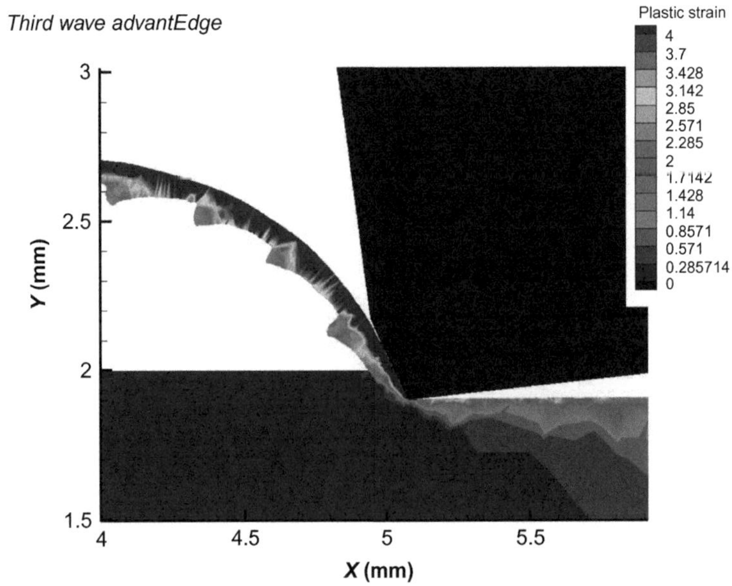

FIG. 11.33

Finite element modeling of serrated chip formation.

The latter is of utmost importance regarding simulations involving materials with low thermal conductivity, such as titanium and its alloys. These materials tend to produce a serrated chip type, as it can be seen in Fig. 11.33, typical of the chip segmentation appearing in these cases; this chip type is connected to poor surface quality and low tool life [24]. Hua and Shivpuri [25] made use of a fracture criterion in order to simulate chip formation. In another approach, a modification of the Johnson–Cook model was presented by Calamaz et al. [26] at first and Özel et al. [27] later, for simulating serrated chip formation during machining of Ti–6Al–4V alloy. The careful integration of the material constitutive model of the workpiece is more commonly used for the prediction of chip formation in machining of titanium and its alloys, in the relevant literature. However, in the case of NC titanium there are peculiarities that need special attention and cannot be adequately faced with methods so far discussed, due to the structure of the material.

Gleiter [28] introduced UFG and NC materials as a significant field in contemporary materials science. Since then, many researchers have dealt with many aspects of this field, including modeling of these materials. A review on the available micromechanical models of nanostructured metallic materials was recently presented by Mishnaevsky and Levashov [29], including methods based on composite/phase mixture models, self-consistent and multiscale composite models of nanomaterials, quasi-real Voronoi, polycrystal plasticity model, micromechanical models of grain boundary sliding, nonequilibrium grain boundaries, and damage initiation in nanomaterials. As pointed out by the authors, the main challenge of the computational modeling of NC metals is the limited knowledge of the real mechanisms of deformation and strength for given loading conditions.

Liu et al. [30] made use of the constitutive model developed by Farrokh and Khan [31] for the determination of the mechanical behavior of NC titanium. Additionally, the various proposed constitutive models can be incorporated to numerical models covering the simulation of UFG materials, including NC titanium, in continuum methods such as FEM or atomistic methods such as molecular dynamics (MD); one application being the machining of NC titanium. It is true that the numerical modeling works pertaining to the simulation of machining of NC titanium are rather limited. Ding and Shin [32] presented a coupled Eulerian–Lagrangian finite element model supplemented with a dislocation density-based material plasticity model for the simulation of orthogonal cutting of commercially pure titanium. Several tests under various strains, strain rates, and temperatures were performed for the simulation of chip formation, grain sizes, and dislocation densities, which were compared to experimental results and proven successful. In another approach, Melkote et al. [33] implemented in FEM software AdvantEdge™ by Third Wave Systems, USA, a suitable user-defined subroutine for a physically based constitutive model of commercially pure titanium. The model considers the mechanics of interaction of mobile dislocations with microstructural barriers and incorporates an additional deformation mechanism for the accurate simulation of serrated chip formation in orthogonal cutting of titanium. More specifically, the model incorporates the inverse Hall–Petch effect to describe the softening of UFG structure in the shear band on the material flow strength, which is an explicit function of the grain size and dislocation density, evolving with deformation. The model was validated against experimental results on cutting forces and chip geometry with good results. Although there are no MD models applied for the machining of NC titanium in the relevant literature, this meshless modeling method could work as an alternative to FEM, as it can capture the fine details of the material microstructure. The MD method has been commonly employed in atomistic simulation of machining; it pertains to the numerical time integration of the Newton equations of motion, which are solved for each atom in the studied system, using an appropriate time integration scheme. An example of modeling machining at nanolevel with MD can be seen in Fig. 11.34.

Besides numerical modeling, statistical methods are used for the modeling and optimization of machining processes. Optimization involves a suitable mathematical procedure which can provide the optimum set of characteristics, related to the optimum performance of a system. More specifically, an optimization problem consists of an objective function that describes the goal of the process that needs to be minimized or maximized. By using suitable heuristic algorithms, the area of possible solutions is searched in order to determine the region when the optimum point lies in an ordered and efficient way. In the work of Markopoulos et al. [34] a review of some common statistical and soft computing methods used in machining are presented, including factorial design method, Taguchi method, response surface methodology (RSM), Grey relational analysis, and statistical regression methods, among others.

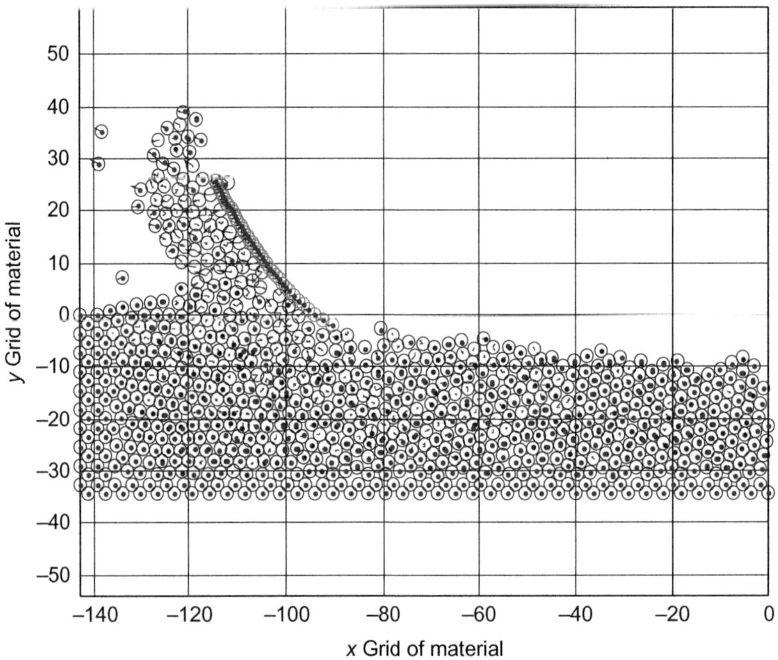

FIG. 11.34

Modeling of machining with molecular dynamics.

Among the aforementioned methods, RSM has been employed for the analysis of surface roughness when turning titanium alloys [35] and for the modeling of end milling of titanium alloy Ti–6Al–4V and the analysis of results with analysis of variance (ANOVA) [34]. In the work of Habrat et al. [4], RSM and ANOVA are employed to evaluate whether the grain size of pure titanium has influence on the cutting force components and the surface roughness in the milling process. Models of cutting force components for MC and NC titanium were prepared and examined. It was concluded that the tangential component of the cutting force in milling of NC titanium is 40% lower than that of MC titanium. Furthermore, roughness parameter S_a was up to 24% higher in the machined NC workpiece, in comparison to MC titanium.

It may be concluded that the available works on modeling and simulation of NC titanium are quite limited. This is due to the fact that the use of this material is still recent in technological applications. However, due to its advantages, its use is expected to increase and more sophisticated models to appear in the literature. Furthermore, more published experimental works will assist in the validation of proposed models and the refinement of the simulations.

11.7 SUMMARY

It may be concluded that the properties of NC titanium are decisive, regarding its machinability. The modulus of elasticity for titanium is only about half of that of steel, a titanium part may deflect several times as much as a similar steel part during machining. For NC pure titanium, significantly increased yield strength and the work hardening behavior compared to coarse-grained samples is limited. In the case of cutting, the work hardening results mainly from the roundness or sharpness of the cutting edge. The coefficient of thermal conductivity of pure titanium is reduced with the grain size, as the increased number of grain boundaries create additional impediment of phonon and electron heat transport through the crystal lattice. The examples show that the specific properties of NC titanium affect machinability indicators. Moreover, it was determined that the properties of the blank made from NC titanium depend on method of fabrication.

The basic quality indicators (cutting force components, chip geometry, and surface roughness) have shown the difference in machinability of NC titanium when compared to MC Ti. The test results show that the structure of titanium has little effect on the main (tangential) force in the case of the turning process. A significant influence was observed in the case of thrust and feed forces. For the NC titanium, these forces are significantly lower than for MC Ti.

For the milling process of NC titanium, the obtained results show that the values of tangential component of cutting force during the milling process are lower (in the range from 5% to 35%) than in the case of MC titanium. No significant differences in the range of radial component values F_r were found.

The profiles and geometric structure of the surface after turning for MC and NC titanium are similar. The visual comparison showed no significant differences. The obtained results of roughness parameters after turning of NC titanium are greater (about 10% for R_a) than for MC Ti. The roughness significantly depends on the feed rate.

The vibrating effects during machining can be observed on the images of surface topography after side milling of NC titanium. This shows the differences in the mechanics of the cutting process of MC and NC titanium.

The obtained roughness parameters after milling of NC titanium are greater than for MC Ti. The significant impact of feed rate on the roughness parameters for side milling was not determined.

Based on images of the chip morphology after turning and milling processes it can be concluded that there are significant differences in the chip forming process. For NC titanium segmented chips and continuous (micro-segmented) were observed. This indicates a change in the material's properties along with grain refinement.

In issue of modeling and simulation of machining of NC titanium, it may be concluded that the available works on modeling and simulation of NC titanium are quite limited. However, significant development in this scientific area has been observed.

ACKNOWLEDGMENTS

Financial support of Structural Funds in the Operational Program—Innovative Economy (IE OP) financed by the European regional Development Fund—Project No POIG.01.03.01-00-015/08 (NANOMET) is gratefully acknowledged.

REFERENCES

[1] R.Z. Valiev, I.P. Semenova, V.V. Latysh, H. Rack, T.C. Lowe, J. Petruzelka, et al., Nanostructured titanium for biomedical applications. Adv. Eng. Mater. 10 (2008) 8–11, https://doi.org/10.1002/adem.200800026.

[2] P.D. Lund, Nanostructured materials for energy applications. Microelectron. Eng. 108 (2013) 84–85, https://doi.org/10.1016/j.mee.2013.04.002.

[3] American Machinist. Machinability of metals, (2018). http://www.americanmachinist.com/cutting-tools/chapter-3-machinability-metals (Chapter 3) (accessed January 21, 2018).

[4] W. Habrat, M. Motyka, K. Topolski, J. Sieniawski, Evaluation of the cutting force components and the surface roughness in the milling process of micro- and nanocrystalline titanium, Arch. Metall. Mater. 61 (2016) 1033–1038 https://doi.org/10.1515/amm-2016-0226.

[5] R. Naseri, M. Kadkhodayan, M. Shariati, Static mechanical properties and ductility of biomedical ultrafine-grained commercially pure titanium produced by ECAP process. Trans. Nonferrous Met. Soc. China (English Ed.) 27 (2017) 1964–1975, https://doi.org/10.1016/S1003-6326(17)60221-8.

[6] J.Y. Chang, J.S. Yoon, G.H. Kim, Development of submicron sized grain during cyclic equal channel angular pressing. Scr. Mater. 45 (2001) 347–354, https://doi.org/10.1016/S1359-6462(01)01040-5.

[7] Y.T. Zhu, T.C. Lowe, T.G. Langdon, Performance and applications of nanostructured materials produced by severe plastic deformation. Scr. Mater. 51 (2004) 825–830, https://doi.org/10.1016/j.scriptamat.2004.05.006.

[8] C.T. Olofson, F.W. Boulger, J.A. Gurklis, Machining and grinding of titanium and its alloys, NASA Technical Memorandum X-53312 (1965) 1–131. http://www.dtic.mil/dtic/tr/fulltext/u2/620508.pdf.

[9] Y.G. Ko, D.H. Shin, K.T. Park, C.S. Lee, An analysis of the strain hardening behavior of ultra-fine grain pure titanium. Scr. Mater. 54 (2006) 1785–1789, https://doi.org/10.1016/j.scriptamat.2006.01.034.

[10] X.C. Cai, J. Song, T.T. Yang, Q.M. Peng, J.Y. Huang, T.D. Shen, A bulk nanocrystalline Mg–Ti alloy with high thermal stability and strength. Mater. Lett. 210 (2018) 121–123, https://doi.org/10.1016/j.matlet.2017.09.021.

[11] R. Lapovok, A. Molotnikov, Y. Levin, A. Bandaranayake, Y. Estrin, Machining of coarse grained and ultra fine grained titanium. J. Mater. Sci. 47 (2012) 4589–4594, https://doi.org/10.1007/s10853-012-6320-7.

[12] American Machinist, Cool tips for cutting titanium, nforma USA, Inc., 2018. http://www.americanmachinist.com/features/cool-tips-cutting-titanium (accessed January 24, 2018).

[13] H. Sumiya, K. Harano, Innovative ultra-hard materials: Binderless nanopolycrystalline diamond and nano-polycrystalline cubic boron nitride, SEI Tech. Rev. 82 (2016) 21–26.

[14] A. Saini, B.S. Pabla, S.S. Dhami, Developments in cutting tool technology in improving machinability of Ti6Al4V alloy: a review. Proc. Inst. Mech. Eng. B J. Eng. Manuf. 230 (2016) 1977–1989, https://doi.org/10.1177/0954405416640176.
[15] M.D. Morehead, Machinability and microstructure stability during the machining of pure copper and titanium processed by equal channel angular pressing, Master Science, Thesis (2007).
[16] A. Khan, K. Maity, Comparative study of some machinability aspects in turning of pure titanium with untreated and cryogenically treated carbide inserts. J. Manuf. Process. 28 (2017) 272–284, https://doi.org/10.1016/j.jmapro.2017.05.018.
[17] J. Kumar, J.S. Khamba, An experimental study on ultrasonic machining of pure titanium using designed experiments. J. Brazilian Soc. Mech. Sci. Eng. 30 (2008) 231–238, https://doi.org/10.1590/S1678-58782008000300008.
[18] E.O. Ezugwu, J. Bonney, Y. Yamane, An overview of the machinability of aeroengine alloys. J. Mater. Process. Technol. 134 (2003) 233–253, https://doi.org/10.1016/S0924-0136(02)01042-7.
[19] W. Habrat, M. Motyka, K. Topolski, J. Sieniawski, A. Nowotnik, Machinability of micro- and nanocrystalline titanium in milling process. in: Proceedings of the 13th World Conference on Titanium, 2015, pp. 999–1005, https://doi.org/10.1002/9781119296126.ch170.
[20] B. Davis, D. Dabrow, R. Newell, A. Miller, J. Schueller, G. Xiao, et al., Chip morphology and chip formation mechanisms during machining of ECAE- processed titanium. J. Manuf. Sci. Eng. 140 (3) (2017) 031008, https://doi.org/10.1115/1.4038442.
[21] S. Swaminathan, M. Ravi Shankar, B.C. Rao, W.D. Compton, S. Chandrasekar, A.H. King, et al., Severe plastic deformation (SPD) and nanostructured materials by machining. J. Mater. Sci. 42 (2007) 1529–1541, https://doi.org/10.1007/s10853-006-0745-9.
[22] A.P. Markopoulos, Machining processes. in: A.P. Markopoulos (Ed.), Finite Element Method in Machining Processes, Springer, 2013, pp. 1–9, https://doi.org/10.1007/978-1-4471-4330-7_1.
[23] A.P. Markopoulos, Finite element modeling. in: A.P. Markopoulos (Ed.), Finite Element Method in Machining Processes, Springer, 2013, pp. 29–57, https://doi.org/10.1007/978-1-4471-4330-7_3.
[24] P.J. Arrazola, T. Özel, D. Umbrello, M. Davies, I.S. Jawahir, Recent advances in modelling of metal machining processes. CIRP Ann. Manuf. Technol. 62 (2013) 695–718, https://doi.org/10.1016/j.cirp.2013.05.006.
[25] J. Hua, R. Shivpuri, Prediction of chip morphology and segmentation during the machining of titanium alloys. J. Mater. Process. Technol. 150 (2004) 124–133, https://doi.org/10.1016/j.jmatprotec.2004.01.028.
[26] M. Calamaz, D. Coupard, F. Girot, A new material model for 2D numerical simulation of serrated chip formation when machining titanium alloy Ti-6Al-4V. Int. J. Mach. Tools Manuf. 48 (2008) 275–288, https://doi.org/10.1016/j.ijmachtools.2007.10.014.
[27] T. Özel, T. Thepsonthi, D. Ulutan, B. Kaftanolu, Experiments and finite element simulations on micro-milling of Ti-6Al-4V alloy with uncoated and cBN coated micro-tools. CIRP Ann. Manuf. Technol. 60 (2011) 85–88, https://doi.org/10.1016/j.cirp.2011.03.087.
[28] H. Gleiter, Nanocrystalline materials. Prog. Mater. Sci. 33 (1989) 223–315, https://doi.org/10.1016/0079-6425(89)90001-7.
[29] L. Mishnaevsky, E. Levashov, Micromechanical modelling of nanocrystalline and ultrafine grained metals: a short overview. Comput. Mater. Sci. 96 (2015) 365–373, https://doi.org/10.1016/j.commatsci.2014.09.024.

[30] J. Liu, A.S. Khan, L. Takacs, C.S. Meredith, Mechanical behavior of ultrafine-grained/nanocrystalline titanium synthesized by mechanical milling plus consolidation: experiments, modeling and simulation. Int. J. Plast. 64 (2015) 151–163, https://doi.org/10.1016/j.ijplas.2014.08.007.

[31] B. Farrokh, A.S. Khan, Grain size, strain rate, and temperature dependence of flow stress in ultra-fine grained and nanocrystalline Cu and Al: synthesis, experiment, and constitutive modeling. Int. J. Plast. 25 (2009) 715–732, https://doi.org/10.1016/j.ijplas.2008.08.001.

[32] H. Ding, Y.C. Shin, Dislocation density-based grain refinement modeling of orthogonal cutting of titanium, J. Manuf. Sci. Eng. 136 (2014) 41003–41011.

[33] S.N. Melkote, R. Liu, P. Fernandez-Zelaia, T. Marusich, A physically based constitutive model for simulation of segmented chip formation in orthogonal cutting of commercially pure titanium. CIRP Ann. Manuf. Technol. 64 (2015) 65–68, https://doi.org/10.1016/j.cirp.2015.04.060.

[34] A.P. Markopoulos, W. Habrat, N.I. Galanis, N.E. Karkalos, Modelling and optimization of machining with the use of statistical methods and soft computing. in: J.P. Davim (Ed.), Design of Experiments in Production Engineering, Springer International Publishing, Cham, 2016, pp. 39–88, https://doi.org/10.1007/978-3-319-23838-8_2.

[35] S. Ramesh, L. Karunamoorthy, K. Palanikumar, Measurement and analysis of surface roughness in turning of aerospace titanium alloy (Gr5). Meas. J. Int. Meas. Confed. 45 (2012) 1266–1276, https://doi.org/10.1016/j.measurement.2012.01.010.

CHAPTER

Dental application 12

Carlos N. Elias, Daniel J. Fernandes, José A. Galiza, Emília dos Santos Monteiro, Ana C. Crema de Almeida

Biomaterials Laboratory, Instituto Militar de Engenharia, Rio de Janeiro, Brazil

12.1 BIOMATERIALS

Biomaterials are part of a class of biologically compatible materials used to restore the activity of a damaged body part. They are used to repair, replace, or treat a part of the musculoskeletal system. The term "implant" refers to biomaterials that remain in the body for an indeterminate time, whereas a "prosthesis" is inserted in the body to perform the function for a certain time. It is very important that the implant when inserted into the body cannot induce adverse reactions or have any inappropriate effect on tissues. Additionally, the biomaterial must withstand the adverse effects of the medium and withstand the imposed loads. Among the various definitions of biomaterials found in the literature, the most accepted definition is currently the one employed by the American National Institute of Health that describes biomaterial as:

> *Any substance or combination of substances, other than drugs, synthetic or natural in origin, which can be used for any period of time, which augments or replaces partially or totally any tissue, organ or function of the body, in order to maintain or improve the quality of life of the individual.*

The most important property of biomaterials is the biocompatibility. Biocompatibility was defined by Williams [1] as:

> *Biocompatibility is the ability of a material to perform with a specific tissue response a specific application.*

Based on this definition it is concluded that the biomaterial does not present unlimited biocompatibility. One example is stainless steel and the Cr–Co alloy that are used in orthopedic implants but are not used in dental implants. This is because the reaction of the biomaterial is different with the type of application. The biomaterial's behavior and interaction with the organism depend on the application. The biomaterial needs excellent corrosion resistance for dental and orthopedic implants.

Nanocrystalline Titanium. https://doi.org/10.1016/B978-0-12-814599-9.00012-2

However, in some applications the material needs to exhibit total degradation when in contact with the body, for example the use of magnesium alloys (Mg–Ca–Zn) for plate fractured bone stability.

When considering the interaction with biological tissues, biomaterials can be classified into the categories of bioinert, bioreactive, or bioactive [2]. In the first class are Ti alloys, stainless steels, Cr–Co–Mo alloys, polymers, and ceramics (alumina, zirconia), which when in contact with the hard tissues of the body induce the formation of a fibrous tissue capsule. In the second group are commercial pure (CP) Ti, Nb and Ta, which allow osseointegration (the implant is attached to the bone and there is contact of bone cells with the surface of the implant). In the third group are bioglass, Ca phosphates (hydroxyapatites) and autogenous bones, which react and exchange ions with the organism.

In this chapter we will look closely at dental implants, due to their association with a variety of procedures and implanted devices. There are different diameters and designs of dental implants. Some models of them are shown in Fig. 12.1. Although the majority are cylindrical with an external hexagon for dental prosthesis fixation, Fig. 12.1 shows conical dental implant with internal prosthesis connection. The angles and designs of the screws threads in the implant body also vary among manufacturers.

Dental implants must be capable of osseointegration. Osseointegration occurs when a material such as CP Ti becomes firmly bonded to the bone. The first criterion to select the material to be used in the manufacture of these implants is that it does not present degradation of the compact film of the oxides and hydroxides on its surface when exposed to the body fluids.

Regardless of the function, application, and region of insertion of the implants into maxillary bone, the materials must have good corrosion resistance,

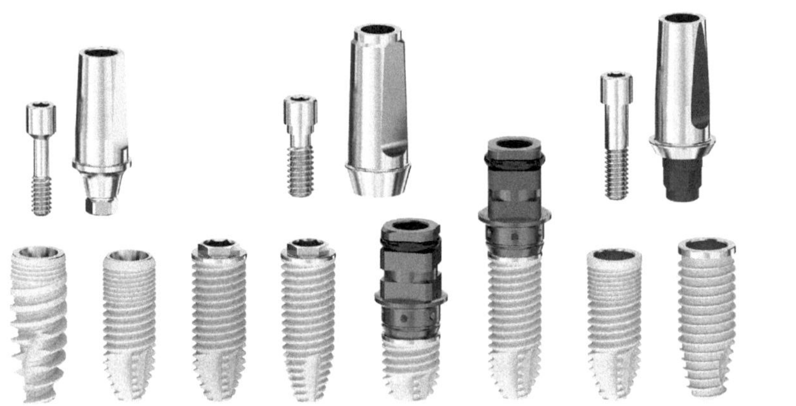

FIG. 12.1

Example of different shapes of commercial dental implant made with Ti Hard.

Courtesy of Conexão Sistemas de Prótese Co, Brazil.

biocompatibility, and not release toxic elements [3]. Such requirements reduce the range of materials that can be used. In addition to the biocompatibility, the selection of materials for application as an implant must take into account mechanical strength, market availability, cost, and machinability. Based on these requirements, the biomaterial for use in the manufacture of dental implants is restricted to CP Ti and Ti–6Al–4V ELI alloy [2].

Commercially pure titanium has an excellent interaction with bone-forming cells and osseointegration. This metal is the best biomaterial to manufacture dental implants. In addition, the surface of the passive cover film is slightly negatively charged at physiological pH, facilitating interaction with cells and proteins. Titanium has a dielectric constant comparable to that of water, so that the Coulomb interaction is similar to water [4–6].

12.2 DENTAL IMPLANTS

Currently, dental implant therapy constitutes a viable treatment option for both completely and partially edentulous patients. The selection of material for dental implant applications is based on requirements related to mechanical properties, workability, chemical properties, and biocompatibility. The choice of the material for dental implant is critical because it must have osseointegration, corrosion resistance, and sufficient mechanical strength.

It is very important that the biomaterials exposed to body fluids, i.e. in contact with an aggressive environment, have no degradation of the compact oxide that covers their surface. In other words, regardless of the role and place of application of the implants, the materials should have good corrosion resistance, biocompatibility, and absence of toxic elements. These requirements reduce the number of materials available for implants.

Over 100 Ti alloys are found on the market. The market is concentrated in three commercial alloys: Ti–6Al–4V (Grade 5—Ti G5) (60%), CP Ti Grades 1–4 (G1–G4) (13%) and Ti–3Al–2.5V (7.2%).

CP Ti (ASTM F67), Ti G5 (ASTM F 136 and F1472) and Ti–Zr alloy are used for dentistry application. These applications for dentistry devices include implants, crowns, bridges, overdentures, and dental implant prosthesis components (screws and abutment). The ASTM F136 alloy has lower impurities concentration than Ti–6Al–4V for engineering applications.

The titanium oxide into the CP Ti dental implant surface induces the osseointegration. Other important physical properties of CP Ti responsible for the biocompatibility are: a low level of electronic conductivity, high corrosion resistance, thermodynamic state at physiological pH values, low ion formation in aqueous environments, an isoelectric point of the oxide of 5–6, capacity to influence redox reactions at the tissue interface in a way that can modulate cell behavior, and no allergic reactions in patient.

The Technical Standards ASTM F-67 classifies CP Ti for biomaterial application into four grades based on the chemical composition (Table 12.1). The main difference among CP Ti grades is the maximum allowable content of iron and oxygen. This technical standards specification covers the chemical composition, metallurgical requirements, testing procedures, dimensional tolerances between implant and prosthesis, surface characteristics, cleaning procedures, surface inspection, handling, packaging, sterilization, labeling, and information about dental implants.

Commercially pure titanium is free of alloying elements and has lower concentration of impurities that may be harmful to the organism, but has low mechanical strength. Due to a relatively small mechanical resistance, CP Ti is not used in applications that involve high stresses, such as small dental implants and orthopedic prostheses. Fig. 12.2 shows a fractured dental implant that was installed in the posterior region of the mandible. The fracture occurred due to the small thickness of the implant wall. In the posterior mandible region the oral loads are high and narrow dental implant is not adequate. In the posterior region of mandible the most suitable choice is the installation of CP Ti implants with diameters above 3.75 mm or Ti G5 (F136). Some companies use a Ti–Zr alloy.

12.3 LIMITATION OF CP Ti FOR DENTAL IMPLANT APPLICATIONS

The disadvantages of CP Ti for dental implants include a higher Young's modulus, low mechanical strength, and a low wear resistance for some applications. It is difficult to improve the Ti mechanical properties without reducing biocompatibility. Table 12.2 shows that Ti G1 has lower mechanical strength than Ti G2, G3, and G4. Ti G4, due to higher iron content, has higher mechanical strength than Ti G1, G2, and G3. The tensile strength of CP Ti is significantly lower than that of Ti G5 and has higher elastic modulus than natural bone. The elastic moduli of CP Ti

Table 12.1 Chemical Composition (wt%) of CP Ti for Biomaterials Applications

	CP Ti Grade			
Element	**1**	**2**	**3**	**4**
Nitrogen (max)	0.03	0.03	0.05	0.05
Carbon (max)	0.08	0.08	0.08	0.08
Hydrogen (max)	0.015	0.015	0.015	0.015
Iron (max)	0.20	0.30	0.30	0.50
Oxygen (max)	0.18	0.25	0.35	0.40
Titanium	Balance	Balance	Balance	Balance

Based on ASTM F67.

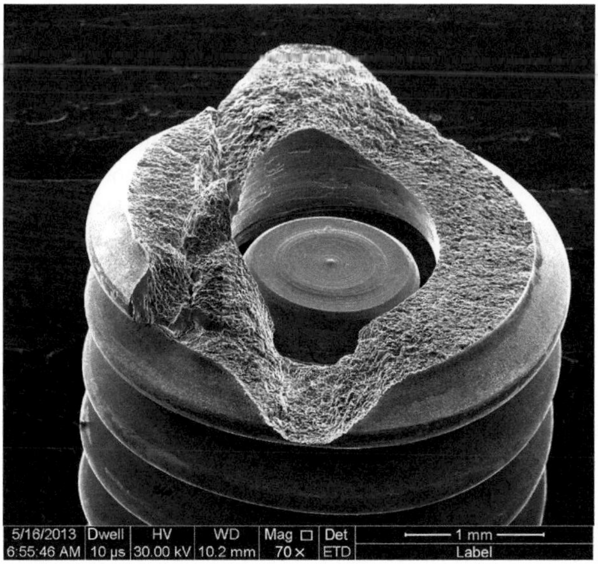

FIG. 12.2

Fractured dental implant due to small wall thickness and loads of intense chewing. The Technical Standard ASTM F136 (Ti–6Al–4V ELI) and ASTM F1472 (standard Ti–6Al–4V) specify the requirements of the Ti–6Al–4V alloy for biomedical applications. ASTM F136 alloy has long been favored for biomedical applications, but it has a possible toxic effect resulting from released vanadium and aluminum [7]. For this reason, vanadium and aluminum-free alloys have been researched for implant applications.

Table 12.2 Mechanical Requirements of Annealed-Bar, Billet, Forgings, and Other Forms of CP Ti, Ti G5 Alloy

Grade	Tensile Strength (min, MPa)	Yield Strength 0.2% (min, MPa)	Elongation[a] in 4*D* (min, %)	Reduction of Area[b] (min, %)
1	240	170	24	30
2	345	275	20	30
3	450	380	18	30
4	550	483	15	25
5	860	795	10	—

[a]*Elongation of material 0.063 in. (1.6 mm) or greater in diameter (D) or width (W) shall be measured using a gauge length of 2 in. or 4D or 4W. The gauge length must be reported with the test results.*
[b]*Reduction of area not required for tubing.*
Adapted from ASTM F67 (Grade 1–4) and F136 (Grade 5).

(110–130 GPa) is about three to six times higher than human cortical bone (20–30 GPa). Finite element simulations show that materials with lower elastic modulus have better stress distribution at the implant–bone interface and lead to less bone atrophy. A high difference between the modulus of elasticity of the implant material and bone can induce the phenomenon of stress shielding.

Although the mechanical strength of implants is important, the elongation in tensile test must be analyzed. The implant must also be somewhat flexible to avoid the shielding of bones from stress. The human body tends to reduce or eliminate their own parts when they are not used. Stress shielding is a mechanism that occurs when the forces exerted on the bone are different from the forces applied to a normal limb. This difference reduce bone density at the site (osteopenia), leading to bone atrophy. A common site for stress shielding is the proximal femoral diaphysis after placement of a femoral prosthesis. The more tightly the stem of the prosthesis fits into the distal medullary canal, the greater the shift of body weight to the prosthetic stem from the proximal femoral cortex. This causes loss of the normal remodeling forces above the level at which the stem is fixated against the endosteal surface of the medullary canal, resulting in osteopenia of the proximal femoral diaphysis [8,9].

During oral activities, implants are required to support axial loads around 120 N. Overloads and cyclic forces are also expected during chewing and biting. Shemtov-Yona and Rittel [10,11] showed that fatigue fracture is the main failure mechanism of dental implants. Fractographic analysis showed that implants may fracture after a relatively small number of cycles when subjected to forces of the kind that matches the load levels generated during mastication. The fracture mechanisms of Ti G5 and Ti G4 implants in an oral environment were identical to those found on specimens fractured in laboratory test conditions.

Choe et al. [12] analyzed the origins of mechanical failure in dental implants. They observed that most problems arose from fatigue failure and traumatic overload. Relevant implications for implant life are related to the ability of the implant–abutment junction to transmit the load to the structure during mastication efforts. The stress distribution upon implantable devices relies on the interface between hard tissue and the implant–abutment junction, but an overall approach seems reasonable, because the fatigue endurance of all components has a significant influence on the long-term performance of the devices.

One limitation of CP Ti and Ti G5 for dental applications is the impossibility to make an adequate implant for replacing a single missing tooth, especially for the maxillary and mandibular anterior areas. The replacement of a tooth by implant in the anterior region is critical because it is a place that demands aesthetics and a mechanical integrity of the implant–abutment joint, where biological requirements are amplified. New surgical techniques and applications have been developed and has become necessary to use implants with higher mechanical strengths. In the anterior maxilla the thickness of the bone plate is small and the implants must have narrow diameters. Yet there are still limitations for reducing the diameter of dental implants due to the relatively low mechanical strength of CP Ti.

12.4 LIMITATIONS OF Ti G5 FOR MEDICAL APPLICATIONS

The Ti–6Al–4V ELI alloy (F136) has excellent mechanical properties for use in hip prostheses and dental implants, but releases metal ions (Al and V), which may cause local and systemic carcinogenic effects. Aluminum and vanadium ions may inhibit the formation of apatite, making it difficult to obtain mineralization at the interface. Aluminum may affect cell function, cell proliferation, and extracellular matrix synthesis. Aluminum release has also been associated with neurological disorders. On the other hand, 4% of vanadium and 6% of aluminum may be sufficient to provoke systemic and local adverse reactions. Cytotoxicity, adverse tissue reactions, and general pathologies have been associated with vanadium and aluminum. Vanadium may cause local or systemic reactions and inhibit cell proliferation. Aluminum may be associated with osteomalacia, pulmonary granulomatosis, and neurotoxicity.

The ions liberated by the degradation and corrosion process of Ti–6Al–4V ELI (F136) implants do not remain in the vicinity of the implanted site. Blood or lymph vessels, cells, and cellular fluids transport metal particles to distant tissues such as brain, skin, hair, nails, and the gastrointestinal system. Metal elements released by the process of corrosion of oral implants were detected in adjacent tissues, saliva, and distant tissues. Variable amounts of titanium particles were detected in macrophages, fibrocytes, and giant cells in tissues adjacent to titanium implants [7,13].

Titanium can also be found in the liver, lymph nodes, and soft tissue close to the implant. Hematogenous transport may explain the accumulation of metal in parenchymal organs such as the kidneys and lungs. Concentrations of titanium, vanadium, and aluminum can be found in the brain, kidney, liver, and spleen of rabbits [7,13]. The accumulation of metals in some tissues and the diffusion of these by the organism can cause unfavorable reactions, ranging from a slight tissue discoloration to aseptic osteolysis or sterile necrosis.

One of the stimuli to develop the CP Ti with better mechanical properties is related to the release of toxic ions by titanium alloy elements, especially the release of aluminum and vanadium by F136 alloy.

12.5 ULTRAFINE-GRAINED (UFG) Ti FOR DENTISTRY

New surgical techniques and applications require dental implants stronger than the old ones, but the mechanical strength of CP Ti Grades 1–4 is not suitable for the new uses.

To minimize some disadvantages of CP Ti and Ti G5 for dental implant applications, a new CP Ti G4 was developed. This new Ti is cold hardening and is nominated "Ti Hard" [14,15].

The Ti Hard was processed by equal-channel angular pressing (ECAP) [16]. This processing resulted in a large reduction in grain size, from 25 μm to 150 nm. ECAP has been used extensively as a way to refine microstructures of metals and alloys and to improve mechanical behavior. Valiev and Langdon [16] showed that the ultrafine grains contained predominantly high-angle nonequilibrium grain boundaries with increased grain-to-grain internal stresses. The strength of nanostructured titanium is nearly twice that of conventional titanium.

Elias et al. [15] compared the mechanical properties of dental implants machined from bars of Ti G2, 4, 5, and CP Ti Hard. The implants were subjected to surface treatments with double etching (HCl and H_2SO_4 solution) with the same concentration and for the same temperature and time intervals.

The implants were submitted to compression test and fatigue test according to ISO 14801 (Dentistry–Fatigue test for endosseous dental implants). The mechanical test simulated the oral loading. Fig. 12.3 shows the set up used for mechanical tests. The maximum torque used in the tests is not recommended for surgical insertion but was done to evaluate the strength of Ti Hard under extreme load.

Elias et al. [15] tested the commercially available Ti Hard dental implants known as Easy®, Flash® and Torq®. Fig. 12.4 shows the tested dental implants. The first implant model has an external hexagon for dental prosthesis connection. The Flash®

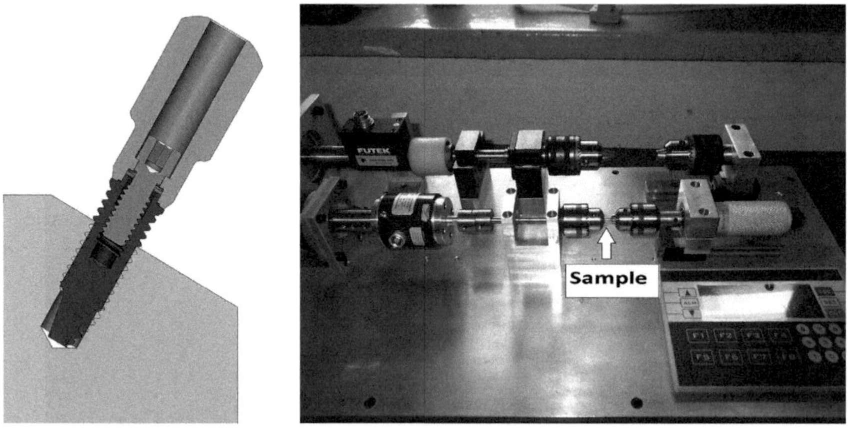

(A) (B)

FIG. 12.3

Setup for mechanical test of dental implants. (A) Compression and fatigue test. (B) Torque test.

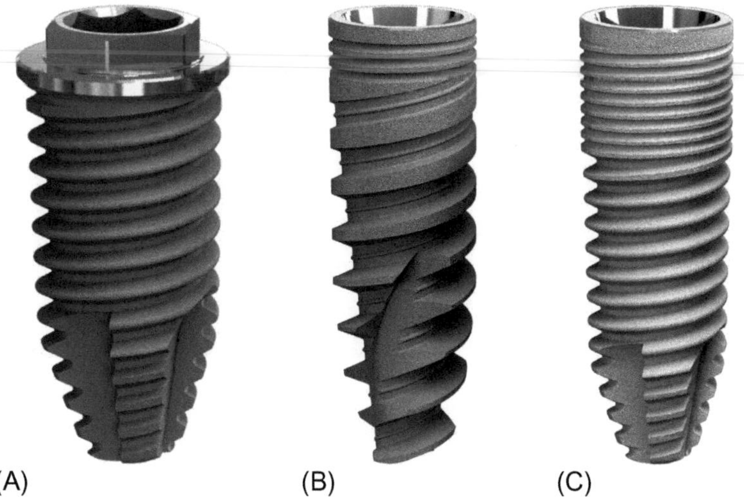

FIG. 12.4

Commercial Ti Hard dental implant available in the Brazilian market. (A) Easy® model with external hexagon. (B) Flash® model with internal cone Morse configuration. (C) Torq® model with internal cone Morse configuration.

Courtesy of Conexão Sistemas de Prótese Co, Brazil.

and Torq® implant models have internal dental prosthesis fixation such as the cone Morse hexagonal configuration.

Fig. 12.5 shows the external hexagon region of the implant before and after the application of torque 120 N cm. The torque test simulated the insertion of dental implant in the maxillary bone. It can be observed that dental implants Ti Hard have a smaller plastic deformation than Ti G4.

Table 12.3 shows the mean fracture peak loads during static compression test of dental implant. The mechanical behavior of implants made from Ti Hard was better than that of implants made of Ti G4 for all implant designs tested. The uniaxial compressive forces necessary to cause component failure were 889.97 ± 79.4 N and 588.9 ± 74.7 N for Easy® and Flash® implants made with Ti Hard, respectively. External hexagon implants (Easy®) are stronger than the cone Morse designs (Torq® and Flash®). Fatigue testing of the CP Ti and Ti Hard dental implants was carried out with a unidirectional compressive load [16]. Failures were defined as the fracture of any component as per ISO 14801 recommendations. A series of loads was applied until a lower limit was reached at which at least three specimens survived 5×10^6 cycles and none failed in the specified number of cycles (maximum endured load). Table 12.3 shows the maximum loading of dental implants without fracture during fatigue test.

Table 12.3 also shows that for the same number of cycles, the Ti Hard Easy® implant fractured with the highest load (400 N). The maximum number of cycles

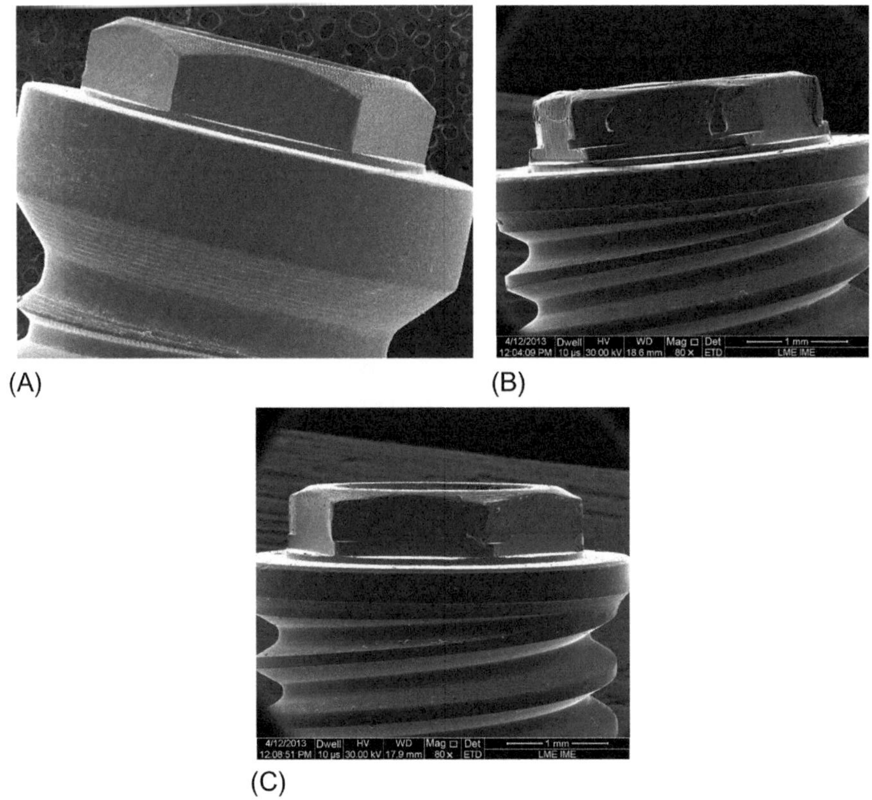

FIG. 12.5

Dental implant with external hexagon before and after application of torque equal to 120 N cm: (A) CP Ti G4 before torque test. (B) CP Ti G4 after torque test. (C) Ti Hard after torque test.

Table 12.3 Maximum Static Compression and Fatigue (5×10^6 Cycles) Forces (N) of Failure for Different External Hexagon (EH) and Cone Morse (CM) Dental Implants

Implant	Alloy	Compression	Fatigue
		Force (SD)	Force (N)
Easy® (EH)	CP Ti G4	776.4 (74.5)	300
	G4 Hard	889.9 (79.4)	400
Flash® (CM)	CP Ti G4	308.8 (15.2)	200
	G4 Hard	588.9 (74.7)	280
Torq® (CM)	CP Ti G4	410.3 (25.2)	200
	G4 Hard	498.3 (54.6)	260

Adapted from C.N. Elias, D.J. Fernandes, R.S. de Biasi, Comparative study of compressive and fatigue strength of dental implants made of nanocrystalline Ti Hard and microcrystalline Ti G4, Fatigue Fract. Eng. Mater. Struct. 40 (2017) 696–705.

used in the experiment (5×10^6) was considered adequate to simulate 5–6 years of service in the oral cavity under masticatory loads, assuming 2000–2400 functional cycles per day [17]. Cyclic plastic deformations play an important role in all aspects of fatigue. After a high number of cycles (in the order of 10^6 cycles), fatigue life is governed by crack nucleation, being controlled by dislocation–dislocation and dislocation–grain boundary interactions [18]. As in the static compression test, a dental implant made with Ti Hard is more resistant than CP Ti G4.

12.6 SURFACE MORPHOLOGY OF UFG CP Ti DENTAL IMPLANT

The surface morphology is an important parameter used to compare the success of dental implants. The surface treatment or surface modification is considered the main concern in recent developments of dental implants. The objective of surface dental implant treatment is to improve the biological response of materials by changing the surface properties without changing the bulk properties.

Surface topography and bone–implant interactions determine the success of dental implant osseointegration. Surface roughness has been shown to affect osteoblast proliferation and differentiation [19]. Alkaline phosphatase activity was enhanced by surface roughness, and cells grown on rougher surfaces produced a higher quantity of alkaline phosphatase activity than cells grown on smoother surfaces.

Titanium is normally covered with a thin protective oxide layer, which largely determines the surface properties of an implant. This thin oxide, naturally formed on a titanium substrate, is presumably responsible for the excellent biocompatibility of titanium implants due to a low level of electronic conductivity, a high corrosion resistance, and a thermodynamically stable state at physiological pH values [6]. In addition, titanium and its oxide have a low ion formation tendency in aqueous environments.

Surface modifications of dental implants include a change in the morphology and chemical composition and they are useful because biological tissues interact only through the outermost atomic layer. Surface morphology changes such as micro- and nanoroughness, texture, and porosity affect the ability of cells to adhere to solid substrate [20].

In the last two decades, many efforts have undertaken to improve the surface properties of titanium and titanium alloys in dental applications. The literature shows that some surface treatments enhance osseointegration and the success of dental implants [19,21–23]. Several techniques including acid etching, particle blasting, and anodizing have been tried in order to obtain titanium dental implant surfaces with better biological properties. Nowadays, emphasis is placed on the design of proactive surface and implant design for a specific bone quality and timely responses from surrounding cells and tissues [19–23].

Fig. 12.6 shows the surface morphology of dental implants after acid etching. It can be seen that the surface is homogeneous with uniformly distributed

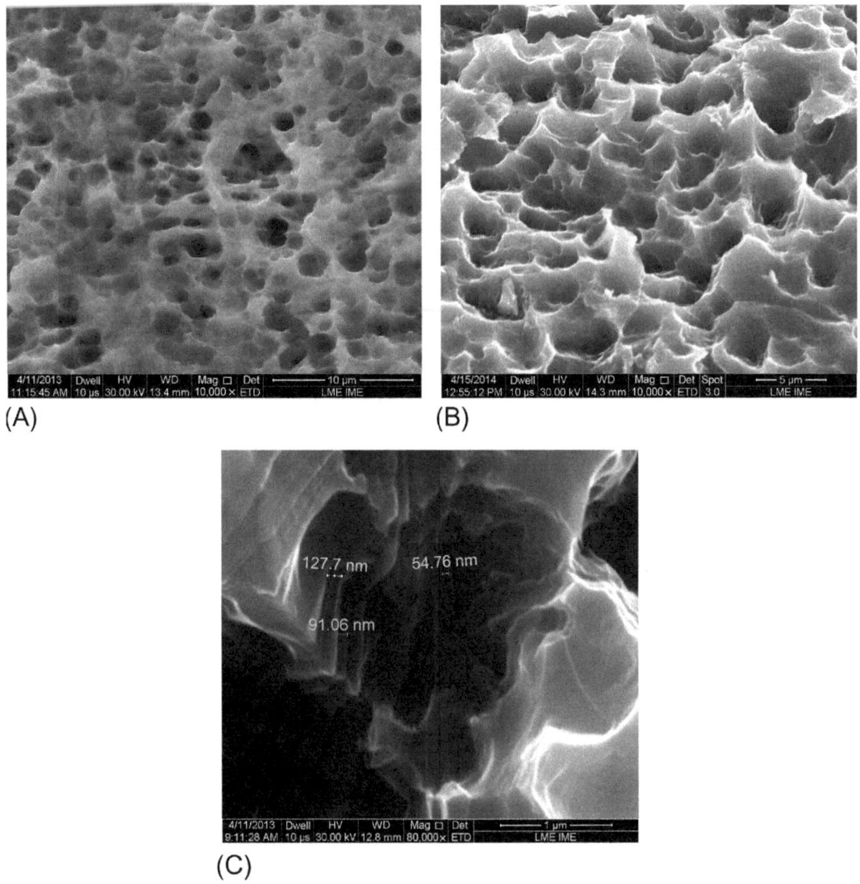

FIG. 12.6

Dental implant surface morphology after acid etching. (A) Ti alloy G5 dental implant. (B) CP Ti G4 dental implant. (C) CP Ti Hard dental implant.

microcavities. Fig. 12.6 does not show a significant difference between the macro- and micromorphology of samples Ti G4 and Ti Hard after treatment with acid. This result suggests that the use of dental implant made with Ti Hard can be subjected to the same machining processes and surface chemical treatment as the older materials without significant changes in biocompatibility. Only at high magnification it is possible to observe that Ti Hard has a higher number of nanoscale surface features than Ti G4 and G5 and thus it is expected that the interactions between the cells and the implant surface are improved in the early stages after implantation. According to Meirelles et al., the nanoscale surface features improve the osseointegration mechanism [23].

Fig. 12.7 shows the sandblasted dental implant surface. It is possible to observe particle contamination into the implant surface. Some dental implants use

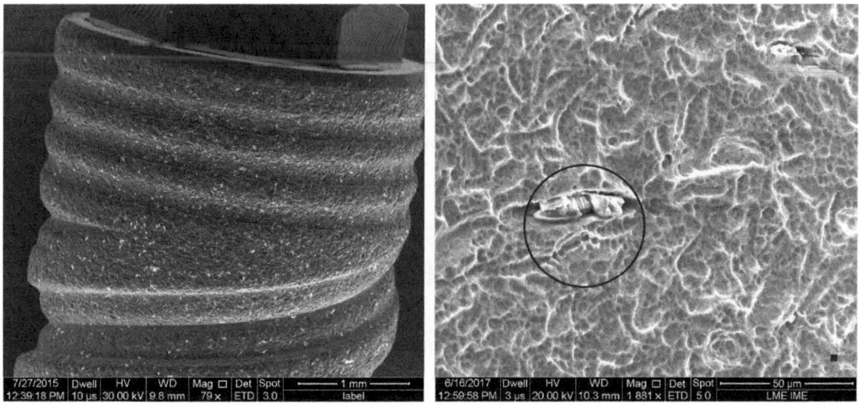

(A) (B)

FIG. 12.7

(A) Sandblasted dental implant surface with alumina particles. (B) Detail of (A), it is possible to observe particle contamination.

sandblasting of the dental implant with titanium particles following with acid treatment. Sandblasting dental implant surface with titanium particles is better than the treatment with alumina particles. The titanium dioxide particles have no negative effects on the cells' adhesion. However, the alumina contamination reduces the cell proliferation.

Fig. 12.8 shows a dental implant surface after electrochemical treatment using the spark methodology and nanoparticles of hydroxyapatite deposition. The surface morphology after Ca and P deposition was very rough and had rounded topographic

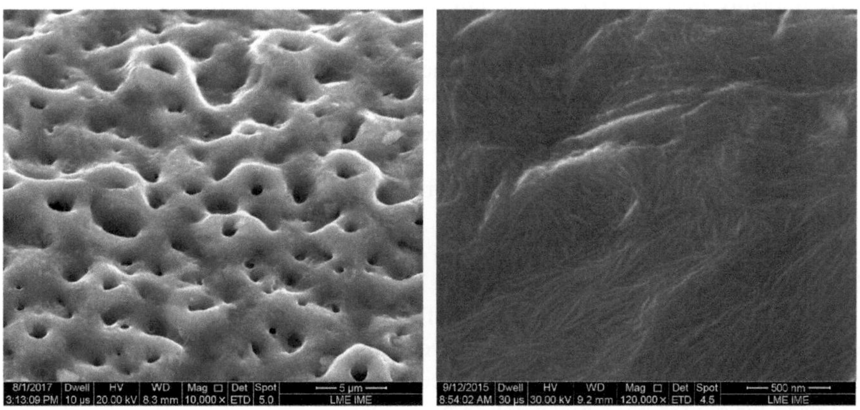

(A) (B)

FIG. 12.8

Dental implant after surface treatment. (A) Anodic spark deposition of Ca and P. (B) Chemical deposition of hydroxyapatite nanoparticles.

features. A lot of open pores were homogeneously distributed over the surface. The pores present a shape like a volcano. In the case of chemical modification, the purpose of the change is to provide ions or nanoparticles (hydroxyapatite or calcium phosphate) with specific biological responses on the metallic surface, in order to increase the attachment and stability of biomolecules.

12.7 DESIGN AND NARROW DENTAL IMPLANT SIZES

Unquestionably, important changes have been made in the surgical techniques for dental implant installation, contributing to the increase of the success rate of the treatment. Likewise, advances have been made in the manufacturing process that improved the quality of the implants. With these advances, new applications, planning, and surgical techniques have emerged that require implant osseointegration with greater mechanical resistance. As mentioned previously, CP Ti of Grades 1–4 have limited mechanical properties that do not allow the manufacture of some implants, and if used, present fracture risks. Therefore, there is a need to increase the mechanical strength of CP Ti without reducing the biocompatibility. One of the limitations of the use of Ti G2 and G4 is the low yield strength and tensile strength. To prevent any fracture, the wall thickness has to be larger than 0.4 mm. Because of the limitations of the mechanical strength of CP Ti G4, it is not possible to reduce the external implant diameter to less than 3.3 mm. However, the use of dental implants with small external diameter for insertion in places with high oral loads is critical. To overcome some limitations of CP Ti, three procedures are adopted by dental implant manufacturers are:

(i) implants are restricted to medium and large diameters (higher than 3.5 mm);
(ii) narrow dental implants are made with Ti G5;
(iii) implants and the prosthetic component are integrated in the same solid body, instead of attaching the prosthetic component by means of screws. The implant does not have a hole for abutment screw and prosthesis connection.

The use of Ti Hard allows the manufacture of dental implants with small diameters and internal connections. As shown in Fig. 12.9, a narrow dental implant with 3.5 mm external diameter and internal connection can be manufactured with walls as thin as 0.4 mm.

Cold worked CP Ti allows the use of narrow dental implants in the maxillary region. Smaller implants are less invasive solutions for patients who have narrow bone ridges and limited space between teeth. A smaller diameter can also decrease the need for guided bone regeneration procedures.

Another advantage of using Ti Hard is its higher hardness, which prevents deformation of the hexagon during insertion and increases the stability of the implants (Fig. 12.5). The use of implants with smaller diameters benefits the insertion of the implant with minimal trauma, and offers less invasive solutions, especially in cases where the bone site has a small thickness and where there is a limited space

FIG. 12.9

Narrow dental implant with internal prosthesis connection and 0.4 mm wall thickness.

between the teeth. The smaller diameter reduces the need for procedures requiring grafts to increase bone volume.

12.8 SUMMARY

UFG titanium obtained by the severe plastic deformation method was analyzed for dental implants because it provides the strength of titanium alloys without the toxic alloying elements. The results of this work showed that the mechanical behavior of dental implants made of Ti Hard processed by ECAP was better than that of dental implants made of Ti G4 for all implant designs tested. The surface morphology of dental implants Ti Hard did not show a significant difference between the macro and micro morphology of samples Ti G4 after treatment with acid. This result suggests that the use of dental implants made with Ti Hard can be subjected to the same machining processes and surface chemical treatment as the older materials without significant changes in biocompatibility. At high magnification it was possible to observe that Ti Hard has higher number of nanoscale surface features than Ti G4 and G5 and thus it is expected that the interactions between the cells and the implant surface are improved in the early stages after implantation. The use of Ti Hard allows the manufacture of dental implants with small diameters and internal connections. The clinical tests results showed that dental implants made with Ti Hard have adequate performance.

ACKNOWLEDGMENTS

This work was carried out with financial support from the Brazilian Agencies CNPq (Process 302252-2009-6 e 476757/2012-6) and FAPERJ (Process E-224/102.766/2012 e E26/111.829/2011). The authors are grateful to Conexão Sistemas de Prótese for supplying the dental implant images.

REFERENCES

[1] D. Williams, Essential Biomaterials Science, Cambridge University Press, 2014 D.F. Williams, The Williams Dictionary of Biomaterials, 1999, ISBN 0-85323-921-5.

[2] C.N. Elias, J.H.C. Lima, R. Valiev, M.A. Meyers, Biomedical applications of titanium and its alloys, JOM 60 (2008) 46–49.

[3] A. Remes, D.F. Williams, Immune response in biocompatibility: review, Biomaterials 13 (1992) 731–743.

[4] M. Ozcan, C. Hammerle, Titanium as a reconstruction and implant material in dentistry: advantages and pitfalls, Materials (Basel) 5 (2012) 1528–1545.

[5] M.A.-H. Gepreel, M. Niinomi, Biocompatibility of Ti-alloy for long term implantation, J. Mech. Behav. Biomed. Mater. 20 (2013) 407–415.

[6] P. Tengvall, Cap 14 Proteins at titanium interface, in: D.M. Brunett, P. Tengvall, M. Textor, P. Thomsen (Eds.), Titanium in Medicine, Springer Verlag, 2001, pp. 458–483.

[7] L.S. de Morais, G.G. Serra, E.F.A. Palermo, L.R. Andrade, C.A. Müller, M.A. Meyers, et al., Systemic levels of metallic ions released from orthodontic mini-implants, Am. J. Orthod. Dentofacial Orthop. 135 (4) (2009) 522–529.

[8] M. Niinomi, M. Nakai, Titanium-based biomaterials for preventing stress shielding between implant devices and bone, Int. J. Biomater. 2011 (2011) 10 Article ID 836587.

[9] L. Cristofolini, M. Viceconti, In vitro stress shielding measurement can be affected by large errors, J. Arthroplasty 14 (2) (1999) 215–220.

[10] K. Shemtov-Yona, D. Rittel, Identification of failure mechanisms in retrieved fractured dental implants, Eng. Fail. Anal. 38 (2014) 58–65.

[11] K. Shemtov-Yona, D. Rittel, On the mechanical integrity of retrieved dental implants, J. Mech. Behav. Biomed. Mater. 49 (2015) 290–299.

[12] H. Choe, J. Lee, C. Chung, Analyses of fractured implant fixture after prolonged implantation, Met. Mater. Int. 10 (2004) 327–335.

[13] G.C. Serra, L.S. de Morais, C.N. Elias, C.A. Müller, Sequential bone healing of immediately loaded mini-implants: histomorphometric and fluorescence analysis, Am. J. Orthod. Dentofacial Orthop. 137 (2010) 80–90.

[14] D.J. Fernandes, C.N. Elias, R.Z. Valiev, Properties and performance of ultrafine grained Ti for biomedical applications, Mater. Res. 18 (6) (2015) 1163–1169.

[15] C.N. Elias, D.J. Fernandes, R.S. de Biasi, Comparative study of compressive and fatigue strength of dental implants made of nanocrystalline Ti Hard and microcrystalline Ti G4, Fatigue Fract. Eng. Mater. Struct. 40 (2017) 696–705.

[16] R.Z. Valiev, T.G. Langdon, Principles of equal-channel angular pressing as a processing tool for grain refinement, Prog. Mater. Sci. 51 (2006) 881–981.

[17] I.S. Park, S.Y. Won, T.S. Bae, K.Y. Song, C.W. Park, T.G. Eom, C.M. Jeong, Fatigue characteristics of five types of implant-abutment joint designs, Met. Mater. Int. 14 (2008) 133–138.

[18] Y. Estrin, A. Vinogradov, Fatigue behaviour of light alloys with ultrafine grain structure produced by severe plastic deformation: an overview, Int. J. Fatigue 32 (2010) 898–907.

[19] A. Wennerberg, The importance of surface roughness for implant incorporation, Int. J. Mach. Tool Manuf. 0890-695538 (1998) 657–662.

[20] S. Komasa, Y. Taguchi, H. Nishida, M. Tanaka, T. Kawazoe, Bioactivity of nanostructure on titanium surface modified by chemical processing at room temperature, J. Prosthodont. Res. 56 (2012) 170–177.

[21] M. Suzuki, M.V. Guimaraes, C. Marin, R. Granato, J.N. Gil, P.G. Coelho, Histomorphometric evaluation of alumina-blasted/acid-etched and thin ion beam-deposited bioceramic surfaces: an experimental study in dogs, Int. J. Oral Maxillofac. Surg. 0278-239167 (2009) 602–607.

[22] K.H. Schuckert, S. Jopp, U. Muller, De novo grown bone on exposed implant surfaces using photodynamic therapy and recombinant human bone morphogenetic protein-2: case report, Implant Dent. 1056-616315 (2006) 361–365.

[23] L. Meirelles, L. Melin, T. Peltola, P. Kjellin, I. Kangasniemi, F. Currie, et al., Effect of hydroxyapatite and titania nanostructures on early in vivo bone response, Clin. Implant Dent. Relat. Res. 10 (4) (2008) 245–254.

Index

Note: Page numbers followed by *f* indicate figures, and *t* indicate tables.

J

L

M

N